Advanced Composite Materials and Structures

This book bridges the gap between theoretical concepts and their implementations, especially for the high-performance structures/components related to advanced composite materials. This material focuses on the prediction of various structural responses, such as deformations, natural frequencies, etc., of advanced composites under complex environments and/or loading conditions. In addition, it discusses micro-mechanical material modeling of various advanced composite materials that involve different structures ranging from basic to advanced, such as beams, flats, curved panels, shells, skewed, corrugated, and other materials, as well as various solution techniques via analytical, semi-analytical, and numerical approaches.

This book:

- Covers micro-mechanical material modeling of advanced composite materials
- Describes constitutive models of different composite materials and kinematic models of different structural configuration
- Discusses pertinent analytical, semi-analytical, and numerical techniques
- Focuses on structural responses relating to deformations, natural frequencies, and critical loads under complex environments
- Presents actual demonstration of theoretical concepts as applied to real examples using Ansys APDL scripts

This book is aimed at researchers, professionals, and graduate students in mechanical engineering, material science, material engineering, structural engineering, aerospace engineering, and composite materials.

Advanced Composite Materials and Structures

Modeling and Analysis

Edited by
Mohamed Thariq Hameed Sultan
Vishesh Ranjan Kar
Subrata Kumar Panda
Kandaswamy Jayakrishna

CRC Press
Taylor & Francis Group
Boca Raton London New York

CRC Press is an imprint of the
Taylor & Francis Group, an **informa** business

First edition published 2023
by CRC Press
6000 Broken Sound Parkway NW, Suite 300, Boca Raton, FL 33487-2742

and by CRC Press
4 Park Square, Milton Park, Abingdon, Oxon, OX14 4RN

CRC Press is an imprint of Taylor & Francis Group, LLC

ISBN: 978-0-367-74631-5 (hbk)
ISBN: 978-0-367-74634-6 (pbk)
ISBN: 978-1-003-15881-3 (ebk)

DOI: 10.1201/9781003158813

Typeset in Times
by SPi Technologies India Pvt Ltd (Straive)

Contents

Preface

In past few decades, composite materials have been utilized in various applications, namely aerospace, biomedical, defense, sports, architecture sectors, etc., mainly due to their light weight and flexibility in design. But the performance of traditional composites is limited in hostile conditions. The development of advanced composite materials can be exciting for the development of high-performance components. In this context, to confirm the viability of such materials and their structures, in-depth structural analysis is highly desirable.

This book focuses on the prediction of various structural responses such as deformations, natural frequencies, critical loads, damage, etc., of advanced composites under complex environment and/or loading conditions. In addition, this book discusses:

- The micro-mechanical material modeling of various advanced composite materials such as 2D/3D FGMs, nanocomposites, sandwich composites, biocomposites, hybrid composites, etc.
- Different structures raging from basic to advanced, such as flat and curved panel, shell, corrugated, perforated panels, etc.
- Various solution techniques via analytical, semi-analytical, and numerical approaches.

This book aims to allow engineers and researchers to extend their basic knowledge of composite materials to predict the structural behavior of novel composite materials and their structures.

Editors

Prof. Ir. Ts. Dr. Mohamed Thariq Bin Haji Hameed Sultan holds a PhD in mechanical engineering from the Department of Mechanical Engineering, University of Sheffield, United Kingdom. He specializes in the fields of hybrid composites, advanced materials, structural health monitoring, and impact studies. He is also a Professional Engineer (PEng) registered with the Board of Engineers Malaysia (BEM), and a Chartered Engineer (CEng) registered with the Institution of Mechanical Engineers (IMechE) United Kingdom. Recently, he was also awarded the Professional Technologist (PTech) from Malaysian Board of Technologists (MBOT). He has published more than 220 journal articles and 15 books internationally.

Dr. Vishesh Ranjan Kar is presently working as an assistant professor in the Department of Mechanical Engineering, NIT Jamshedpur, India. He completed his PhD from NIT Rourkela in 2015 as a full-time research scholar (2012–2015) in the field of computational solid mechanics. His research interests are the nonlinear finite element method, advanced composite structures, computational mechanics, and shape optimization. He has authored (and coauthored) over 60 research articles in peer-reviewed journals, books, and conference proceedings in the field of modeling and analysis of composite structures. Presently, he is supervising six PhD students in the area of advanced composite structures. He is the recipient of the Early Career Research Award 2017 from DST, Government of India; Young Scientist 2019 from Venus International Foundation, India; and the Preeminent Researcher Award 2019 from the International Institute of Organized Research, India in association with Western Sydney University, Australia. He is also recognized as one of the Top 2% Scientists in the World by Elsevier BV, Stanford University for the year 2020. He is a life time member of Indian Society for Applied Mechanics and a number of other international engineering societies. His h-index is more than 23 in Google Scholar and Scopus.

Dr. Subrata Kumar Panda is an associate professor in the Department of Mechanical Engineering, NIT Rourkela, India. He holds a PhD in smart composite structures from the Department of Aerospace Engineering, IIT Kharagpur. His research area comprises of multiscale multi-physics material modeling and experimental mechanics. He has published more than 200 articles in peer-reviewed international journals and conference proceedings. He has handled three sponsored projects from different government agencies (DRDO, DST, and AICTE) as principal investigator (PI) and two as co-PI (BRNS, IMPRINT). He has also worked as one of the mechanical engineering collaborators for the Centre of Excellence Orthopaedic Tissue Engineering and Rehabilitation, NIT Rourkela, India. He has already guided eight PhD scholars (six as main/sole supervisor) in different areas relevant to the fields of smart composite structural modeling and analysis, acoustics, and biomechanics. He has received an award in recognition for his contribution in mechanical sciences from NIT Rourkela. He is also recognized as one of the Top 2% Scientists in the World by Elsevier BV, Stanford University for the year 2020. His h-index is more than 40 in Google Scholar and Scopus.

Dr. Kandaswamy Jayakrishna is an associate professor in the School of Mechanical Engineering at the Vellore Institute of Technology University, India. His research is focused on biocomposites for biomedical and ballistic applications. He has published 66 journal publications in leading SCI/SCOPUS indexed journals, 26 book chapters, 91 refereed conference proceedings, and 7 books in CRC/Springer Series. He has been awarded the Young Engineer Award from the Institution of Engineers, India, in 2019, and the Global Engineering Education Award from the Industrial Engineering and Operations Management (IEOM) Society International, USA, in 2021.

Contributors

Mehmet Avcar
Suleyman Demirel University
Isparta, Turkey

Kamran Behdinan
University of Toronto
Toronto, Canada

Sunkesula Mohammad Bilal
National Institute of Technology
  Jamshedpur
Jamshedpur, India

Madan Lal Chandravanshi
Indian Institute of Technology (ISM)
  Dhanbad
Dhanbad, India

Mrinal Chaudhury
National Institute of Technology
  Rourkela
Rourkela, India

Bipin Kumar Chaurasia
National Institute of Technology
  Jamshedpur
Jamshedpur, India

Ömer Civalek
China Medical University
Taichung, Taiwan

Dhiman K. Das
Gaighata Government ITI
Ramchandrapur, India

Balakrishnan Devarajan
Applied Materials, Santa Clara
CA, USA

Pankaj S. Ghatage
Vellore Institute of Technology
Vellore, India
Rajarambapu Institute of Technology,
  Rajaramnagar,
    Affiliated to Shivaji University
Kolhapur, Maharashtra, India

Mayank K. Ghosh
National Institute of Technology
  Jamshedpur
Jamshedpur, India

Lazreg Hadji
University of Tiaret
Tiaret, Algeri

Chetan K. Hirwani
National Institute of Technology Patna
Patna, India

K. Jayakrishna
School of Mechanical Engineering
  VIT Vellore
Vellore, India

Kamal Kishore Joshi
Kalinga Institute of Industrial
  Technology Bhubaneswar,
Bhubaneswar, India
National Institute of Technology
  Jamshedpur
Jamshedpur, India

Rahul Kalyankar
Railroad Manufacture,
  Loram Technologies Inc.,
Georgetown, TX, USA

Vishesh Ranjan Kar
National Institute of Technology
  Jamshedpur
Jamshedpur, India

Abhilash Karakoti
National Institute of Technology
  Jamshedpur
Jamshedpur, India

Deepak Kumar
National Institute of Technology
  Jamshedpur
Jamshedpur, India

Mrityunjay Kumar
Indian Institute of Technology (ISM)
  Dhanbad
Dhanbad, India

Samarjeet Kumar
National Institute of Technology
  Jamshedpur
Jamshedpur, India

Shyam Kumar Chaudhary
National Institute of Technology
  Jamshedpur
Jamshedpur, India

Subrata Kumar Panda
National Institute of Technology Rourkela
Rourkela, India

Bakhtiyar Alimovich Khudayarov
Tashkent Institute of Irrigation and
  Agricultural Mechanization Engineers
Tashkent, Uzbekistan

Rama Kanta Layek
LUT University
Lahti, Finland

Rishabh Pal
National Institute of Technology Rourkela
Rourkela, India

Rene Roy
Research Center for Aircraft Parts
  Technology,
  Gyeongsang National University
Jinju, South Korea

S. Sahoo
National Institute of Technology
Durgapur, India

Jit Sarkar
Boldink Technologies Private Limited
Howrah, India

Hamid M. Sedighi
Shahid Chamran University of Ahvaz
Ahvaz, Iran

Taha Sheikh
University of Toronto
Toronto, Canada

Nitin Sharma
Kalinga Institute of Industrial
  Technology Bhubaneswar,
Bhubaneswar, India

Karunesh Kumar Shukla
National Institute of Technology
  Jamshedpur
Jamshedpur, India
MN National Institute of Technology
Allahabad, India

P. Edwin Sudhagar
Vellore Institute of Technology
Vellore, India

Mohamed Thariq Hameed Sultan
University Putra Malaysia,
  UPM Serdang
Selangor Darul Ehsan, Malaysia

1 Multi-Directional Graded Composites
An Introduction

Kamal Kishore Joshi

Kalinga Institute of Industrial Technology Bhubaneswar,
Bhubaneswar, India

National Institute of Technology Jamshedpur,
Jamshedpur, India

Vishesh Ranjan Kar

National Institute of Technology Jamshedpur,
Jamshedpur, India

Pankaj S. Ghatage

Vellore Institute of Technology,
Vellore, India

Rajarambapu Institute of Technology, Rajaramnagar,
Affiliated to Shivaji University,
Kolhapur, Maharashtra, India

Rama Kanta Layek

LUT University,
Lahti, Finland

CONTENTS

DOI: 10.1201/9781003158813-1

1

1.1   INTRODUCTION

Because of the high strength-to-weight ratio, lightweight, and stiffness-to-weight ratio, traditional composite materials find application in the field of biomedical, defense, nuclear, and aerospace industries. The main limitation of conventional composites is delamination caused by interlaminar stresses at high-temperature applications. This problem was overcome by using a special class of heterogeneous composite materials called functionally graded materials (FGMs) processed by the continuous gradation of two or more constituent materials, usually metal and ceramic, in more than one direction [1]. In FGM, the desired mechanical properties in the preferred direction are obtained by varying the volume fraction of the constituent materials [2]. Alla [3] implemented the rule of mixture to evaluate material properties of FGM to determine the thermal stresses in a 2D-FGM plate. Free vibration behavior of thick circular FG plate is analyzed using FSDT by varying the material properties continuously in radial and thickness direction according to exponential law [4]. Chi and Chung [5] investigated the mechanical behavior of simply supported FG rectangular plates subjected to transverse loading conditions. Three different gradation schemes viz. P-FGM (power-law), S-FGM (sigmoid law), and E-FGM (exponential law) were employed to estimate material properties. Shariyat et al. [6–12] employed various modeling schemes like Voigt's model, exponential law, and Mori-Tanaka law to determine effective material properties of bidirectional circular FG plate for carrying out low-velocity impact analysis with/without elastic foundation. Few authors [13, 14] reported the effect of porosity on vibration and buckling behavior of imperfect bidirectional FG micro/nanobeams using the GDQ method. The multivariable power law is used to describe material properties along the axial and thickness direction of the beam.

1.2   GRADATION SCHEMES IN FGMS

In FGMs, the material properties are assumed to be graded continuously from one surface to another surface. Various schemes are used to obtain gradation in FGMs namely power law, exponential law, and sigmoidal law. Among all of them, power law is the most commonly used scheme. In the aforementioned gradation schemes, the FGM structure consists of two constituents: metal and ceramic.

1.2.1 POWER LAW

In the open literature, the power-law behavior is well recognized by the researchers. If h is the thickness of the FGM plate, then according to multivariable power-law function, the volume fraction of ceramic can be evaluated as [15]:

For 1D-FGM plate:

$$
\left.\begin{array}{l}
x - fgm : V_c(x) = \left(\dfrac{x}{a}\right)^{q_x}, V_m(x) = 1 - V_c(x) \\[3mm]
y - fgm : V_c(y) = \left(\dfrac{y}{b}\right)^{q_y}, V_m(y) = 1 - V_c(y) \\[3mm]
z - fgm : V_c(z) = \left(0.5 + \dfrac{z}{h}\right)^{q_z}, V_m(z) = 1 - V_c(z)
\end{array}\right\}
\tag{1.1a}
$$

For 2D-FGM plate:

$$
\left.\begin{array}{l}
xy - fgm : V_c(x,y) = \left(\dfrac{x}{a}\right)^{q_x}\left(\dfrac{y}{b}\right)^{q_y}, V_m(x,y) = 1 - V_c(x,y) \\[3mm]
yz - fgm : V_c(y,z) = \left(\dfrac{y}{b}\right)^{q_y}\left(0.5 + \dfrac{z}{h}\right)^{q_z}, V_m(y,z) = 1 - V_c(y,z) \\[3mm]
xz - fgm : V_c(x,z) = \left(\dfrac{x}{a}\right)^{q_x}\left(0.5 + \dfrac{z}{h}\right)^{q_z}, V_m(y,z) = 1 - V_c(x,z)
\end{array}\right\}
\tag{1.1b}
$$

For 3D-FGM plate:

$$
xyz - fgm : V_c(x,y,z) = \left(\dfrac{x}{a}\right)^{q_x}\left(\dfrac{y}{b}\right)^{q_y}\left(0.5 + \dfrac{z}{h}\right)^{q_z}, V_m(y,z) = 1 - V_c(x,y,z)
\tag{1.1c}
$$

where V_c and, V_m denote the volume fraction of ceramic and metal.

1.2.2 EXPONENTIAL LAW

The material properties of FGMs can also be described by using exponential laws as a function of volume fraction of constituent materials as [16]

$$
P(z) = P_t e^{(V_{fc})}
\tag{1.2}
$$

$$
V_{fc} = (1/h)\ln(P_b/P_t)(z + h/2)
\tag{1.3}
$$

where P_t and P_b represent properties at the top and bottom surface of the material and V_{fc} denotes volume fraction of ceramic.

1.2.3 SIGMOID LAW

In the literature, it is revealed that using power-law and exponential law, stress concentration between the adjacent layer changes rapidly. To overcome this problem, sigmoidal law [17] is introduced, which is the combination of two power laws as described below.

$$G_1(z) = 1 - \frac{1}{2}\left(\frac{\frac{h}{2} - z}{\frac{h}{2}}\right)^q \qquad 0 \le z \le \frac{h}{2} \qquad (1.4a)$$

$$G_2(z) = 1 - \frac{1}{2}\left(\frac{\frac{h}{2} + z}{\frac{h}{2}}\right)^q \qquad -\frac{h}{2} \le z \le 0 \qquad (1.4b)$$

1.3 HOMOGENIZATION SCHEMES IN FGMs

The overall mechanical properties of FGMs structures are obtained from the microscopic heterogeneous structure of FGMs by employing a homogenization scheme. In literature, various micromechanical modeling schemes are available, such as Voigt's model, Mori-Tanaka model, Ruess model, Tamura model, Wakashima-Tsukamoto model, and Self-Consistent Method, among others. Some of the models are discussed below.

1.3.1 VOIGT'S SCHEME

This model is used by most researchers because it is simple and easy to apply. The overall material properties of FGMs are either unidirectional and multidirectional and determined by the rule of mixtures. Here material properties of composite structure are obtained as a function of volume fractions of their constituent phases, and in the form of [18]

$$P_{eff} = \sum_{i=1}^{j}(P_iV_i) \qquad (1.5)$$

where P_i and V_i represent material property and volume fraction of constituent phases, respectively.

By using Equation (1.8), the effective mechanical properties like young's modulus, Poisson's ratio, and density for two-phase constituents is given by

$$\left.\begin{aligned} E_{eff} &= E_1V_1 + E_2V_2 \\ v_{eff} &= v_1V_1 + v_2V_2 \\ \rho_{eff} &= \rho_1V_1 + \rho_2V_2 \end{aligned}\right\} \qquad (1.6)$$

where subscripts '1' and '2' denote material properties of constituent elements. Also, by using rule of mixtures, the volume fraction of constituents can be expressed as

$$V_1 + V_2 = 1 \qquad (1.7)$$

1.3.2 REUSS SCHEME

In this model, the uniform stress distribution is assumed throughout the FGM and the mechanical properties are derived from the rule of mixture, which is the harmonic mean of material properties of constituents and their volume fraction. The effective material property by using Reuss [19, 20] estimate can be expressed as

$$P_{eff} = \frac{P_{\alpha 1} P_{\alpha 2}}{V_{\alpha 1} P_{\alpha 2} + V_{\alpha 2} P_{\alpha 1}} \qquad (1.8)$$

1.3.3 MORI-TANAKA SCHEME

This scheme is suitable for composites having a discontinuous particulate phase in the graded structure. According to this scheme, the overall young's modulus (E) and Poisson's ratio (v) [21, 22] can be expressed as a function of overall local bulk (K) and shear (G) modulus as

$$E = \frac{9K + G}{3K + G}, \quad v = \frac{3K - 2G}{2(3K + G)} \qquad (1.9)$$

Here, K and G can be further expressed in terms of bulk (K_1, K_2) and shear (G_1, G_2) modulus of constituent phases and volume fraction (V_1, V_2) as

$$\frac{K - K_1}{K_2 - K_1} = \frac{V_2}{1 + V_1 \left[\dfrac{3(K_2 - K_1)}{3K_1 + 4G_1} \right]} \qquad (1.10a)$$

$$\frac{G - G_1}{G_2 - G_1} = \frac{V_2}{1 + V_1 \left[\dfrac{G_2 - G_1}{G_1 + f_1} \right]} \qquad (1.10b)$$

where

$$f_1 = \frac{G_1 (9K_1 + 8G_1)}{6(K_1 + 2G_1)} \qquad (1.10c)$$

1.3.4 SELF-CONSISTENT METHOD (SCM)

This method assumes that each reinforcement inclusion embedded into a continuum structure has the same material property as that of composite [18]. In other words, both reinforcements and matrix are indistinguishable. Here the effective bulk (K_e) and shear (G_e) modulus are derived by the following relationship:

$$\frac{\chi}{K_e} = \frac{V_1}{K_e - K_2} + \frac{V_2}{K_e - K_1} \qquad (1.11a)$$

$$\frac{\varrho}{G_e} = \frac{V_1}{G_e - G_2} + \frac{V_2}{G_e - G_1} \tag{1.11b}$$

where

$$\chi = 3 - 5\varrho = \frac{K_e}{K_e + \frac{4}{3}G_e} \tag{1.11c}$$

Finally, effective material properties like elastic modulus and Poisson's ratio can be evaluated by using Equation (1.9).

1.3.5 TAMURA SCHEME

In this model, the linear rule of mixture is used to estimate the effective mechanical properties of two-phase constituents with the help of the empirical fitting parameter (p) in the stress-strain relationship of the constituent phases. The fitting parameter [23, 28] is given by the following relation:

$$p = \frac{\sigma_1 - \sigma_2}{\varepsilon_1 - \varepsilon_2} \tag{1.12}$$

In uniaxial loading, overall effective and local stress in composite and each phase is given by

$$\sigma_{eff} = V_1\sigma_1 + V_2\sigma_2 = E_{eff}\varepsilon \tag{1.13a}$$

$$\sigma_1 = E_1\varepsilon_1, \quad \sigma_2 = E_2\varepsilon_2 \tag{1.13b}$$

By combining Equations (1.12), (1.13a), and (1.13b), the effective young's modulus is expressed by the following relation:

$$E_{eff} = \frac{V_1E_1(p - E_2) + V_2E_2(p - E_1)}{V_1(p - E_2) + V_2(p - E_1)} \tag{1.14}$$

1.3.6 GASIK-UEDA MODEL

In this model, it is assumed that the composite structure is made of several sub-cells that are orthotropic. Each sub-cell forms local representative volume elements (LRVE). In addition to this, inclusions are considered as equivalent to cube in LRVE phase, and for each cell mechanical properties are relying on the volume fraction of inclusions. The young's modulus is using LRVE is given as [24, 25]

$$E_{jj} = E_1\left(1 - \sqrt[3]{V_2^2}\left(1 - \frac{1}{1 - \sqrt[3]{V_2^2}\left(1 - \frac{E_1}{E_2}\right)}\right)\right) \tag{1.15}$$

where j denotes one of the coordinate axes.

1.3.7 COHERENT POTENTIAL APPROXIMATION (CPA)

Here, it is assumed that effective material properties obtained using this method are independent of constituent phases in FGM structures [23]. The approximation used in this method to calculate bulk and shear modulus is given below.

$$\frac{V_1(K_1 - K_e)}{3K_1 + 4G_e} + \frac{V_2(K_2 - K_e)}{3K_2 + 4G_e} = 0 \tag{1.16a}$$

$$\frac{V_1(G_1 - G_e)}{G_1 + \dfrac{G_e(9K_e + 8G_e)}{6K_e + 12G_e}} + \frac{V_2(G_2 - G_e)}{G_2 + \dfrac{G_e(9K_e + 8G_e)}{6K_e + 12G_e}} = 0 \tag{1.16b}$$

1.3.8 KERNER MODEL

In this approach, it is assumed that spherical particles are reinforced in the isotropic phase [23] and both phases are perfectly bonded. Due to its predictive capability and ease of application, it is more frequently used over the years [9]. The set of implicit relations in the Kerner model is in the form

$$K_f = \frac{K_1 V_1 + K_2 V_2 \left(\dfrac{3K_1 + 4G_1}{3K_2 + 4G_1}\right)}{V_1 + V_2 \left(\dfrac{3K_1 + 4G_1}{3K_2 + 4G_1}\right)} \tag{1.17a}$$

$$G_f = G_1 \left\{ \frac{\dfrac{V_2 G_2}{G_1(7 - 5v_1) + G_2(8 - 10v_1)} + \dfrac{V_1}{15(1 - v_1)}}{\dfrac{V_2 G_1}{G_1(7 - 5v_1) + G_2(8 - 10v_1)} + \dfrac{V_1}{15(1 - v_1)}} \right\} \tag{1.17b}$$

where (K_f) and (G_f) are effective bulk and shear modulus, respectively.

1.3.9 HIRANO MODEL

In this scheme, effective material properties at intermediate volume fraction are the weighted average of the Kerner model estimate. So, the effective bulk modulus becomes [23, 26, 27]:

$$K_e = \lambda(V_2) K_1'(V_2) + (1 - \lambda(V_2)) K_2'(V_2) \tag{1.18}$$

where $K_1'(V_2)$ and $K_2'(V_2)$, respectively denote the effective bulk modulus estimated using the Kerner model considering phases 1 and 2 as matrix, and $\lambda(V_2)$ denotes interpolating function.

1.4   COMPARISON OF MICROMECHANICAL MODELING SCHEMES

In this section, effective material properties obtained from different modeling schemes are compared. Zuiker [23] founds that for the SiC/C system (refer to Figure 1.1(a)), the bulk modulus estimated from CPA and SCM schemes are almost identical. The fuzzy logic approximation used by the Hirano model resembles that of SCM and CPA schemes due to the choice of interpolation function and limits. A similar comparison can be shown in Figure 1.1(b) for Poisson's ratio.

In another comparative study, Akbarzadeh et. al [28] analyzed various homogenization schemes for aluminum matrix and ceramic inclusions. It is clear from Figure 1.2(a) that at a volume fraction of 0.5, Voigt's estimate of young's modulus is

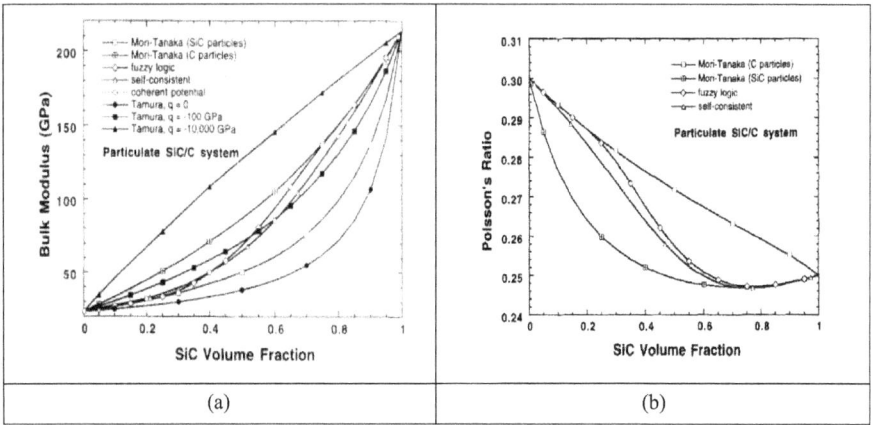

FIGURE 1.1   Overall (a) Bulk modulus and (b) Poisson's ratio versus SiC volume fraction for various modeling schemes [13].

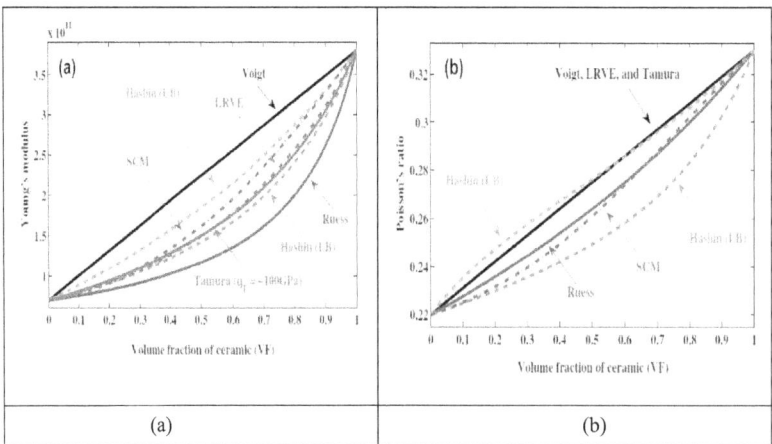

FIGURE 1.2   Overall (a) Young's modulus and (b) Poisson's ratio versus ceramic volume fraction for various modeling schemes [15].

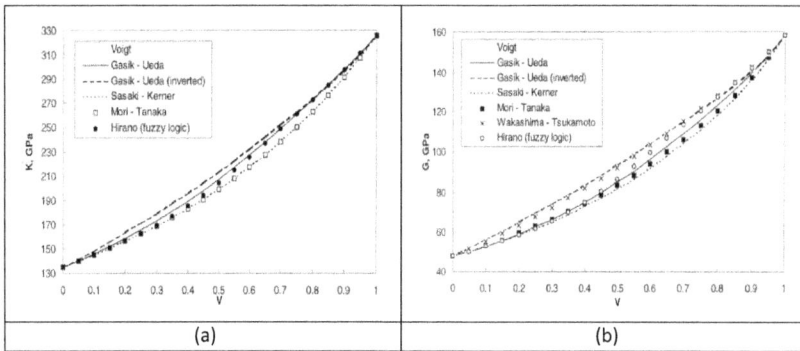

FIGURE 1.3 Overall (a) Bulk modulus and (b) Shear modulus versus volume fraction for W-Cu composite system [16].

approximately 48% higher compared to Reuss model. The effective Young's modulus estimated by the upper bound of Hashin–Shtrikman is 30% higher compared to its lower-bound counterpart. The estimation done by other modeling schemes falls in between Hashin–Shtrikman bounds. Figure 1.2(b) shows that Poisson's ratios estimated from Voigt's, LRVE, and Tamura's model are identical while Reuss and SCM models predict properties in between Hashin–Shtrikman bounds.

For W-Cu composite system, effective properties like bulk and shear modulus for the different micromechanical models are compared by Gasik [24]. Figure 1.3 shows that the Gasik model resembles closely to Hirano model with an appropriate interpolating function.

1.5 SUMMARY

In summary, various micromechanical modeling schemes are studied to evaluate the effective mechanical properties of FGMs with different reinforcements and volume fractions. It was found that the SCM method is most accurate, as it also satisfies the upper and lower bounds of the Hashin–Shtrikman domain. Apart from the SCM model, the Gasik-Ueda model also predicts the material properties accurately, but the major limitation is that it cannot be applied to fiber/particulate reinforced composite with a higher aspect ratio.

REFERENCES

[1] M. Zaidi, K.K. Joshi, A. Shukla, B. Cherinet, A review of the various modelling schemes of unidirectional functionally graded material structures, *AIP Conf. Proc.* 2341 (2021). https://doi.org/10.1063/5.0050306

[2] A. Gupta, M. Talha, Recent development in modeling and analysis of functionally graded materials and structures, *Prog. Aerosp. Sci.* 79 (2015) 1–14. https://doi.org/10.1016/j.paerosci.2015.07.001

[3] M. Nemat-Alla, Reduction of thermal stresses by developing two-dimensional functionally graded materials, *International Journal of Solids and Structures*, 40 (2003) 7339–7356. https://doi.org/10.1016/j.ijsolstr.2003.08.017

[4] M.M. Alipour, M. Shariyat, M. Shaban, A semi-analytical solution for free vibration of variable thickness two-directional-functionally graded plates on elastic foundations, *International Journal of Mechanics and Materials in Design*, (2010) 293–304. https://doi.org/10.1007/s10999-010-9134-2

[5] S.H. Chi, Y.L. Chung, Mechanical behavior of functionally graded material plates under transverse load-Part I: Analysis, *Int. J. Solids Struct.* 43 (2006) 3657–3674. https://doi.org/10.1016/j.ijsolstr.2005.04.011

[6] K. Asemi, H. Ashrafi, M. Shariyat, Three-dimensional stress and free vibration analyses of functionally graded plates with circular holes by the use of the graded finite element method, *J. Appl. Mech. Tech. Phys.* 57 (2016) 690–700. https://doi.org/10.1134/S0021894416040131

[7] M. Shariyat, R. Jafari, Nonlinear low-velocity impact response analysis of a radially preloaded two-directional-functionally graded circular plate : A refined contact stiffness approach, *Compos. Part B.* 45 (2013) 981–994. https://doi.org/10.1016/j.compositesb.2012.05.014

[8] M.M. Alipour, M. Shariyat, Stress analysis of two-directional FGM moderately thick constrained circular plates with non-uniform load and substrate stiffness distributions, *J. Solid Mech.* 2 (2010) 316–331.

[9] M.M. Alipour, M. Shariyat, M. Shaban, A semi-analytical solution for free vibration of variable thickness two-directional-functionally graded plates on elastic foundations, (2010) 293–304. https://doi.org/10.1007/s10999-010-9134-2

[10] H. Ashrafi, K. Asemi, M. Shariyat, A three-dimensional boundary element stress and bending analysis of transversely/longitudinally graded plates with circular cutouts under biaxial loading, *Eur. J. Mech. A/Solids.* 42 (2013) 344–357. https://doi.org/10.1016/j.euromechsol.2013.07.009

[11] M. Shariyat, R. Jafari, A micromechanical approach for semi-analytical low-velocity impact analysis of a bidirectional functionally graded circular plate resting on an elastic foundation, *An Int. J. Theor. Appl. Mech. AIMETA.* 48 (2013) 2127–2148. https://doi.org/10.1007/s11012-013-9729-4

[12] M. Shariyat, M.M. Alipour, A power series solution for vibration and complex modal stress analyses of variable thickness viscoelastic two-directional FGM circular plates on elastic foundations, *Appl. Math. Model.* 37 (2013) 3063–3076. https://doi.org/10.1016/j.apm.2012.07.037

[13] N. Shafiei, M. Kazemi, Buckling analysis on the bi-dimensional functionally graded porous tapered nano- / micro-scale beams, *Aerosp. Sci. Technol.* 66 (2017) 1–11. https://doi.org/10.1016/j.ast.2017.02.019

[14] N. Shafiei, S.S. Mirjavadi, B.M. Afshari, S. Rabby, M. Kazemi, Vibration of two-dimensional imperfect functionally graded (2D-FG) porous nano-/micro-beams, *Comput. Methods Appl. Mech. Engrg.* 322 (2017) 615–632. https://doi.org/10.1016/j.cma.2017.05.007

[15] K.K. Joshi, V.R. Kar, Effect of material heterogeneity on the deformation behaviour of multidirectional (1D/2D/3D) functionally graded composite panels, *Eng. Comput. (Swansea, Wales).* 38 (2021) 3325–3350. https://doi.org/10.1108/EC-06-2020-0301

[16] F. Delale, F. Erdogan, The crack problem for a nonhomogeneous plane, *J. Appl. Mech.* 50 (1983) 609–614.

[17] S.H. Chi, Y.L. Chung, Mechanical behavior of functionally graded material plates under transverse load-Part I: Analysis, *Int. J. Solids Struct.* 43 (2006) 3657–3674. https://doi.org/10.1016/j.ijsolstr.2005.04.011

[18] H.S. Shen, *Functionally graded materials: Nonlinear analysis of plates and shells*, CRC Press, Boca Raton, 2009.

[19] Y. Miyamoto, W.A. Kaysser, B.H. Rabin, A. Kawasaki, R.G. Ford, *Functionally graded materials: Design, processing and applications* (Springer, New York, NY, 1999), pp. 63–88.

[20] L. Hadji, F. Bernard, Bending and free vibration analysis of functionally graded beams on elastic foundations with analytical validation, *Advances in Materials Research*, 9(1) 63–98 (2020).

[21] T. Mori, K. Tanaka, Average stress in matrix and average elastic energy of materials with misfitting inclusions, *Acta Metall.* 2 (1973) 1571–1574.

[22] Y. Benveniste, A new approach to the application of Mori–Tanaka's theory in composite materials, *Mech. Mater.* 6 (1987) 147–157.

[23] J. R. Zuiker, Functionally graded materials: Choice of micromechanics model and limitations in property variation. *Composites Engineering*, 5(7) (1995), pp. 807–819

[24] M. M. Gasik, R. R. Lilius, Evaluation of properties of W-Cu functional gradient materials by micromechanical model, *Computational Materials Science* 3(1) (1994) 41–49.

[25] M. M. Gasik, Micromechanical modelling of functionally graded materials, *Computational Materials Science*, 13 (1–3) (1998) 42–55.

[26] T. Hirano, J. Teraki, T. Yamada, Application of fuzzy theory to the design of functionally gradient materials, *SMiRT 11 Transactions*, 23(21) (1991) 49–54.

[27] T. Hirano, J. Teraki, T. Yamada, On the design of functionally gradient materials, in *Proc. 1st lnt. Syrup. Functionally Gradient Materials*, pp. 5–10. Functionally Gradient Materials Forum, Sendai, Japan, (1990).

[28] A. H. Akbarzadeh, A. Abedini, Z. T. Chen, Effect of micromechanical models on structural responses of functionally graded plates. *Composite Structures*, 119 (2015) 598–609.

2 Free Vibration Characteristics of Bi-Directional Functionally Graded Composite Panels

Pankaj S. Ghatage

Vellore Institute of Technology,
Vellore, India

Rajarambapu Institute of Technology, Rajaramnagar,
Affiliated to Shivaji University,
Kolhapur, India

Vishesh Ranjan Kar

National Institute of Technology Jamshedpur,
Jamshedpur, India

Kamal Kishore Joshi

National Institute of Technology Jamshedpur,
Jamshedpur, India

Kalinga Institute of Industrial Technology Bhubaneswar,
Bhubaneswar, India

P. Edwin Sudhagar

Vellore Institute of Technology,
Vellore, India

Rahul Kalyankar

Railroad Manufacture, Loram Technologies Inc.,
Georgetown, TX, USA

DOI: 10.1201/9781003158813-2

CONTENTS

2.1   INTRODUCTION

The typical functionally graded material (FGM) structures are known for their tailor-made material properties in which metal and ceramic materials are graded smoothly in thickness direction [1]. Due to this, FGM structures are being adopted in different engineering applications where the structures/components demand lightweight material with high strength and stiffness [2–4] and also in biomedical applications [5–7]. However, material gradation in two or more directions may facilitate more promising FGMs under critical and unlike environmental conditions [8]. Analysis of any novel materials is always vital to exhibit the mechanical behavior. Therefore, in recent years, the analyses of these types of heterogeneous materials have been encountered by the group of researchers to examine various mechanical responses.

Alipour et al. [9] studied free vibration behavior of 2D functionally graded circular plate by varying thickness using first-order shear deformation theory (FSDT). Nie and Zhong [10] analyzed free vibration behavior of multidirectional FG annular plates using state space-based differential quadrature method. Shariyat and Alipour [11] developed power series solution in order to analyze free vibration and damping behavior of bidirectional viscoelastic functionally graded plates. Buckling and vibration analysis of circular plate constituted with bidirectional functionally graded materials under the action of hydrostatic in-plane force using classical plate theory are carried out by Lal and Ahlawat [12]. Nejati et al. [13] applied differential quadrature method (DQM) to analyze free vibration behavior of 2D functionally graded annular plates. Yas and Moloudi [14] analyzed multidirectional functionally graded piezoelectric annular plate using DQM. Ahlawat and Lal [15] investigated buckling and vibration behavior of FGM circular plate in which material properties are assumed to be varied in both radial and transverse directions under uniform plane load resting on elastic foundation using generalized differential quadrature method. Van Do et al. [16] carried out buckling and bending behaviors of bidirectional FGM plates using finite element method (FEM) and also they have proposed new third-order shear deformation plate theory in order to analyze the multidirectional FGMs. Mahinzare et al. [17] proposed numerical method to analyze free vibration of bidirectional FG micro circular plate under thermal load using the FSDT. The thermomechanical properties of the plate were assumed to be varied in thickness and radial direction. The same work without considering thermal load extended by Shojaeefard et al. [18] and investigated free vibration as well as angular velocity of the plate. Lieu et al. [19] have presented first-time bending and vibration analysis of in-plane bidirectional FG plate with variable thickness using isogeometric analysis frame-work. Wang et al. [20] analyzed free vibration behavior multidirectional FG parallelepipeds using exiting quadrature element method under general boundary condition. Asgari and Akhlaghi [21] examined free vibration behavior of bidirectional FG thick hollow cylinder using three-dimensional theory of elasticity. On the basis of Love's first approximation classical shell theory, the free vibration analysis of 2D FG cylindrical shell was carried out by Ebrahimi and Najafizadeh [22]. Voigt and Mori–Tanaka models were used in order to describe the material properties and results of both models were compared. Zafarmand et al. [23] investigated free vibration and transient response of 2D FG cylindrical panel on the basis of three-dimensional equations of elasticity.

It is evident from the above literature that very limited works have been reported on the analyses of bidirectional FGM structures. Here, an attempt has been made to compute the eigenfrequencies of biaxial functionally graded composite (B-FGC) panel with two different homogenization schemes, i.e., Mori-Tanaka and Voigt's rule-of-mixture, in conjunction with power-law function. In this study, the strain field of the structure is described with equivalent single-layer higher-order displacement field. The eigenfrequencies of B-FGC panel structures are computed by considering different sets of conditions using finite element approximations.

2.2 MICROMECHANICAL MATERIAL MODELING OF B-FGC STRUCTURE

In this work, B-FGC flat panel structure of sides a and b and uniform thickness h is modeled with in-plane material gradation (refer Figure 2.1). Here, the B-FGC panel is assumed to be constituted with two materials, i.e., metal and ceramic, and the material properties are mentioned in Table 2.1. Here, the constituent materials are assumed to be varying smoothly from one edge $(0, 0, z)$ of the panel to another edge (a, b, z) by following the power-law functions of their respective volume fractions [22], as

$$\left. \begin{aligned} V_{fc}(x,y) &= \left(\frac{x}{a}\right)^{n_x}\left(\frac{y}{b}\right)^{n_y} \\ V_{fm}(x,y) &= 1 - V_{fc}(x,y) \end{aligned} \right\} \begin{aligned} 0 \le n_x < \infty \\ 0 \le n_y < \infty \end{aligned} \tag{2.1}$$

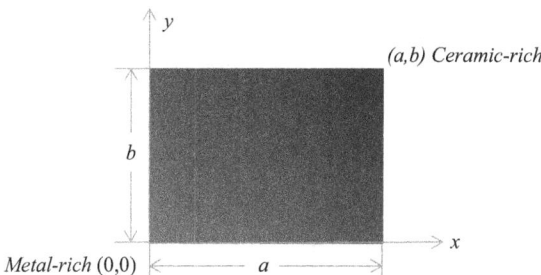

FIGURE 2.1 Geometry and dimensions of B-FGC structure with gradation in x and y directions.

TABLE 2.1
Properties of FGM Constituents [26]

Materials	Properties		
	Young's Modulus $E(GPa)$	Poisson's Ratio v	Density ρ (kg/m^3)
Aluminum (Al)	70	0.3	2707
Alumina (Al_2O_3)	380	0.3	3000
Stainless Steel (SUS304)	207.78	0.3177	8166
Silicon Nitride (Si_3N_4)	322.27	0.24	2370

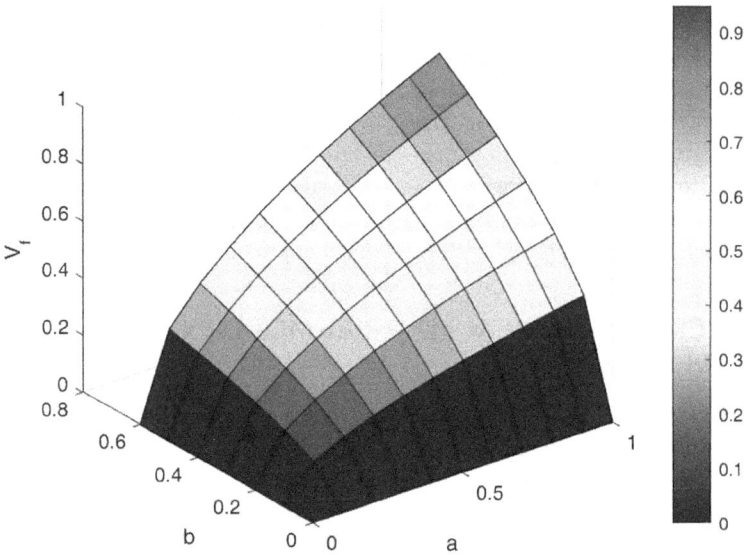

FIGURE 2.2   Variation of volume fraction of ceramic in B-FGC panel with $n_x = 0.5$, $n_y = 0.5$.

where, V_{fc} and V_{fm} are the volume fractions of ceramic and metal materials, respectively. n_x and n_y are the power-law indices in x and y directions, respectively. Figures 2.2–2.5 exhibit the material distribution profile in x–y plane of B-FGC panel structure at different values of n_x and n_y.

To evaluate the instantaneous material properties of B-FGC panel with respect to the in-plane coordinates, Voigt's rule-of-mixture and Mori-Tanaka micromechanical material schemes are adopted.

2.2.1   VOIGT MODEL

The Voigt's rule-of-mixture material model is simple and can be adopted to compute the overall material properties of B-FGC panel [24]. To evaluate various material

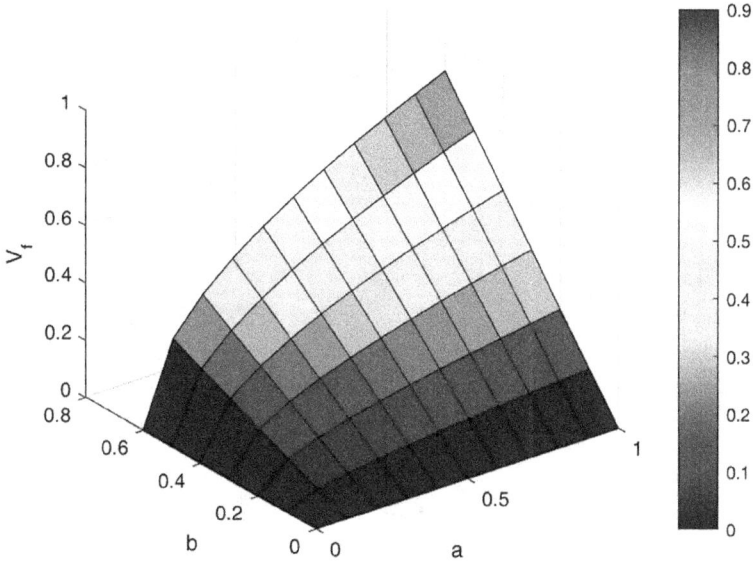

FIGURE 2.3   Variation of volume fraction of ceramic in B-FGC panel with $n_x = 0.5, n_y = 1$.

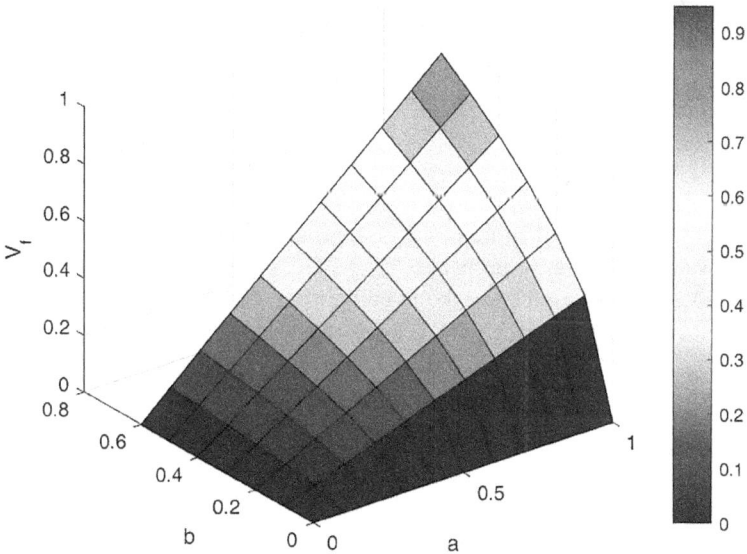

FIGURE 2.4   Variation of volume fraction of ceramic in B-FGC panel with $n_x = 1, n_y = 0.5$.

properties, such as Young's modulus ($E_{\text{B-FGC}}$), mass density ($\rho_{\text{B-FGC}}$) and Poisson's ratio ($\nu_{\text{B-FGC}}$) of B-FGC panel, Voigt's model can be implemented, as [24]

$$E_{\text{B-FGC}}(x,y) = (E_c - E_m)V_{fc}(x,y) + E_m \qquad (2.2)$$

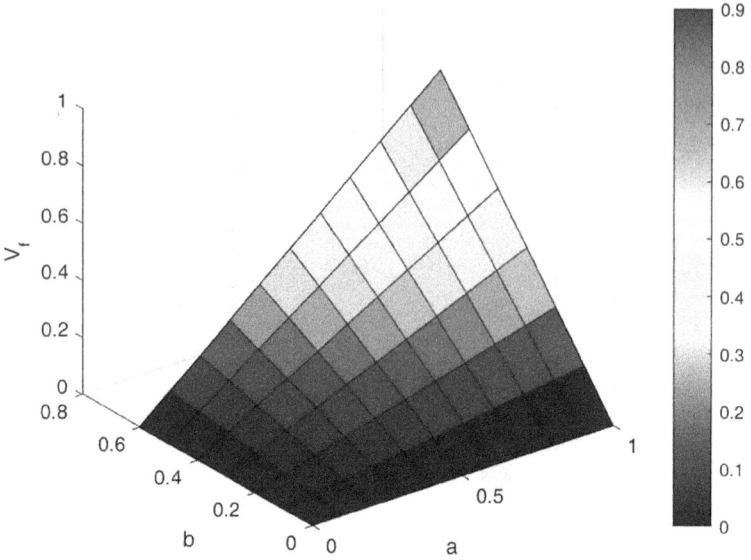

FIGURE 2.5 Variation of volume fraction of ceramic in B-FGC panel with $n_x = 1$, $n_y = 1$.

$$\rho_{\text{B-FGC}}(x,y) = (\rho_c - \rho_m)V_{fc}(x,y) + \rho_m \tag{2.3}$$

$$v_{\text{B-FGC}}(x,y) = (v_c - v_m)V_{fc}(x,y) + v_m \tag{2.4}$$

2.2.2 Mori–Tanaka Scheme

Mori–Tanaka scheme, which accounts discontinuous particulate phase, can be utilized to evaluate the elastic properties of B-FGC panel. $K_{\text{B-FGC}}$ and $G_{\text{B-FGC}}$ are the effective local bulk modulus and the effective shear modulus, respectively, and can be expressed in Equations (2.7) and (2.8), as

$$\frac{K_{\text{B-FGC}} - K_m}{K_c - K_m} = \frac{V_{fc}}{1 + \left(\dfrac{(1 - V_{fc})(K_c - K_m)}{3K_m + 4G_m} \right)} \tag{2.7}$$

$$\frac{G_{\text{B-FGC}} - G_m}{G_c - G_m} = \frac{V_{fc}}{1 + \dfrac{(1 - V_{fc})(G_c - G_m)}{G_m + f_1}} \tag{2.8}$$

where,

$$f_1 = G_m \left(\frac{9K_m + 8G_m}{6(K_m + 2G_m)} \right)$$

In Mori–Tanaka scheme, the effective Young's Modulus ($E_{\text{B-FGC}}$) and Poisson's ratio ($\nu_{\text{B-FGC}}$) can be evaluated as

$$E_{\text{B-FGC}} = \left(\frac{9 K_{\text{B-FGC}} G_{\text{B-FGC}}}{3 K_{\text{B-FGC}} + G_{\text{B-FGC}}} \right) \tag{2.9}$$

$$\nu_{\text{FGBC}} = \left(\frac{3 K_{\text{FGBC}} - 2 G_{\text{FGBC}}}{2 \left(3 K_{\text{FGBC}} + G_{\text{FGBC}} \right)} \right) \tag{2.10}$$

However, to evaluate the effective mass density ($\rho_{\text{B-FGC}}$), Equation (2.3) is utilized.

2.3 FINITE ELEMENT FORMULATIONS

2.3.1 Higher-Order Kinematic Model

To model the displacement field of B-FGC panel, HSDT with nine degree of freedom is adopted [25]. The global displacements $\{\lambda\} = \lfloor u\ v\ w \rfloor^T$ (u, v, w) at any point along (x, y, z) directions in the structure are presented in terms of mid-plane displacements $\{\lambda_0\} = \lfloor u_0\ v_0\ w_0\ \theta_x\ \theta_y\ u_0^*\ v_0^*\ \theta_x^*\ \theta_y^* \rfloor^T$ Equation (2.10), as

$$\begin{pmatrix} u \\ v \\ w \end{pmatrix} = \begin{bmatrix} 1 & 0 & 0 & z & 0 & z^2 & 0 & z^3 & 0 \\ 0 & 1 & 0 & 0 & z & 0 & z^2 & 0 & z^3 \\ 0 & 0 & 1 & 0 & 0 & 0 & 0 & 0 & 0 \end{bmatrix} \begin{pmatrix} u_0 \\ v_0 \\ w_0 \\ \theta_x \\ \theta_y \\ u_0^* \\ v_0^* \\ \theta_x^* \\ \theta_y^* \end{pmatrix} \tag{2.10}$$

$$\{\lambda\} = \left[f \right] \{\lambda_0\} \tag{2.11}$$

The strain tensor $\varepsilon = \lfloor \varepsilon_{xx}\ \varepsilon_{yy}\ \varepsilon_{xy}\ \varepsilon_{xz}\ \varepsilon_{yz} \rfloor^T$ of B-FGC panel is given as

$$\begin{Bmatrix} \varepsilon_{xx} \\ \varepsilon_{yy} \\ \varepsilon_{xy} \\ \varepsilon_{xz} \\ \varepsilon_{yz} \end{Bmatrix} = \begin{bmatrix} \dfrac{\partial u}{\partial x} \\[2mm] \dfrac{\partial v}{\partial y} \\[2mm] \dfrac{1}{2}\left(\dfrac{\partial u}{\partial y} + \dfrac{\partial v}{\partial x} \right) \\[2mm] \dfrac{1}{2}\left(\dfrac{\partial u}{\partial z} + \dfrac{\partial w}{\partial x} \right) \\[2mm] \dfrac{1}{2}\left(\dfrac{\partial v}{\partial z} + \dfrac{\partial w}{\partial y} \right) \end{bmatrix} = \begin{Bmatrix} \varepsilon_x^0 + z k_x^1 + z^2 k_x^2 + z^3 k_x^3 \\ \varepsilon_y^0 + z k_y^1 + z^2 k_y^2 + z^3 k_y^3 \\ \varepsilon_{xy}^0 + z k_{xy}^1 + z^2 k_{xy}^2 + z^3 k_{xy}^3 \\ \varepsilon_{xz}^0 + z k_{xz}^1 + z^2 k_{xz}^2 + z^3 k_{xz}^3 \\ \varepsilon_{yz}^0 + z k_{yz}^1 + z^2 k_{yz}^2 + z^3 k_{yz}^3 \end{Bmatrix} \tag{2.13}$$

Now, $\{\bar{\varepsilon}\} = \left\lfloor \varepsilon_x^0\, \varepsilon_y^0\, \varepsilon_{xy}^0\, \varepsilon_{xz}^0\, \varepsilon_{yz}^0\, k_x^1\, k_y^1\, k_{xy}^1\, k_{xz}^1\, k_{yz}^1\, k_x^2\, k_y^2\, k_{xy}^2\, k_{xz}^2\, k_{yz}^2\, k_x^3\, k_y^3\, k_{xy}^3\, k_{xz}^3\, k_{yz}^3 \right\rfloor^T$ includes membrane, curvature, and higher-order strain terms, and can be expressed as:

$$\varepsilon_x^0 = u_{,x}, \quad \varepsilon_y^0 = v_{,y}, \quad \varepsilon_{xy}^0 = u_{,y} + v_{,x}, \quad \varepsilon_{xz}^0 = w_{,x} + \theta_x, \quad \varepsilon_{yz}^0 = w_{,y} + \theta_y;$$

$$k_x^1 = \theta_{x,x}, \quad k_y^1 = \theta_{y,y}, \quad k_{xy}^1 = \theta_{x,y} + \theta_{y,x}, \quad k_{xz}^1 = 2u_0^*, \quad k_{yz}^1 = 2v_0^*;$$

$$k_x^2 = u_{0,x}^*, \quad k_y^2 = v_{0,y}^*, \quad k_{xy}^2 = u_{0,y}^* + v_{0,x}^*, \quad k_{xz}^2 = 3\theta_x^*, \quad k_{yz}^2 = 3\theta_y^*;$$

$$k_x^3 = \theta_{x,x}^*, \quad k_y^3 = \theta_{y,y}^*, \quad k_{xy}^3 = \theta_{x,y}^* + \theta_{y,x}^*.$$

2.3.2 CONSTITUTIVE RELATIONS

The stress tensors $\sigma = \lfloor \sigma_{xx}\, \sigma_{yy}\, \sigma_{xy}\, \sigma_{xz}\, \sigma_{yz} \rfloor^T$ at any point of B-FGC panel can be expressed as

$$\begin{Bmatrix} \sigma_{xx} \\ \sigma_{yy} \\ \sigma_{xy} \\ \sigma_{xz} \\ \sigma_{yz} \end{Bmatrix} = \begin{bmatrix} \dfrac{E(x,y)}{1-\upsilon^2(x,y)} & \dfrac{\upsilon E(x,y)}{1-\upsilon^2(x,y)} & 0 & 0 & 0 \\ \dfrac{\upsilon E(x,y)}{1-\upsilon^2(x,y)} & \dfrac{E(x,y)}{1-\upsilon^2(x,y)} & 0 & 0 & 0 \\ 0 & 0 & \dfrac{E(x,y)}{2(1+\upsilon(x,y))} & 0 & 0 \\ 0 & 0 & 0 & \dfrac{E(x,y)}{2(1+\upsilon(x,y))} & 0 \\ 0 & 0 & 0 & 0 & \dfrac{E(x,y)}{2(1+\upsilon(x,y))} \end{bmatrix} \begin{Bmatrix} \varepsilon_{xx} \\ \varepsilon_{yy} \\ \varepsilon_{xy} \\ \varepsilon_{xz} \\ \varepsilon_{yz} \end{Bmatrix}$$

$$(2.14)$$

$$\sigma = C(x,y)\varepsilon \tag{2.15}$$

where $[C]$ is the elastic material matrix.

2.3.3 ENERGY EQUATIONS

The governing equations of present B-FGC model are obtained using Hamilton's principle as

$$\delta \int_{t_1}^{t_2} (U_k - U_\varepsilon)\, dt = 0 \tag{2.16}$$

Here, strain energy (U_ε) and kinetic energy (U_k) of B-FGC panel are given as

$$U_\varepsilon = \frac{1}{2} \int \left(\int_{-h/2}^{+h/2} \{\varepsilon\}^T [C] \{\varepsilon\} dz \right) dA$$

$$= \frac{1}{2} \int_A \left(\{\bar{\varepsilon}\}^T [R] \{\bar{\varepsilon}\} \right) dA$$

(2.17)

$$U_k = \frac{1}{2} \int_A \left(\int_{-h/2}^{+h/2} \rho(x,y) \left(\dot{u}^2 + \dot{v}^2 + \dot{w}^2 \right) dz \right) dA$$

$$= \frac{1}{2} \int_A \left(\int_{-h/2}^{+h/2} \{\dot{\lambda}_0\}^T [f]^T \rho [f] \{\dot{\lambda}_0\} dz \right) dA$$

(2.18)

$$= \frac{1}{2} \int_A \{\dot{\lambda}_0\}^T [m] \{\dot{\lambda}_0\} dA$$

where

$$[R] = \begin{bmatrix} R_1 & R_2 & R_3 & R_4 \\ R_2 & R_3 & R_4 & R_5 \\ R_3 & R_4 & R_5 & R_6 \\ R_4 & R_5 & R_6 & R_7 \end{bmatrix}$$

where

$$\left(R_1, R_2, R_3, R_4, R_5, R_6, R_7 \right)_{ij} = \int_{-h/2}^{h/2} C(x,y) \left(1, z, z^2, z^3, z^4, z^5, z^6 \right) dz \quad \{i,j = 1,2,6,5,4\}$$

$$[m] = \begin{bmatrix} I_1 & 0 & 0 & I_2 & 0 & I_3 & 0 & I_4 & 0 \\ 0 & I_1 & 0 & 0 & I_2 & 0 & I_3 & 0 & I_4 \\ 0 & 0 & I_1 & 0 & 0 & 0 & 0 & 0 & 0 \\ I_2 & 0 & 0 & I_3 & 0 & I_4 & 0 & I_5 & 0 \\ 0 & I_2 & 0 & 0 & I_3 & 0 & I_4 & 0 & I_5 \\ I_3 & 0 & 0 & I_4 & 0 & I_5 & 0 & I_6 & 0 \\ 0 & I_3 & 0 & 0 & I_4 & 0 & I_5 & 0 & I_6 \\ I_4 & 0 & 0 & I_5 & 0 & I_6 & 0 & I_7 & 0 \\ 0 & I_4 & 0 & 0 & I_5 & 0 & I_6 & 0 & I_7 \end{bmatrix}$$

where

$$\left(I_1, I_2, I_3, I_4, I_5, I_6, I_7\right) = \int\limits_{-h/2}^{h/2} \rho(x,y)\left(1, z, z^2, z^3, z^4, z^5, z^6\right) dz$$

Further, 2D isoparmatric finite element approach is adopted to obtain the system of algebraic equations from Equation (2.16). For this purpose, nine-node quadrilateral Lagrangian elements are utilized. For any element, the displacement vector can now expressed in nodal ($i = 1, 2, 3, \ldots$ NNE) form using shape function N as

$$\begin{Bmatrix} u_0 \\ v_0 \\ w_0 \\ \theta_x \\ \theta_y \\ u_0^* \\ v_0^* \\ \theta_x^* \\ \theta_y^* \end{Bmatrix} = \sum_{i=1}^{NNE} N_i \begin{Bmatrix} u_{0_i} \\ v_{0_i} \\ w_{0_i} \\ \theta_{x_i} \\ \theta_{y_i} \\ u_{0_i}^* \\ v_{0_i}^* \\ \theta_{x_i}^* \\ \theta_{y_i}^* \end{Bmatrix} \tag{2.19}$$

where $\{\lambda_{0_i}\} = \begin{bmatrix} u_{0_i} & v_{0_i} & w_{0_i} & \theta_{x_i} & \theta_{y_i} & u_{0_i}^* & v_{0_i}^* & \theta_{x_i}^* & \theta_{y_i}^* \end{bmatrix}^T$ is the nodal displacement vector at node i.

Now, by imposing Equation (2.19) into Equation (2.16), the finite element equations are obtained as

$$\left[K\right]\lambda + \left[M\right]\ddot{\lambda} = 0 \tag{2.20}$$

where $\left[M\right]^e = \int_{-1}^{1}\int_{-1}^{1} \left[N\right]^T \left[m\right]\left[N\right]|J| d\xi d\eta$ is the elemental mass matrix and $\left[K\right]^e = \int_{-1}^{1}\int_{-1}^{1}\left[L\right]^T\left[R\right]\left[L\right]|J| d\xi d\eta$ is the elemental stiffness matrix. Here [L] comprises differential operators and shape functions.

By substituting $\lambda(x, y, z, t) = \bar{\lambda}(x, y, z, t) e^{i\omega t}$ in Equation (2.20), eigenequation of the present B-FGC model is given, as

$$\left(\left[K\right] - \omega^2\left[M\right]\right)\{\bar{\lambda}\} = 0 \tag{2.21}$$

where ω is the natural frequency and $\bar{\lambda}$ is the mode shape.

2.3.4 BOUNDARY CONDITIONS

The following essential boundary conditions are considered to restrict the rigid body motion of B-FGC panel:

Fully clamped (CCCC):

$$u_0 = v_0 = w_0 = \theta_x = \theta_y = u_0^* = v_0^* = \theta_x^* = \theta_y^* = 0 \text{ at } x = 0, a \text{ and } y = 0, b$$

Fully simply supported (SSSS):

$$v_0 = w_0 = \theta_y = v_0^* = \theta_y^* = 0 \text{ at } x = 0, a$$

$$u_0 = w_0 = \theta_x = u_0^* = \theta_x^* = 0 \text{ at } y = 0, b$$

Clamped/Simply supported (SCSC):

$$v_0 = w_0 = \theta_y = v_0^* = \theta_y^* = 0 \text{ at } x = 0, a$$

$$u_0 = v_0 = w_0 = \theta_x = \theta_y = u_0^* = v_0^* = \theta_x^* = \theta_y^* = 0 \text{ at } y = 0, b$$

Cantilever (CFFF):

$$u_0 = v_0 = w_0 = \theta_x = \theta_y = u_0^* = v_0^* = \theta_x^* = \theta_y^* = 0 \text{ at } y = 0$$

2.4 CONVERGENCE AND VALIDATION STUDY

The free vibration behavior of B-FGC panels is obtained using the proposed mathematical model. MATLAB environment is used to develop the computer program. In order to ensure the accuracy and efficacy of present computational model, convergence and validation study are carried out with two different examples, which are discussed in detail below.

Example 2.1

In this example free vibration behavior of bidirectional FG square panel is studied with $n_x = 1$, $n_y = 1$ and thickness ratio $a/h = 5$. The panel is considered to be constituted with Al (Aluminium) and Al_2O_3 (Alumina); the material properties are depicted in Table 2.1. The nondimensional frequency parameters are obtained by $\bar{\omega} = (\omega a^2/h)\left(\sqrt{\rho_c/E_c}\right)$. To the best knowledge of author's, similar kind of study is not reported, so the obtained nondimensional frequency parameters are compared with 3D-FEM solutions of commercially available tool (ANSYS APDL), which are reported in Table 2.2. Both results show good agreement with each other. From this convergence study it is concluded that the mesh size (5 × 5) is adequate to analyze free vibration behavior of B-FGC panels.

TABLE 2.2

Convergence and Comparison of Linear Frequency $\bar{\omega} = (\omega a^2/h)(\sqrt{\rho_c/E_c})$ for Clamped Bi-Directional FG (Al/Al_2O_3) Square Panel for $n_x = 1$, $n_y = 1$ and $a/h = 5$

| Mesh | VRM/MT | \multicolumn{5}{c}{Frequency Modes} |
		1	2	3	4	5
2×2	VRM	6.2483	11.1505	11.1872	12.5513	13.0159
	M-T	4.9397	8.7179	8.9252	9.8405	10.2649
3×3	VRM	5.6074	10.0792	10.1960	12.1573	12.7517
	M-T	4.4404	7.9546	8.1756	9.6683	10.1524
4×4	VRM	5.4519	9.5564	9.7154	12.0135	12.6644
	M-T	4.339	7.5519	7.7992	9.6155	10.1268
5×5	VRM	5.3935	9.3663	9.5523	11.9484	12.6298
	M-T	4.3056	7.4229	7.6849	9.5939	10.1191
6×6	VRM	5.3662	9.2803	9.4829	11.9135	12.6135
	M-T	4.2915	7.3691	7.6393	9.583	10.116
APDL (3D)	VRM	5.6841	9.7719	9.9566	12.8871	13.5060
	M-T	5.1739	8.6796	9.2551	11.4668	12.2843

Example 2.2

To prove the accuracy of present finite element formulation, the results are compared with available results of Talha and Singh [26] as shown in Table 2.3. In this example the free vibration analysis of clamped square FG panel with power law indices ($n_z = 0, 0.5, 1, 5, 10, \infty$) and thickness ratio $a/h = 10$ in conjunction with Voigt model and Mori–Tanaka scheme of material distribution is carried out, the nondimensional frequency parameters using equation $\bar{\omega} = (\omega a^2/h)(\sqrt{\rho_c/E_c})$ are presented. The panel is considered to be comprised of stainless steel (SUS304) and silicon nitride (Si_3N_4) material, the properties of the same is reported in Table 2.1. It is observed that the percentage difference between computed results and the results of Talha and Singh [26] is within 4%.

TABLE 2.3

Nondimensional Frequency Parameter for Clamped Square FG ($SUS304/Si_3N_4$) Panel

| n | Mode | \multicolumn{2}{c}{APDL (3D)} | \multicolumn{2}{c}{Present (HSDT)} | Talha and Singh [26] |
		VRM	M-T	VRM	M-T	
Ceramic	1	9.9526	9.8518	10.0232	10.013	10.1599
	2	19.7359	18.7677	19.4779	19.4608	19.9367
	3	19.7359	18.8970	19.4779	19.4608	19.9367
	4	27.6808	26.4405	27.2849	27.2627	28.1367
	5	34.5443	31.1330	33.3335	33.3098	34.6017

(Continued)

TABLE 2.3 (Continued)
Nondimensional Frequency Parameter for Clamped Square FG
(SUS304/Si$_3$N$_4$) Panel

n	Mode	APDL (3D) VRM	APDL (3D) M-T	Present (HSDT) VRM	Present (HSDT) M-T	Talha and Singh [26]
0.5	1	6.9691	6.9896	6.9127	6.8565	7.0202
	2	13.8374	13.2210	13.4397	13.3292	13.7978
	3	13.8374	13.3514	13.4397	13.3292	13.7978
	4	19.4207	18.9007	18.8249	18.6692	19.4845
	5	24.2593	21.4310	23.0112	22.8201	23.9945
1	1	6.1135	5.9421	6.0533	6.0108	6.1489
	2	12.1267	11.1735	11.7633	11.677	12.0812
	3	12.1267	11.5135	11.7633	11.677	12.0812
	4	17.0111	16.2740	16.4743	16.351	17.0625
	5	21.2344	18.0553	20.1308	19.9762	20.9992
5	1	4.9671	4.5445	4.9114	4.8862	4.9816
	2	9.8006	8.5834	9.5137	9.4671	9.7440
	3	9.8006	8.9186	9.5137	9.4671	9.7440
	4	13.7103	12.4742	13.3081	13.2446	13.0873
	5	17.0504	14.1607	16.2234	16.1479	16.0556
10	1	4.7387	4.3964	4.681	4.6599	4.7457
	2	9.3534	8.3188	9.0701	9.0327	9.2841
	3	9.3534	8.5193	9.0701	9.0327	9.2841
	4	13.0874	11.8993	12.6901	12.6401	13.0873
	5	16.2788	13.7760	15.4713	15.4142	16.0556
Metal	1	4.4486	4.3283	4.3811	4.3314	4.4410
	2	8.8097	8.2202	8.5099	8 4183	8.7107
	3	8.8097	8.2973	8.5099	8.4183	8.7107
	4	12.3476	11.5900	11.9187	11.7931	12.2919
	5	15.3941	13.6203	14.5554	14.409	15.1084

2.5   RESULTS AND DISCUSSION

In this section, numerical solution for free vibration behavior of B-FGC panel using higher-order kinematics are presented and discussed in detail. Some new illustrations have been analyzed to explore the effectiveness of the developed model. The influence of thickness ratio, aspect ratio, and boundary conditions on nondimensional frequency of B-FGC panel has been presented. The panel is assumed to be comprised with steel (*SUS304*) and silicon nitride (*Si$_3$N$_4$*) throughout analysis otherwise stated.

Table 2.4 shows the variation of nondimensional frequency parameter $\bar{\omega} = \left(\omega a^2/h\right)\left(\sqrt{\rho_c/E_c}\right)$ of fully clamped (CCCC) 2D-FG (SUS304/Si$_3$N$_4$) rectangular panels having aspect ratio equal to 1.5 and different volume fraction indices (n_x and n_y) with variable thickness ratio (a/h) ranging from 10 to 100 and different frequency modes. It is observed that the nondimensional frequency parameter increases

TABLE 2.4

The Nondimensional Frequency Parameters of Clamped (CCCC) B-FGC Panels at Different Thickness Ratios (a/h)

(n_x, n_y)	a/h	VRM/M-T	Frequency Modes									
			1	2	3	4	5	6	7	8	9	10
(0.5, 0.5)	10	VRM	9.3481	14.0038	19.9848	21.4643	23.9527	25.1522	29.7146	29.8606	31.3235	33.2120
		M-T	9.2407	13.8438	19.7581	21.2181	23.689	24.8838	29.393	29.5232	30.9881	32.867
	20	VRM	10.7038	16.8564	25.4472	28.2044	30.98	40.4936	46.8704	47.2522	50.3044	53.1886
		M-T	10.5773	16.6568	25.1452	27.8662	30.6236	40.0263	46.3343	46.7423	49.7676	52.5798
	50	VRM	11.5671	18.9476	30.0453	34.5454	36.9929	51.0144	62.4738	71.1515	83.5571	90.1992
		M-T	11.4281	18.7162	29.6746	34.1059	36.5486	50.3914	61.6967	70.2812	82.554	89.3243
	100	VRM	11.8218	19.5375	31.6763	36.4082	39.0666	54.6557	68.7722	78.5037	93.2531	166.4704
		M-T	11.6788	19.2963	31.2789	35.9359	38.5879	53.9701	67.8844	77.5075	92.0829	164.9607
(0.5, 1)	10	VRM	8.5147	12.6918	18.4628	19.4126	21.8768	22.9054	27.2619	27.4351	28.5469	30.3462
		M-T	8.4329	12.567	18.2898	19.2217	21.6727	22.7035	26.9965	27.1963	28.296	30.0947
	20	VRM	9.738	15.256	23.4383	25.4389	28.2232	36.9017	43.2088	43.8093	45.8109	48.6068
		M-T	9.6404	15.0982	23.2032	25.1729	27.9413	36.5242	42.8296	43.3824	45.407	48.1238
	50	VRM	10.5143	17.1272	27.5961	31.0573	33.6044	46.2854	57.9496	64.6018	75.6653	83.4152
		M-T	10.4061	16.9424	27.3025	30.7088	33.2455	45.7718	57.3122	63.8684	74.8051	82.7864
	100	VRM	10.7436	17.6546	29.0681	32.7121	35.4512	49.5247	63.6032	71.0589	84.1654	154.0604
		M-T	10.6318	17.4613	28.7513	32.3357	35.0617	48.9557	62.8646	70.2009	83.1404	153.0212

(1, 0.5)	10	VRM	8.4675	12.8309	18.0229	19.7694	21.9568	22.9523	26.8475	27.4214	28.9391	29.9199
		M-T	8.3825	12.7042	17.8549	19.5754	21.742	22.7518	26.6137	27.1541	28.6964	29.6723
	20	VRM	9.6897	15.4178	22.9201	25.9066	28.3576	37.1402	42.1224	43.3762	45.9046	48.5205
		M-T	9.5884	15.2571	22.6914	25.6307	28.0616	36.7542	41.7464	43.0444	45.5037	48.0161
	50	VRM	10.4663	17.3025	27.0241	31.6028	33.7979	46.6111	55.8353	64.7006	76.6480	82.7565
		M-T	10.354	17.1136	26.7388	31.2297	33.4223	46.0698	55.282	63.9514	75.7482	82.2827
	100	VRM	10.695	17.8319	28.476	33.2575	35.6635	49.8461	61.3656	71.3000	85.3220	153.7585
		M-T	10.5792	17.6342	28.1589	32.8531	35.2567	49.2445	60.7266	70.4351	84.2564	152.9976
(1, 1)	10	VRM	7.9787	11.9855	17.2020	18.4082	20.6214	21.6102	25.6586	25.6591	27.2483	28.6018
		M-T	7.9123	11.886	17.069	18.2565	20.4547	21.4525	25.4515	25.4717	27.0564	28.3994
	20	VRM	9.1188	14.3852	21.8172	24.0659	26.5660	34.6967	40.5258	41.1628	43.2204	45.6447
		M-T	9.0393	14.2583	21.6351	23.8508	26.3339	34.39	40.2232	40.8933	42.905	45.2445
	50	VRM	9.8407	16.1280	25.6613	29.2882	31.5813	43.3932	53.5079	60.4752	70.9916	79.5226
		M-T	9.7523	15.9786	25.4335	29.0008	31.2843	42.9631	53.0563	59.866	70.2543	79.1398
	100	VRM	10.0535	16.6174	27.0202	30.8151	33.2936	46.3696	58.6821	66.4592	78.8193	147.8806
		M-T	9.9623	16.4609	26.7745	30.5037	32.9711	45.8911	58.1592	65.7488	77.94	147.2818

TABLE 2.5
The Nondimensional Frequency Parameters of Clamped B-FGC panels at Different Aspect Ratios

							Frequency Modes					
(n_x, n_y)	a/b	VRM/M-T	1	2	3	4	5	6	7	8	9	10
(0.5, 0.5)	1	VRM	6.9428	15.1568	15.3893	22.7946	31.321	31.3937	38.3314	39.0039	53.1375	87.4835
		M-T	6.8591	14.9619	15.2055	22.5151	30.9204	30.9955	37.8394	38.5391	52.4846	86.6661
	1.5	VRM	11.5671	18.9476	30.0453	34.5454	36.9929	51.0144	62.4738	71.1515	83.5571	90.1992
		M-T	11.4281	18.7162	29.6746	34.1059	36.5486	50.3914	61.6967	70.2812	82.554	89.3243
	2	VRM	18.3513	24.8965	39.4804	49.3447	56.8422	69.3907	92.6911	100.573	107.7045	112.2936
		M-T	18.1312	24.5978	38.9895	48.7421	56.1591	68.5576	91.768	99.3612	106.5592	110.9338
	2.5	VRM	26.9753	33.1137	46.4556	72.4174	81.464	93.0704	95.8651	121.4791	126.1683	144.047
		M-T	26.6524	32.7215	45.8884	71.5451	80.4866	91.9667	94.903	120.1479	124.7659	142.3747
(0.5, 1)	1	VRM	6.2956	13.713	14.005E	20.753	28.1019	28.8209	34.7791	35.5508	48.3557	80.7366
		M-T	6.2289	13.5583	13.8816	20.5223	27.7967	28.4863	34.3581	35.1547	47.7843	80.1886
	1.5	VRM	10.5143	17.1272	27.5961	31.0573	33.6044	46.2854	57.9496	64.6018	75.6653	83.4152
		M-T	10.4061	16.9424	27.3025	30.7088	33.2455	45.7718	57.3122	63.8684	74.8051	82.7864
	2	VRM	16.7271	22.5042	35.5448	45.5956	51.5297	62.7924	86.1182	93.9561	101.0565	102.3091
		M-T	16.5576	22.2688	35.1536	45.122	50.985	62.1155	85.4275	92.978	100.0991	101.1896
	2.5	VRM	24.674	29.9021	41.8279	67.2792	73.8608	84.0029	89.3262	109.685	115.0119	135.3362
		M-T	24.4268	29.5968	41.3784	66.5972	73.0893	83.114	88.5884	108.5467	114.0028	133.997

(1, 0.5)	1	VRM	6.2956	13.713	14.(005E	20.753	28.1019	28.8209	34.7791	35.5508	48.3557	80.7366
		M-T	6.2289	13.5583	13.8816	20.5223	27.7967	28.4863	34.3581	35.1547	47.7843	80.1886
	1.5	VRM	10.4663	17.3025	27.0241	31.6028	33.7979	46.6111	55.8353	64.7006	76.6480	82.7565
		M-T	10.354	17.1136	26.7388	31.2297	33.4223	46.0698	55.282	63.9514	75.7482	82.2827
	2	VRM	16.5677	22.7541	36.1285	44.2596	51.9136	63.6971	84.5215	89.9114	97.3366	101.7538
		M-T	16.3919	22.5051	35.7089	43.8173	51.3254	62.9598	84.0252	89.0923	96.6922	100.6101
	2.5	VRM	24.2929	30.2948	42.5645	64.8722	74.2315	85.6612	87.323	110.1316	120.0116	129.0123
		M-T	24.0416	29.9611	42.0741	64.2574	73.3921	84.7143	86.7588	109.3523	118.7028	127.9265
(1, 1)	1	VRM	5.903	12.8434	13.1703	19.4215	26.637	26.7315	32.3745	33.3277	45.0722	77.5953
		M-T	5.8495	12.7213	13.0508	19.2359	26.3861	26.4803	32.0329	33.0006	44.5971	77.2529
	1.5	VRM	9.8407	16.1280	25.6613	29.2882	31.5813	43.3932	53.5079	60.4752	70.9916	79.5226
		M-T	9.7523	15.9786	25.4335	29.0008	31.2843	42.9631	53.0563	59.866	70.2543	79.1398
	2	VRM	15.6321	21.1942	33.5079	42.2979	48.4413	59.0645	81.4468	86.6841	93.9564	95.6563
		M-T	15.4942	21.0001	33.1826	41.944	47.9816	58.4855	81.039	86.0216	93.3962	94.7381
	2.5	VRM	23.02	28.1809	39.4394	62.307	69.4071	79.2218	84.1506	104.9954	108.5275	124.9241
		M-T	22.8221	27.924	39.0606	61.8156	68.7519	78.4524	83.7142	104.1574	107.7465	124.0042

TABLE 2.6
The Nondimensional Frequency Parameters of B-FGC Panels at Different Support Conditions

(n_x, n_y)	Support Conditions	VRM/ M-T	Frequency Modes									
			1	2	3	4	5	6	7	8	9	10
(0.5, 0.5)	SSSS	VRM	5.7681	11.2616	18.0093	21.3424	23.556	33.0021	36.7661	39.5024	45.9679	47.6444
		M-T	5.6973	11.1276	17.8049	21.0955	23.294	32.6275	36.3717	39.0856	45.4812	47.1042
	CCCC	VRM	11.5671	18.9476	30.0453	34.5454	36.9929	51.0144	62.4738	71.1515	83.5571	90.1992
		M-T	11.4281	18.7162	29.6746	34.1059	36.5486	48.6332	50.3914	61.6967	70.2812	71.9599
	SCSC	VRM	10.5265	14.8287	23.9221	28.8623	34.1073	38.6569	42.127	53.8921	55.3074	60.1521
		M-T	10.4022	14.6552	23.6398	28.5172	33.7056	38.2034	41.6458	53.2023	54.6862	59.4478
	CFFF	VRM	2.2256	4.0194	7.4894	10.5245	11.9003	15.2447	15.8301	16.4605	24.6535	25.9527
		M-T	2.2018	3.9745	7.4084	10.3977	11.7655	15.0931	15.648	16.2794	24.3903	25.6537
(0.5, 1)	SSSS	VRM	5.2527	10.2258	16.6474	19.3472	21.5814	30.2603	33.555	36.9653	42.1827	43.8868
		M-T	5.196	10.1194	16.4874	19.1539	21.372	29.959	33.2574	36.6435	41.8045	43.4468
	CCCC	VRM	10.5143	17.1272	27.5961	31.0573	33.6044	46.2854	57.9496	64.6018	75.6653	83.4152
		M-T	10.4061	16.9424	27.3025	30.7088	33.2455	45.7718	48.6332	57.3122	63.8684	71.9599
	SCSC	VRM	9.6208	13.4213	21.6641	26.7777	30.9901	35.2419	38.3839	47.8897	50.7376	56.5074
		M-T	9.5245	13.2835	21.4361	26.5031	30.6666	34.8813	37.9898	47.3481	50.2313	55.9277
	CFFF	VRM	2.0439	3.6558	6.9133	9.5078	10.7744	13.8136	14.5287	15.2404	22.2811	23.6831
		M-T	2.0253	3.622	6.8469	9.4007	10.6797	13.703	14.3806	15.0842	22.1182	23.4507

(1, 0.5)	SSSS	VRM	5.2385	10.3523	16.2742	19.6952	21.6623	30.3582	33.9912	35.5912	42.1533	43.5572
		M-T	5.1825	10.2451	16.1212	19.5005	21.4488	30.0607	33.7137	35.307	41.7581	43.1604
	CCCC	VRM	10.4663	17.3025	27.0241	31.6028	33.7979	46.6111	55.8353	64.7006	76.6480	82.7565
		M-T	10.354	17.1136	26.7388	31.2297	33.4223	46.0698	48.6332	55.282	63.9514	71.9599
	SCSC	VRM	9.5205	13.634	22.0476	25.9446	31.2651	35.5071	38.8652	49.7954	51.4145	53.941
		M-T	9.4238	13.4924	21.8242	25.7015	30.9319	35.1854	38.4751	49.2571	50.8757	53.4932
	CFFF	VRM	2.0505	3.7115	6.8868	9.7582	11.1276	14.1282	14.4685	15.1254	23.2025	24.0675
		M-T	2.0331	3.6765	6.8199	9.6477	11.017	14.0399	14.315	14.9725	22.9823	23.8153
(1, 1)	SSSS	VRM	4.9264	9.6673	15.5092	18.3466	20.3571	28.4698	32.021	34.3009	39.724	41.1161
		M-T	4.8817	9.5819	15.3873	18.1932	20.1867	28.2329	31.7988	34.0734	39.4165	40.7946
	CCCC	VRM	9.8407	16.1280	25.6613	29.2882	31.5813	43.3932	53.5079	60.4752	70.9916	79.5226
		M-T	9.7523	15.9786	25.4335	29.0008	31.2843	42.9631	48.6332	53.0563	59.866	70.2543
	SCSC	VRM	9.001	12.6891	20.5228	24.8701	29.2239	33.3612	36.309	45.6443	47.9489	52.2162
		M-T	8.9242	12.5771	20.342	24.6707	28.9576	33.0982	35.9904	45.1847	47.5702	51.8396
	CFFF	VRM	1.9213	3.4647	6.4737	9.0232	10.358	13.1048	13.5418	14.2623	21.5848	22.3613
		M-T	1.9079	3.4381	6.4193	8.9311	10.2786	13.0389	13.4185	14.1331	21.4342	22.1784

considerably with increase in thickness ratios, and that the frequency parameter increases with increases in frequency modes. However, the eigenfrequencies computed using the Voigt model and the Mori–Tanaka model are well attuned with each other.

Table 2.5 show the effect of aspect ratio (a/b) on nondimensional frequency parameter of fully clamped (CCCC) B-FGC ($a/h = 50$) panels with different volume fraction indices (n_x and n_y). The nondimensional frequency parameter of B-FGC panel is presented for the first 10 frequency modes. It is noticed that as aspect ratio increases from lower value to higher value, the nondimensional frequency parameter increases remarkably. Also, the frequency parameter of B-FGC panel decreases approximately 15% with increase in volume faction indices from $n_x = n_y = 0.5$ to $n_x = n_y = 1$.

Table 2.6 depicts the nondimensional frequency parameters of B-FGC panels at different support conditions such as fully simply supported (SSSS), fully clamped (CCCC), simply supported and clamped (SCSC), and one-edge clamped (CFFF). The aspect ratio and thickness ratio of the panel are kept at 1.5 and 50, respectively, for all the cases. Here, it can be observed that the frequency parameters of B-FGC panels are diminishing in the order of CCCC, SCSC, SSSS, and CFFF due to the number of degrees-of-freedom.

2.6 CONCLUSIONS

The eigenfrequencies of B-FGC panels are examined using two different homogenization schemes, i.e., Voigt and Mori–Tanaka. The panel is assumed to be graded in longitudinal as well as transverse directions using power-law function. The energy equations are governed using Hamilton's principle and solved subsequently using 2D higher-order finite element method. To ensure the stability and accuracy of the present model, convergence and validation studies are carried out. The obtained results show good agreement with those reported in published literature and with the 3D FEM results obtained via commercially available tool (APDL). Parametric studies have been carried out in order to present the effects of power-law indices, thickness ratios, aspect ratios, and support conditions on the eigenfrequencies of B-FGC panels, and noted the following observations:

- Thin and high aspect ratio B-FGC panels demonstrate higher eigenfrequencies in all the considered cases.
- Fully clamped B-FGC panels demonstrate have maximum frequency values whereas minimum in case of one edge clamed B-FGC panels.
- The effects of power-law indices on the frequency behavior of B-FGC panels are significant.
- The computed eigenfrequencies of B-FGC panels using Voigt model and Mori–Tanaka model are within the expected line and well attuned. In addition, Voigt scheme is simple but overestimates the final results, whereas Mori–Tanaka scheme is adequate for high-performance structures.

REFERENCES

[1] Ahlawat N and Lal R (2016) Buckling and vibrations of multi-directional functionally graded circular plate resting on elastic foundation. *Procedia Engineering* 144: 85–93.

[2] Alipour MM, Shariyat M and Shaban M (2010) A semi-analytical solution for free vibration of variable thickness two-directional-functionally graded plates on elastic foundations. *International Journal of Mechanics and Materials in Design* 6(4): 293–304.

[3] Carrera E and Soave M (2011) Use of functionally graded material layers in a two-layered pressure vessel. *Journal of Pressure vessel Technology* 133(5): 051202–051212.

[4] Tehrani PH and Talebi M (2012) Stress and Temperature Distribution Study in a Functionally Graded Brake Disk. *International Journal of Automobile Engineering* 2: 172–179.

[5] Pompe W, Worch H, Epple M, et al. (2003) Functionally graded materials for biomedical applications. *Materials Science and Engineering: A*, 362(1–2): 40–60.

[6] Lin D, Li Q, Li W, et al. (2009) Design optimization of functionally graded dental implant for bone remodeling. *Composites Part B: Engineering* 40(7): 668–675.

[7] Sola A, Bellucci D and Cannillo V (2016) Functionally graded materials for orthopedic applications–an update on design and manufacturing. *Biotechnology advances* 34(5): 504–531.

[8] Ghatage PS, Kar VR and Sudhagar PE (2020) On the numerical modelling and analysis of multi-directional functionally graded composite structures: A review. *Composite Structures* 236: 111837.

[9] Chin ES (1999) Army focused research team on functionally graded armor composites. *Materials Science and Engineering: A* 259(2): 155–161.

[10] Nie G and Zhong Z (2010) Dynamic analysis of multi-directional functionally graded annular plates. *Applied Mathematical Modelling* 34(3): 608–616.

[11] Shariyat M and Alipour MM (2013) A power series solution for vibration and complex modal stress analyses of variable thickness viscoelastic two-directional FGM circular plates on elastic foundations. *Applied Mathematical Modelling* 37(5): 3063–3076.

[12] Lal R and Ahlawat N (2017) Buckling and vibrations of two-directional functionally graded circular plates subjected to hydrostatic in-plane force. *Journal of Vibration and Control* 23(13): 2111–2127.

[13] Nejati M, Mohsenimonfared H and Asanjarani A (2015) Free vibration analysis of 2D functionally graded annular plate considering the effect of material composition via 2D differential quadrature method. *Mechanics of Advanced Composite Structures* 2(2): 95–111.

[14] Yas MH and Moloudi N (2015) Three-dimensional free vibration analysis of multi-directional functionally graded piezoelectric annular plates on elastic foundations via state space based differential quadrature method. *Applied Mathematics and Mechanics* 36(4): 439–464.

[15] Koizumi M (1997) FGM activities in Japan. *Composites Part B: Engineering*, 28(1–2): 1–4.

[16] Van Do T, Nguyen DK, Duc ND, et al. (2017) Analysis of bi-directional functionally graded plates by FEM and a new third-order shear deformation plate theory. *Thin-Walled Structures* 119: 687–699.

[17] Mahinzare M, Barooti MM and Ghadiri M (2018) Vibrational investigation of the spinning bi-dimensional functionally graded (2-FGM) micro plate subjected to thermal load in thermal environment. *Microsystem Technologies* 24(3): 1695–1711.

[18] Shojaeefard MH, Googarchin HS, Mahinzare M, et al. (2018) Free vibration and critical angular velocity of a rotating variable thickness two-directional FG circular microplate. *Microsystem Technologies* 24(3): 1525–1543.

[19] Lieu QX, Lee S, Kang J, et al. (2018) Bending and free vibration analyses of in-plane bi-directional functionally graded plates with variable thickness using isogeometric analysis. *Composite Structures* 192: 434–451.

[20] Wang X, Yuan Z and Jin C (2019) 3D free vibration analysis of multi-directional FGM parallelepipeds using the quadrature element method. *Applied Mathematical Modelling* 68: 383–404.

[21] Asgari M and Akhlaghi M (2011) Natural frequency analysis of 2D-FGM thick hollow cylinder based on three-dimensional elasticity equations. *European Journal of Mechanics-A/Solids* 30(2): 72–81.

[22] Ebrahimi MJ and Najafizadeh MM (2014) Free vibration analysis of two-dimensional functionally graded cylindrical shells. *Applied Mathematical Modelling*, 38(1): 308–324.

[23] Zafarmand H, Salehi M and Asemi K (2015) Three dimensional free vibration and transient analysis of two directional functionally graded thick cylindrical panels under impact loading. *Latin American Journal of Solids and Structures* 12(2): 205–225.

[24] Gibson LJ, Ashby MF, Karam GN, et al. (1995) The mechanical properties of natural materials. II. Microstructures for mechanical efficiency. *Proceedings of the Royal Society of London. Series A: Mathematical and Physical Sciences* 450(1938): 141–162.

[25] Pandya BN and Kant T (1988) Finite element analysis of laminated composite plates using a higher-order displacement model. *Composites Science and Technology* 32(2): 137–155.

[26] Talha M and Singh BN (2010) Static response and free vibration analysis of FGM plates using higher order shear deformation theory. *Applied Mathematical Modelling* 34(12): 3991–4011.

3 Analytical Solution for the Steady-State Heat Transfer Analysis of Porous Nonhomogeneous Material Structures

Samarjeet Kumar and Vishesh Ranjan Kar

National Institute of Technology Jamshedpur,
Jamshedpur, India

Bakhtiyar Alimovich Khudayarov

Tashkent Institute of Irrigation and Agricultural
Mechanization Engineers,
Tashkent, Uzbekistan

CONTENTS

DOI: 10.1201/9781003158813-3

3.1   INTRODUCTION

The heat transfer phenomena in nature can be categorized as linear (with some assumptions) and nonlinear. These phenomena can be modeled or expressed in the form of partial differential equations. Numerous methods exist for analyzing and expressing these problems. These methods are categorized as either analytical or numerical types. Some of the classical analytical methods used for solving these partial differential equations are the variable separation and homogenization method, integrating factor method, homotopy analysis method (HAM), and adomian decomposition method (ADM), among others [1–5]. The numerical methods include finite element method (FEM), finite difference method (FDM), boundary element method (BEM), meshless method, Heun's method, Euler method, and Runge–Kutta method, among others [6–10]. The main problems associated with the analytical methods are that they can be solved for simple boundary and geometrically conditions. The introduction of complexity in the terms of material, geometry, and nonlinearity makes partial differential equations hard to solve. The solving process contains certain differential terms or integrations factor or some transform function that are not available or known. These terms or their values are fairly approximated to proceed further in numerical methods [11]. Sometimes solutions are referred to as semi-analytical, which means certain steps of partial differential equations are solved via the analytical or algebraic procedure and after that, numerical methods are used to approximate. The differential transform method (DTM) can be introduced as an analytical as well as a numerical method for solving integral equations and ordinary/partial differential equations. This method provides a good sense of continuation and gaining momentum among researchers due to its simplicity and pedagogical benefits [12]. This method provides analytical solutions in the form of a polynomial. It is different from the traditional high-order Taylor series, which is based on the symbolic computation of derivatives of functions. It can be regarded as an iterative procedure for obtaining the series solution of partial differential equations [3.13]. The obtained solution in the form of convergent series has easily computable components. The advantage associated with this method is less computational work along with a high convergence rate providing the accurate series solution.

3.1.1   Differential Transform Method

The differential transform method is an analytical Taylor series-based solution applied for both linear and nonlinear types of problems [14–17]. This method was introduced independently by Zhou and Pukhov in 1986 to solve linear and nonlinear differential equations involved in problems of electrical circuits [18]. The advantage associated with differential transform is that it does not require linearization and discretization of differential equations [3.19]. The error associated with discretization is also nil in this case. This method is widely used by researchers to get a more accurate solution to the differential equation. This analytical method works well with small and large quantities (as in the case of perturbation techniques) along with the complex type of initial and boundary conditions involved in governing equations. There is no need to guess initial values (as in the case of HAM), resulting in a direct solution of the partial differential equation.

For understanding the concept of DTM, Taylor series expansion of function $T(x)$ in the domain D:

$$T(x) = \sum_{r=0}^{\infty} \frac{(x-x_i)^r}{r!} \left[\frac{d^r T(x)}{dx^r} \right]_{x=x_i} \quad \forall x \in D \tag{3.1}$$

where x_i is any point in the domain and $T(x)$ is the original function expressed in form of infinite series whose center is located at x_i.

The Maclaurin series is obtained by setting the value of random point x_i as zero and can be defined as follows [18]:

$$T(x) = \sum_{r=0}^{\infty} \frac{(x)^r}{r!} \left[\frac{d^r T(x)}{dx^r} \right]_{x=0} \quad \forall x \in D \tag{3.2}$$

The differential transform of function $T(x)$ is expressed as follows [18]:

$$U(r) = \sum_{r=0}^{\infty} \frac{H^r}{r!} \left[\frac{d^r T(x)}{dx^r} \right]_{x=0} \quad \forall x \in D \tag{3.3}$$

where $U(r)$ is the transformed function and $T(x)$ is the original function. The differential spectrum of $U(r)$ lies between 0 to H. The right choice of the value of H (constant) reduces the number of discrete spectrums. The discrete spectrum is the values obtained by putting the value of discrete r like X (0), X (1), etc. The original function or solution can be obtained by putting more value of discrete in the inverse differential transform. The inverse differential transform of $U(r)$ can be expressed as

$$T(x) = \sum_{r=0}^{\infty} \left(\frac{x}{H} \right)^r U(r) \tag{3.4}$$

But in real solution, finite series is taken into consideration and the rest of the terms are truncated, and then Equation (3.4) can be written as

$$T(x) = \sum_{r=0}^{N} \left(\frac{x}{H} \right)^r U(r) \tag{3.5}$$

TABLE 3.1

The Fundamental Operation of Differential Transform [18, 19]

Original Function	Transformed Function
$T(x) = o(x) + p(x)$	$U(r) = O(r) + P(r)$
$T(x) = o(x) - p(x)$	$U(r) = O(r) - P(r)$
$T(x) = \lambda p(x)$	$U(r) = \lambda P(r)$
$T(x) = o(x)p(x)$	$U(r) = \sum_{c=0}^{r} O(c)P(r-c)$
$T(x) = o(x)p(x)q(x)$	$U(r) = \sum_{r}^{h} \sum_{c=0}^{r} O(c)P(r-c)Q(h-r)$
$T(x) = e^{nx}$	$U(r) = \dfrac{n^r}{r!}$
$T(x) = x^n$	$U(r) = \delta(r-n) = \begin{cases} 1 & ; r = n \\ 0 & ; \text{otherwise} \end{cases}$
$T(x) = \cos(\omega x + \phi)$	$U(r) = \dfrac{\omega^r}{r!} \cos\left(\dfrac{r\pi}{2} + \phi\right)$
$T(x) = \sin(\omega x + \phi)$	$U(r) = \dfrac{\omega^r}{r!} \sin\left(\dfrac{r\pi}{2} + \phi\right)$
$T(x) = \dfrac{dp(x)}{dx}$	$U(r) = (r+1)P(r+1)$
$T(x) = \dfrac{d^2 p(x)}{dx^2}$	$U(r) = (r+1)(r+2)P(r+2)$
$T(x) = \dfrac{d^m p(x)}{dx^m}$	$U(r) = (r+1)(r+2)......(r+m)P(r+m)$

The value of N is decided by the convergence of the series coefficient. This implies that $\sum_{r=N}^{\infty} \left(\dfrac{x}{H}\right)^r U(r)$ terms are negligibly small and termed as error associated with differential transformation. This method of calculating the solution of the partial differential equation is termed DTM(k).

Equation (3.5) together with value $H = 1$, the standard/ fundamental differential transforms of one-dimensional function is listed in Table 3.1.

The different steps involved in DTM are presented in Figure 3.1.

3.2   MICROMECHANICAL PROPERTY OF EVEN AND UNEVEN POROUS FGM

The FGM is assumed to be made of a mixture of two constituent materials so that the bottom surface is ceramic-rich, whereas the top surface is metal-rich. If the property of FGM is varying along a single direction, then it is called unidirectional FGM. The variation of property in the x-direction is referred to as X FGM. The variation of property in the y-direction results in Y FGM. Similarly, for Z

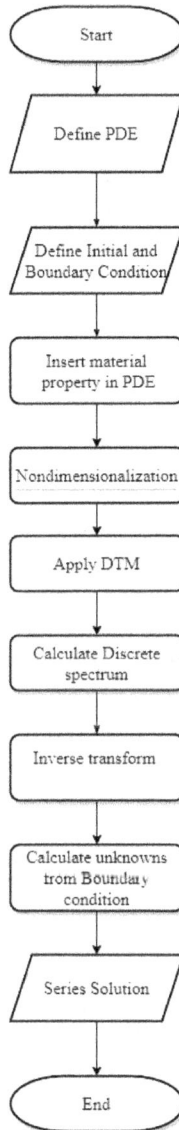

FIGURE 3.1   Flow chart depicting process of DTM.

FGM, variation is found in the z-direction. In the present analysis, the gradation is achieved through power-law scheme of X FGM. The volume fraction of ceramic V_c is defined as

$$V_c(x) = \left(\frac{x}{l}\right)^n \tag{3.6}$$

The volume fraction V_m of metal can be expressed as

$$V_m(x) = 1 - V_c(x) = 1 - \left(\frac{x}{l}\right)^n \tag{3.7}$$

The micromechanical property is based on extended Voigt's law or rule of mixture. The thermal properties such as density, specific heat, and thermal conductivity are functions of coordinate x for nonporous/perfect FGM material and can be expressed as [20, 21]

$$P(x) = P_c V_c(x) + P_m V_m(x) = (P_c - P_m)\left(\frac{x}{l}\right)^n + P_m \tag{3.8}$$

$$= \varepsilon_1 \left(\frac{x}{l}\right)^n + \delta = \frac{\varepsilon_1}{l^n} x^n + \delta = \kappa x^n + \delta \tag{3.9}$$

where

$$\varepsilon_1 = (P_c - P_m)$$

where P_c and P_m are the material properties of ceramic and metal, respectively. n denotes the power-law exponent/coefficient.

The thermal properties of Perfect FGM according to Equation (3.8) can be written as

$$\left.\begin{array}{l}
K(x) = K_c V_c(x) + K_m V_m(x) = (K_c - K_m)\left(\frac{x}{l}\right)^n + K_m = K_1 x^n + K_m \\[3mm]
C(x) = C_c V_c(x) + C_m V_m(x) = (C_c - C_m)\left(\frac{x}{l}\right)^n + C_m = C_1 x^n + C_m \\[3mm]
\rho(x) = \rho_c V_c(x) + \rho_m V_m(x) = (\rho_c - \rho_m)\left(\frac{x}{l}\right)^n + \rho_m = \rho_1 x^n + \rho_m
\end{array}\right\} \tag{3.10}$$

Here, K, C, and ρ denote thermal conductivity, specific heat, and density, respectively. Subscripts c and m refer to ceramic and metal, respectively.

But the assumption of perfect FGM is not accurate. There are always microvoids or porosity present in the FGM fabrication. The porosity in FGM can be modeled two ways, i.e., even and uneven porous. In even porous FGM, the thermal conductivity decreases with an increase in porosity index. Porosity index (α) is defined as a volumetric measure of the number of microvoids present within the total volume of porous FGM specimen.

For even porosity [22–24],

$$P(x) = P_c\left\{V_c(x) - \frac{\alpha}{2}\right\} + P_m\left\{V_m(x) - \frac{\alpha}{2}\right\} = (P_c - P_m)\left(\frac{x}{l}\right)^n + P_m - \frac{\alpha}{2}(P_c + P_m)$$

$$\tag{3.11}$$

The thermal properties of even porous FGM according to Equation (3.11) can be written as

$$
\left.
\begin{aligned}
K(x) &= K_c V_c(x) + K_m V_m(x) = (K_c - K_m)\left(\frac{x}{l}\right)^n + K_m - \frac{\alpha}{2}(K_c + K_m) = K_1 x^n + K_2 \\
C(x) &= C_c V_c(x) + C_m V_m(x) = (C_c - C_m)\left(\frac{x}{l}\right)^n + K_m - \frac{\alpha}{2}(K_c + K_m) = C_1 x^n + C_2 \\
\rho(x) &= \rho_c V_c(x) + \rho_m V_m(x) = (\rho_c - \rho_m)\left(\frac{x}{l}\right)^n + K_m - \frac{\alpha}{2}(K_c + K_m) = \rho_1 x^n + \rho_2
\end{aligned}
\right\}
$$

$$(3.12)$$

3.3   STEADY-STATE HEAT TRANSFER BEHAVIOR OF FGM PLATE

3.3.1   PHYSICAL DERIVATION OF 1 D HEAT EQUATION FOR FGM PLATE

Since the thermal properties of FGM are varying in only one direction, the heat transfer within the plate is assumed or modeled as a unidirectional phenomenon. In the FGM plate, the heat transfer takes place from the region of higher temperature to the one of lower temperature. Three governing principles decide heat transfer within the plate.

3.3.1.1   Thermal Energy Stored within a Body with Nonhomogeneous Material Properties

Thermal energy stored per unit volume can be written as

$$U - \rho(x)C(x)dT \qquad (3.13)$$

where ρ is the density of the body, C is the specific heat, and dT is the temperature change. Specific heat is the amount of energy stored or liberated when temperature changes by one unit.

3.3.1.2   Fourier Law of Heat Transfer

Fourier principle states that conductive heat transfer within the plate is proportional to the negative temperature gradient. Heat transfer rate can be written as

$$Q = -K(x)A\frac{dT}{dx} \qquad (3.14)$$

where $K(x)$ is the thermal conductivity of the plate, A is the cross-sectional area of heat transfer, and $\frac{dT}{dx}$ represents the temperature gradient in the direction of x.

3.3.1.3   Principle of Energy Conservation

Consider a nonhomogeneous plate of length l and width w with nonuniform temperature whose length is along x-axis from $x = 0$ to $x = l$. Nonhomogeneous means the properties density, specific heat, and thermal conductivity is dependent on position. There is no heat source within the plate, and it is assumed that all sides of the plate are insulated except the ends. One-dimensional heat transfer is taken into consideration to simplify the problem, since FGM considered here has a unidirectional variation of property. The temperature distribution within the plate can be fairly assumed to be varied in the x-direction.

Consider an arbitrary thin segment of the plate consist of length dx that lies between x and $x + dx$. The thin segment is assumed to follow uniform temperature at these defined points. Thermal energy of the segment is

$$E = \rho(x)C(x)Adx dT$$

From the principle of energy conservation,

$$Q_{right} - Q_{left} = \Delta E \big|_{dt}$$

$$\left\{ Q + \frac{dQ}{dx} - Q \right\} dt = \rho(x)C(x)Adx \left\{ T(x,t+\Delta t) - T(x,t) \right\}$$

$$\left\{ Q + \frac{dQ}{dx} - Q \right\} = \rho(x)C(x)Adx \left\{ \frac{T(x,t+\Delta t) - T(x,t)}{dt} \right\}$$

$$\left\{ Q + \frac{dQ}{dx} - Q \right\} = \rho(x)C(x)Adx \left\{ \frac{dT}{dt} \right\}$$

For the steady-state heat transfer process, the temperature within the plate is independent of the time. The temperature profile obtained is also called equilibrium temperature. This solution is independent of time, and the rate of change of temperature is zero.

$$\frac{dQ}{dx} = 0$$

From Equation (3.14) we get

$$\frac{d}{dx}\left(-K(x)A\frac{dT}{dx} \right) = 0$$

$$K(x)\frac{d^2T}{dx^2} + \frac{dK(x)}{dx}\frac{dT}{dx} = 0 \qquad 0 \le x \le l$$

(3.15)

3.3.2   BOUNDARY CONDITIONS

Mainly three types of boundary conditions that need to be applying at the end are required in linear problems [25, 26]. However, in reality, boundary conditions have complex nature. Here, three simple cases are considered.

3.3.2.1 Dirichlet Boundary Condition

If the fixed temperature is applied at the end of the plate/rod, this type of condition is called the Dirichlet boundary condition. Sometimes it is called as an essential boundary condition.

$$T(x)\big|_{boundary} = \eta \tag{3.16}$$

where η is constant temperature applied at the boundary.

3.3.2.2 Neuman Boundary Condition

In this case, heat flow or the gradient across the boundaries are taken into consideration. The value of flux may be zero or nonzero depending on the physical situation. Zero value indicates that the boundary is insulated.

$$\frac{dT(x)}{dx}\bigg|_{boundary} = \gamma \tag{3.17}$$

where γ is constant gradient applied at the boundary.

3.3.2.3 Mixed Boundary Condition

In this category, the partial differential is needed for complex conditions at the end of the boundary domain. It is the third type of boundary condition named *robin condition*, which is a weighted combination of Dirichlet and Neuman boundary conditions. It is a linear combination of values of temperature and the value of temperature gradient on the boundary of the plate. Mathematically it is expressed as

$$f\,T(x)\big|_{boundary} + h\frac{dT(x)}{dx}\bigg|_{boundary} = g \tag{3.18}$$

3.3.3 NONDIMENSIONALIZATION OF PARAMETERS

Physical or dimensional terms are present in the ordinary/partial differential equation and boundary conditions such as k, x, l, T. In nondimensional form, many physical terms (as much as possible) are clubbed together to make the solution simple and meaningful [25]. The choice of the dimensionless variable depends on the experience and wisdom of mathematicians. Sometimes it is evident from the problem statement, e.g., the length of the plate/rod can be taken as one instead of arbitrary length l. The dimensionless variable for the plate can be introduced as

$$\text{Dimensionless space variable, } \bar{x} = \frac{x}{\hat{L}}$$

$$\text{Dimensionless temperature, } \bar{T}(\bar{x}) = \frac{T(x)}{\hat{T}}$$

$$\text{Dimensionless function, } \bar{f}(\bar{x}) = \frac{f(x)}{\hat{f}}$$

where $\hat{L}$ and $\hat{T}$ are the characteristic parameter for length and temperature, respectively.

The derivative of space can also be converted into dimensionless form through chain rule

$$
\left.\begin{aligned}
\theta_x &= \frac{dT}{dx} = \hat{T}\frac{d\bar{T}}{d\bar{x}}\frac{d\bar{x}}{dx} = \frac{\hat{T}}{\hat{L}}\frac{d\bar{T}}{d\bar{x}} \\
\theta_{xx} &= \frac{d^2T}{dx^2} = \frac{\hat{T}}{\hat{L}^2}\frac{d^2\bar{T}}{d\bar{x}^2}
\end{aligned}\right\}
\tag{3.19}
$$

Equation (3.15) in dimensionless form can be written as

$$
\begin{aligned}
K(\bar{x})\frac{\hat{T}}{\hat{L}^2}\frac{d^2\bar{T}}{d\bar{x}^2} + \frac{dK(\bar{x})}{\hat{L}d\bar{x}}\frac{\hat{T}}{\hat{L}}\frac{d\bar{T}}{d\bar{x}} = 0 \qquad 0 \le \bar{x} \le 1 \\
K(\bar{x})\frac{d^2\bar{T}}{d\bar{x}^2} + \frac{dK(\bar{x})}{d\bar{x}}\frac{d\bar{T}}{d\bar{x}} = 0 \qquad 0 \le \bar{x} \le 1
\end{aligned}
\tag{3.20}
$$

It can be noted that with change in dimensional to nondimensional form, the range of function also changes (depending on the choice of characteristic parameter). Suppose we are taking characteristic length as length of plate (l), then partial differential equation in dimensionless form should be valid when dimensionless space variable lies between 0 to 1. Along with partial differential equation, initial and boundary conditions must be nondimensional. Equations (3.16), (3.17), and (3.18) can be written in dimensionless form as

$$
\bar{T}(\bar{x})\Big|_{\text{boundary}} = \frac{\eta}{\hat{T}} = \alpha
\tag{3.21}
$$

$$
\frac{d\bar{T}(\bar{x})}{d\bar{x}}\Big|_{\text{boundary}} = \gamma\frac{\hat{L}}{\hat{T}} = \beta
\tag{3.22}
$$

$$
\begin{aligned}
f\hat{T}\,\bar{T}(\bar{x})\Big|_{\text{boundary}} + h\frac{\hat{T}}{\hat{L}}\frac{d\bar{T}(\bar{x})}{d\bar{x}}\Big|_{\text{boundary}} = g \\
f_1\,\bar{T}(\bar{x})\Big|_{\text{boundary}} + h_1\frac{d\bar{T}(\bar{x})}{d\bar{x}}\Big|_{\text{boundary}} = g_1
\end{aligned}
\tag{3.23}
$$

3.4   SOLUTION WITH DIFFERENTIAL TRANSFORM METHOD

Here a one-dimensional heat transfer partial differential equation (PDE) through the DTM method is solved and presented. Two different types of material models, i.e., perfect and even porous FGM, are considered, and their material property is incorporated in PDE.

3.4.1   PERFECT POWER-LAW GRADED X FGM

Equation (3.20) can be expressed in the following ways:

$$\left(K_1\hat{L}^n\left(\overline{x}\right)^n + K_m\right)\frac{d^2\overline{T}}{d\overline{x}^2} + \left(K_1 n\hat{L}^n\left(\overline{x}\right)^{n-1}\right)\frac{d\overline{T}}{d\overline{x}} = 0 \qquad 0 \le \overline{x} \le 1$$

$$\left(a\overline{x}^n + b\right)\frac{d^2\overline{T}}{d\overline{x}^2} + \left(a_1\overline{x}^{n-1}\right)\frac{d\overline{T}}{d\overline{x}} = 0 \qquad 0 \le \overline{x} \le 1$$

(3.24)

Here dimensionless terms are

$$a = 1$$

$$b = \frac{K_m}{K_1\hat{L}^n}$$

$$a_1 = n$$

Rearranging the Equation (3.24) in the following form,

$$a\overline{x}^n\frac{d^2\overline{T}}{d\overline{x}^2} + b\frac{d^2\overline{T}}{d\overline{x}^2} + a_1\overline{x}^{n-1}\frac{d\overline{T}}{d\overline{x}} = 0 \qquad 0 \le \overline{x} \le 1$$

$$DT\left(\frac{d^2\overline{T}}{d\overline{x}^2}\right) = (r+1)(r+2)U(r+2)$$

$$DT\left(\frac{d\overline{T}}{d\overline{x}}\right) = (r+1)U(r+1)$$

(3.25)

$$DT\left(x^n\right) = \delta(r-n) = \begin{cases} 1 & ; r = n \\ 0 & ; \text{otherwise} \end{cases}$$

$$DT\left(x^{n-1}\right) = \delta_1(r-n+1) = \begin{cases} 1 & ; r = n \\ 0 & ; \text{otherwise} \end{cases}$$

Now, applying differential transform method to the Equation (3.25),

$$a\sum_{s=0}^{r}\delta(s-n)(r-s+1)(r-s+2)U(r-s+2) + b(r+1)(r+2)U(r+2)$$

$$+ a_1\sum_{s=0}^{r}\delta_1(s-n+1)(r-s+1)U(r-s+1) = 0$$

(3.26)

Rearranging the Equation (3.15), we get discrete spectrum function as

$$U(r+2) = -\frac{a\sum_{s=0}^{r}\delta(s-n)(r-s+1)(r-s+2)U(r-s+2) + a_1\sum_{s=0}^{r}\delta_1(s-n+1)(r-s+1)U(r-s+1)}{b(r+1)(r+2)}$$

(3.27)

The differential transform of boundary condition can be written as

$$U(0) = \sigma$$
$$U(1) = \lambda$$

(3.28)

$U(0)$ and $U(1)$ represent zero and first differential transform, respectively, and the value of σ, λ is unknown but can be determined through boundary conditions (Equation 3.26). Here, the discrete spectrum for power-law index $n = 1$ is calculated and listed in Table 3.2. For other values of the power-law index, the discrete spectrum can be calculated from Equation (3.27).

The inverse transform of the nondimensional temperature function (Equation 3.5) is stated as

$$\bar{T}(x) = U(0) + U(1)x + U(2)x^2 + U(3)x^3 + \ldots\ldots\ldots$$

$$\bar{T}(x)\Big|_{n=1} = \sigma + \lambda x - \frac{a_1|_{n=1}}{2b}\lambda x^2 + \left(\frac{a+a_1|_{n=1}}{3b}\right)\left(\frac{a_1|_{n=1}}{2b}\right)\lambda x^3 + \ldots\ldots\ldots$$

(3.29)

TABLE 3.2

Value of Discrete Spectrum for Perfect X FGM with Power-Law Index $n = 1$

Value of r ($n = 1$)	Discrete Spectrum Value $U(r + 2)$
0	$-\dfrac{a_1|_{n=1}}{2b}U(1)$
1	$-\left(\dfrac{a+a_1|_{n=1}}{3b}\right)U(2)$
2	$-\left(\dfrac{2a+a_1|_{n=1}}{4b}\right)U(3)$
3	$-\left(\dfrac{3a+a_1|_{n=1}}{5b}\right)U(4)$
4	$-\left(\dfrac{4a+a_1|_{n=1}}{6b}\right)U(5)$
5	$-\left(\dfrac{5a+a_1|_{n=1}}{7b}\right)U(6)$

The final series solution for perfect FGM can be obtained if the value of λ is known.

3.4.2 EVEN POROUS POWER-LAW GRADED X FGM

The PDE for even porous X FGM can be expressed as

$$\left(K_1\hat{L}^n\left(\bar{x}\right)^n + K_2\right)\frac{d^2\bar{T}}{d\bar{x}^2} + \left(K_1 n\hat{L}^n\left(\bar{x}\right)^{n-1}\right)\frac{d\bar{T}}{d\bar{x}} = 0 \qquad 0 \le \bar{x} \le 1$$

$$\left(a\bar{x}^n + b_1\right)\frac{d^2\bar{T}}{d\bar{x}^2} + \left(a_1\bar{x}^{n-1}\right)\frac{d\bar{T}}{d\bar{x}} = 0 \qquad 0 \le \bar{x} \le 1$$

(3.30)

Here dimensionless term is

$$b_1 = \frac{K_2}{K_1\hat{L}^n}$$

Rearranging Equation (3.30) in the following form,

$$a\bar{x}^n\frac{d^2\bar{T}}{d\bar{x}^2} + b_1\frac{d^2\bar{T}}{d\bar{x}^2} + a_1\bar{x}^{n-1}\frac{d\bar{T}}{d\bar{x}} = 0 \qquad 0 \le \bar{x} \le 1 \qquad (3.31)$$

Now, applying differential transform method to the Equation (3.31),

$$a\sum_{s=0}^{r}\delta\left(s-n\right)\left(r-s+1\right)\times\left(r-s+2\right)U\left(r-s+2\right)+b_1\left(r+1\right)$$

$$\times\left(r+2\right)U\left(r+2\right)+a_1\sum_{s=0}^{r}\delta_1\left(s-n+1\right)\times\left(r-s+1\right)U\left(r-s+1\right)=0 \quad (3.32)$$

Rearranging Equation (3.32), we get discrete spectrum function by inserting the value of k zero onward,

$$U\left(r+2\right)=$$

$$-\frac{a\sum_{s=0}^{r}\delta\left(s-n\right)\left(r-s+1\right)\left(r-s+2\right)U\left(r-s+2\right)+a_1\sum_{s=0}^{r}\delta_1\left(s-n+1\right)\left(r-s+1\right)U\left(r-s+1\right)}{b_1\left(r+1\right)\left(r+2\right)}$$

(3.33)

TABLE 3.3

Value of Discrete Spectrum for Perfect X FGM with Power-Law Index $n = 1$

Value of r ($n = 1$)	Discrete Spectrum Value $U(r + 2)$
0	$-\dfrac{a_1\big|_{n=1}}{2b_1\big|_{\alpha=\alpha}} U(1)$
1	$-\left(\dfrac{a + a_1\big|_{n=1}}{3b_1\big|_{\alpha=\alpha}}\right) U(2)$
2	$-\left(\dfrac{2a + a_1\big|_{n=1}}{4b_1\big|_{\alpha=\alpha}}\right) U(3)$
3	$-\left(\dfrac{3a + a_1\big|_{n=1}}{5b_1\big|_{\alpha=\alpha}}\right) U(4)$
4	$-\left(\dfrac{4a + a_1\big|_{n=1}}{6b_1\big|_{\alpha=\alpha}}\right) U(5)$
5	$-\left(\dfrac{5a + a_1\big|_{n=1}}{7b_1\big|_{\alpha=\alpha}}\right) U(6)$

The differential transformed of boundary condition can be written as

$$
\begin{aligned}
U(0) &= \sigma \\
U(1) &= \lambda
\end{aligned}
\tag{3.34}
$$

$U(0)$ and $U(1)$ represent the zero and the first differential of transform, respectively, and value of σ, λ is unknown but can be determined through boundary conditions.

The inverse transform of the nondimensional temperature function (Equation 3.5) is stated as

$$
\bar{T}(x) = U(0) + U(1)x + U(2)x^2 + U(3)x^3 + \dots\dots\dots
$$

$$
\bar{T}(x)\big|_{n=1} = \sigma + \lambda x - \frac{a_1\big|_{n=1}}{2b_1\big|_{\alpha=\alpha}}\lambda x^2 + \left(\frac{a + a_1\big|_{n=1}}{3b_1\big|_{\alpha=\alpha}}\right)\left(\frac{a_1\big|_{n=1}}{2b_1\big|_{\alpha=\alpha}}\right)\lambda x^3 + \dots\dots\dots
\tag{3.35}
$$

Putting the value of λ in Equation (3.35), the final series solution for imperfect FGM can be obtained (Table 3.3).

3.5 RESULTS AND DISCUSSION

3.5.1 VALIDATION STUDY

To check the accuracy of the present method, a typical example of heat transfer in an exponentially graded perfect FGM plate is considered. The result obtained in steady-state heat transfer is compared with analytical and numerical methods. The accuracy of the result can be estimated by the average relative error [29], which is defined as

$$
E\big|_{ar} = \sqrt{\frac{\sum_{i=1}^{n}\left(V^{(num)}(X)-V^{(an)}(X)\right)_{i}^{2}}{\sum_{i=1}^{n}\left(V^{(an)}(X)\right)_{i}^{2}}}
\tag{3.36}
$$

where V represents the value of temperature in the desired method.

Here a square perfect exponentially graded plate graded in the y-direction is considered. Here the side of the square plate is taken as $l = 0.04$ m. The square plate is subjected to zero and finite Dirichlet conditions at the bottom and top, respectively, and the rest of the ends are insulated (zero Neumann condition). The thermal conductivity of the material follows the following rules $K(y) = K_0 e^{2\beta y}$. In this section, the gradation parameter β (25 cm^{-1}) is used for calculation. The obtained DTM result is compared with the analytical solution [27, 28] (see Figure 3.2) and the FEM solution [29] (see Table 3.5). There is good agreement among these methods.

The partial differential equation for perfect exponentially Y graded FGM is

$$
K(y)\frac{d^2T}{dy^2} + \frac{dK(y)}{dy}\frac{dT}{dy} = 0 \qquad 0 \le y \le l
\tag{3.37}
$$

Putting the expression of thermal conductivity in Equation (3.37), we get

$$
\frac{d^2T}{dy^2} + 2\beta\frac{dT}{dy} = 0
\tag{3.38}
$$

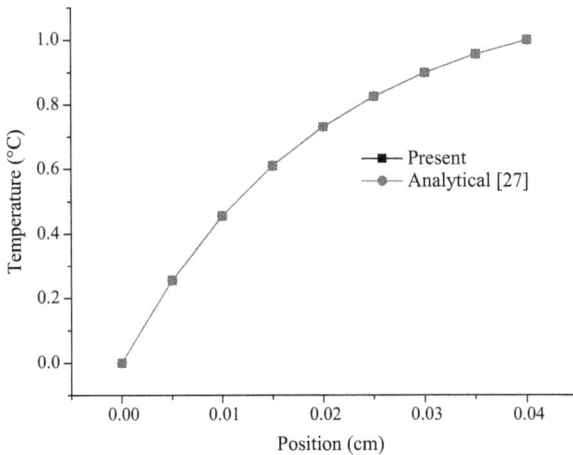

FIGURE 3.2 Comparison of the result obtained from DTM and analytical method.

Boundary condition according to problem statement is

$$T(0) = 0 \text{ and } T(l) = 1$$
$$Q(0) = 0 \text{ and } Q(l) = 0 \tag{3.39}$$

Now, applying differential transform method on the Equation (3.38) and from Table 3.1 of fundamental transformation, we get

$$DT\left(\frac{d^2T}{dy^2}\right) + 2\beta DT\left(\frac{dT}{dy}\right) = 0$$
$$(r+1)(r+2)U(r+2) + 2\beta(r+1)U(r+1) = 0 \tag{3.40}$$

The differential spectrum obtained from Equation (3.40) is

$$U(r+2) = -\frac{2\beta(r+1)U(r+1)}{(r+1)(r+2)} \tag{3.41}$$

The differential transformed of boundary condition can be written as

$$U(0) = 0$$
$$U(1) = \lambda \tag{3.42}$$

$U(1)$ represents the transformation of the first differential, and the value of λ (unknown) can be determined through boundary conditions, i.e., $T(a) = 0$ in the Taylor series solution obtained by inverse transformation.

The inverse transform of the temperature function (see Equation 3.5) can be expressed as

$$T(y) = U(0) + U(1)y + U(2)y^2 + U(3)y^3 + \ldots\ldots\ldots$$

Inserting the value of differential spectrum from Equation (3.42) and Table 3.4, the series solution can be written as

$$T(y) = \lambda \sum_{j=0}^{\infty} \frac{(-1)^j (2\beta)^j}{(j+1)!} y^{j+1} \tag{3.43}$$

The value of λ is unknown but can be determined through boundary conditions, i.e., $T(l) = 1$ in Taylor series solution, and we get

$$\lambda = \frac{1}{l - \frac{(2\beta)}{2!}l^2 + \frac{(2\beta)^2}{3!}l^3 - \frac{(2\beta)^3}{4!}l^4 + \ldots\ldots} = \frac{1}{\sum_{j=0}^{\infty} \frac{(-1)^j (2\beta)^j}{(j+1)!} l^{j+1}} \tag{3.44}$$

TABLE 3.4

Value of Discrete Spectrum for Perfect Y FGM with Gradation Parameter β

Value of r	Discrete Spectrum Value $U(r + 2)$
0	$\dfrac{-(2\beta)}{2!}\lambda$
1	$\dfrac{(2\beta)^2}{3!}\lambda$
2	$\dfrac{-(2\beta)^3}{4!}\lambda$
3	$\dfrac{(2\beta)^4}{5!}\lambda$
4	$\dfrac{-(2\beta)^5}{6!}\lambda$
5	$\dfrac{(2\beta)^6}{7!}\lambda$

TABLE 3.5

Comparison of the Result Obtained from DTM and FEM

Position (cm)	DTM (Present)	HFS- FEM [29]	ANSYS (Plane 77)
0	0	0	0
0.005	0.255821	0.2556	0.2275
0.01	0.455054	0.4553	0.4551
0.015	0.610217	0.6102	0.5931
0.02	0.731059	0.7295	0.7311
0.025	0.82517	0.8252	0.8148
0.03	0.898464	0.8982	0.8984
0.035	0.955545	0.9556	0.94923
0.04	1	1	1
$E|_{ar}$	0	7.92×10^{-4}	1.65×10^{-2}

Putting the value of λ in Equation (3.43), the final series solution for exponentially graded FGM can be written as

$$T(y) = \frac{\displaystyle\sum_{j=0}^{\infty} \frac{(-1)^j (2\beta)^j}{(j+1)!} y^{j+1}}{\displaystyle\sum_{j=0}^{\infty} \frac{(-1)^j (2\beta)^j}{(j+1)!} l^{j+1}} \tag{3.45}$$

3.5.2 NUMERICAL ILLUSTRATION

An X FGM square plate containing dimension $l = 1$ m, subjected to finite Dirichlet and zero Neumann boundary condition is considered (see Figure 3.3). The FGM is made of two different materials: nickel as metal and aluminum oxide as ceramic. The thermal, physical, and mechanical properties of constituents are listed in Table 3.6 [30].

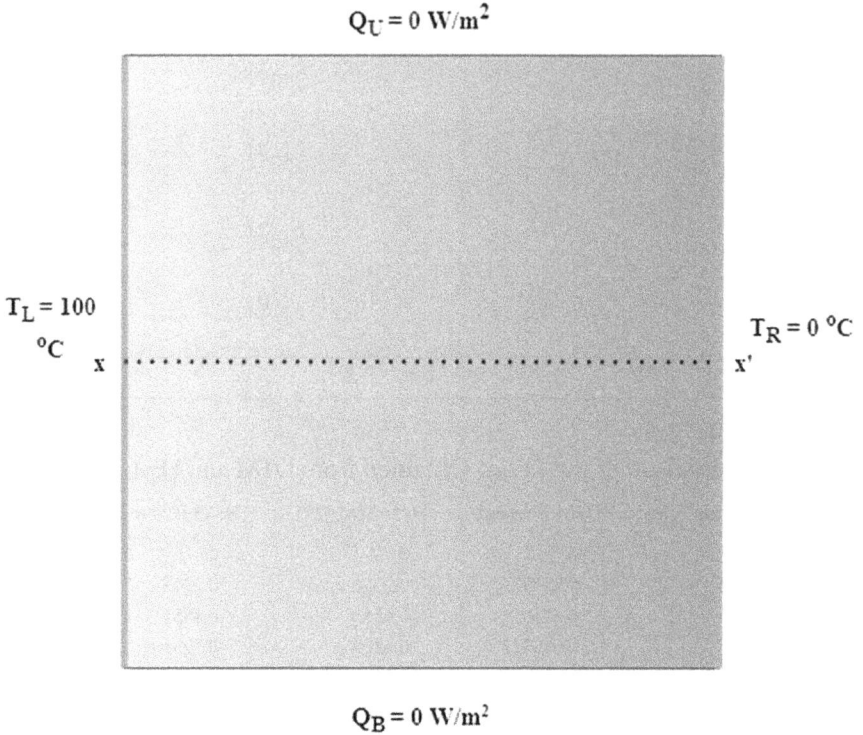

$$Q_U = 0 \text{ W/m}^2$$

$T_L = 100$
$^\circ C$
x ·

$T_R = 0 \,^\circ C$
x'

$$Q_B = 0 \text{ W/m}^2$$

FIGURE 3.3 Illustration of X FGM plate with boundary condition.

TABLE 3.6
The Thermal, Physical, and Mechanical Properties of Ni/Al₂O₃ Composition

Property	Nickel (Ni)	Aluminium Oxide (Al$_2$O$_3$)
Thermal conductivity (K)	60.5 W/mK	46 W/mK
Density (ρ)	8880 kg/m^3	3960 kg/m^3
Specific heat (C)	0.11 Wh/kg K	0.21 Wh/kg K
Shear modulus (G)	76 GPa	150 GPa
Bulk modulus (B)	180 Gpa	172 Gpa
Coefficient of thermal expansion (ξ)	6.6×10^{-6} 1/°C	8.1×10^{-6} 1/°C

A computer program based on MATLAB language is prepared according to the DTM rule stated above. For Ni/Al$_2$O$_3$ FGM plate, the value of the constant is calculated as

$$a = 1; \quad a_{1(n=1)} = 1; \quad b = -4.1714; \quad b_{1(\alpha=\alpha)} = -4.1714 + 5.7241\alpha;$$

The value of boundary condition in the terms of dimensionless temperature can be written as

$$\bar{T}(0) = 1; \bar{T}(L) = 0 \tag{3.46}$$

So DTM transform of first boundary condition is $U(0) = 1$, and from Equations (3.29) and (3.46), value of λ calculated as

$$\lambda = \frac{-1}{1 - \dfrac{a_1|_{n=1}}{2b} + \left(\dfrac{a + a_1|_{n=1}}{3b}\right)\left(\dfrac{a_1|_{n=1}}{2b}\right) + \ldots\ldots} \tag{3.47}$$

The final series solution for power graded X FGM ($n = 1$) plate can be written as

$$\bar{T}(x)\big|_{n=1} = 1 - \frac{x - \dfrac{a_1|_{n=1}}{2b}x^2 + \left(\dfrac{a + a_1|_{n=1}}{3b}\right)\left(\dfrac{a_1|_{n=1}}{2b}\right)x^3 + \ldots\ldots}{1 - \dfrac{a_1|_{n=1}}{2b} + \left(\dfrac{a + a_1|_{n=1}}{3b}\right)\left(\dfrac{a_1|_{n=1}}{2b}\right) + \ldots\ldots} \tag{3.48}$$

Equation (3.48) further reduces for $n = 1$ in infinite series form

$$\bar{T}(x)\big|_{n=1} = 1 - \frac{\displaystyle\sum_{j=0}^{\infty} \dfrac{(-1)^j \, j! \, x^{j+1}}{(j+1)! \, b^j}}{\displaystyle\sum_{j=0}^{\infty} \dfrac{(-1)^j \, j!}{(j+1)! \, b^j}} \tag{3.49}$$

Similarly, in even porous X FGM case with similar boundary and material conditions, λ can be calculated using the described step above (Table 3.7)

$$\lambda = \frac{-1}{1 - \dfrac{a_1|_{n=1}}{2b_1} + \left(\dfrac{a + a_1|_{n=1}}{3b_1}\right)\left(\dfrac{a_1|_{n=1}}{2b_1}\right) + \ldots\ldots} \tag{3.50}$$

TABLE 3.7

Variation of Nondimensional Temperature with Position for Different Porosity Index

Nondimensional Position	$\alpha = 0$	$\alpha = 0.1$	$\alpha = 0.20$
0	1	1	1
0.1	0.9115	0.9127	0.9141
0.2	0.8207	0.8229	0.8256
0.3	0.7277	0.7307	0.7344
0.4	0.6322	0.6357	0.6401
0.5	0.5341	0.5380	0.5428
0.6	0.4334	0.4372	0.4420
0.7	0.3298	0.3332	0.3376
0.8	0.2231	0.2259	0.2294
0.9	0.1133	0.1149	0.1170
1.0	0	0	0

The final series solution for power graded imperfect X FGM ($n = 1$) plate can be written as

$$\bar{T}(x)\Big|_{n=1} = 1 - \frac{x - \dfrac{a_1\big|_{n=1}}{2b_1}x^2 + \left(\dfrac{a + a_1\big|_{n=1}}{3b_1}\right)\left(\dfrac{a_1\big|_{n=1}}{2b_1}\right)x^3 + \ldots\ldots\ldots}{1 - \dfrac{a_1\big|_{n=1}}{2b_1} + \left(\dfrac{a + a_1\big|_{n=1}}{3b_1}\right)\left(\dfrac{a_1\big|_{n=1}}{2b_1}\right) + \ldots\ldots} \tag{3.51}$$

Equation (3.51) further reduces for $n = 1$ in infinite series form

$$\bar{T}(x)\Big|_{n=1} = 1 - \frac{\displaystyle\sum_{j=0}^{\infty} \frac{(-1)^j \, j! \, x^{j+1}}{(j+1)! \, b_1^{\,j}}}{\displaystyle\sum_{j=0}^{\infty} \frac{(-1)^j \, j!}{(j+1)! \, b_1^{\,j}}} \tag{3.52}$$

3.6 CONCLUSIONS

This study is focused on the determination of temperature profile under Dirichlet and Neumann conditions. In this work, DTM is successfully implemented for the steady heat transfer problem that involves material complexity. Two types of the material condition, i.e., perfect and imperfect (even) FGM plate, are considered for the analysis. A validation study is carried out to check the accuracy of the proposed method. This work can be treated as a benchmark for evaluating temperature distribution within FGM plates. Based on numerical illustration, the following conclusion can be drawn:

- DTM can be used for the analytical solution of the FGM plate with material complexity. It can be applied to a one-dimensional heat transfer problem that occurs in unidirectional FGM.
- The effect of porosity index in temperature distribution within the plate is studied. The temperature at spatial position increases with the increase in porosity index.
- The average relative error associated with DTM is very small compared to other numerical methods provided a sufficient number of iterations is reached.

REFERENCES

1. Duan, J. S., Chaolu, T., & Rach, R. (2012). Solutions of the initial value problem for non-linear fractional ordinary differential equations by the Rach–Adomian–Meyers modified decomposition method. *Applied Mathematics and Computation, 218*(17), 8370–8392.
2. Elsaid, A. (2011). Homotopy analysis method for solving a class of fractional partial differential equations. *Communications in Nonlinear Science and Numerical Simulation, 16*(9), 3655–3664.
3. Zhang, X., Zhao, J., Liu, J., & Tang, B. (2014). Homotopy perturbation method for two dimensional time-fractional wave equation. *Applied Mathematical Modelling, 38*(23), 5545–5552.
4. Abdulaziz, O., Hashim, I., & Momani, S. (2008). Application of homotopy-perturbation method to fractional IVPs. *Journal of Computational and Applied Mathematics, 216*(2), 574–584.
5. Sutradhar, A., Paulino, G. H., & Gray, L. J. (2002). Transient heat conduction in homogeneous and non-homogeneous materials by the Laplace transform Galerkin boundary element method. *Engineering Analysis with Boundary Elements, 26*(2), 119–132.
6. Wang, B. L., & Tian, Z. H. (2005). Application of finite element–finite difference method to the determination of transient temperature field in functionally graded materials. *Finite Elements in Analysis and Design, 41*(4), 335–349.
7. Wang, H., Qin, Q. H., & Kang, Y. L. (2006a). A meshless model for transient heat conduction in functionally graded materials. *Computational Mechanics, 38*(1), 51–60.
8. Cui, X. Y., Li, Z. C., Feng, H., & Feng, S. Z. (2016). Steady and transient heat transfer analysis using a stable node-based smoothed finite element method. *International Journal of Thermal Sciences, 110*, 12–25.
9. Sutradhar, A., & Paulino, G. H. (2004). The simple boundary element method for transient heat conduction in functionally graded materials. *Computer Methods in Applied Mechanics and Engineering, 193*(42–44), 4511–4539.
10. Demirbaş, M. D., Ekici, R., & Apalak, M. K. (2020). Thermoelastic analysis of temperature-dependent functionally graded rectangular plates using finite element and finite difference methods. *Mechanics of Advanced Materials and Structures, 27*(9), 707–724.
11. Kenmogne, F. (2015). Generalizing of differential transform method for solving nonlinear differential equations. *Applied & Computational Mathematics, 4*(1), 1000196.
12. Munganga, J. M. W., Mwambakana, J. N., Maritz, R., Batubenge, T. A., & Moremedi, G. M. (2014). Introduction of the differential transform method to solve differential equations at undergraduate level. *International Journal of Mathematical Education in Science and Technology, 45*(5), 781–794.

13. Ayaz, F. (2003). On the two-dimensional differential transform method. *Applied Mathematics and Computation, 143*(2–3), 361–374.

14. Opanuga, A. A., Edeki, S. O., Okagbue, H. I., & Akinlabi, G. O. (2015). Numerical solution of two-point boundary value problems via differential transform method. *Global Journal of Pure and Applied Mathematics, 11*(2), 801–806.

15. Jang, M. J., Chen, C. L., & Liu, Y. C. (2001). Two-dimensional differential transform for partial differential equations. *Applied Mathematics and Computation, 121*(2–3), 261–270.

16. Kangalgil, F., & Ayaz, F. (2007). Solution of linear and nonlinear heat equations by differential transform method. *Selcuk Journal of Applied Mathematics, 8*(1), 75–86.

17. Bildik, N., & Konuralp, A. (2006). The use of variational iteration method, differential transform method and Adomian decomposition method for solving different types of nonlinear partial differential equations. *International Journal of Nonlinear Sciences and Numerical Simulation, 7*(1), 65–70.

18. Zhou, J. K. (1986). Differential transformation and its applications for electrical circuits, Huazhong University Press, Wuhan.

19. Hatami, M., Ganji, D. D., & Sheikholeslami, M. (2016). *Differential transformation method for mechanical engineering problems*. Academic Press, Cambridge, Massachusetts.

20. Kar, V. R., & Panda, S. K. (2015). Thermoelastic analysis of functionally graded doubly curved shell panels using nonlinear finite element method. *Composite Structures, 129*, 202–212.

21. Kar, V. R., Mahapatra, T. R., & Panda, S. K. (2017). Effect of different temperature load on thermal postbuckling behaviour of functionally graded shallow curved shell panels. *Composite Structures, 160*, 1236–1247.

22. Li, S., Zheng, S., & Chen, D. (2020). Porosity-dependent isogeometric analysis of bi-directional functionally graded plates. *Thin-Walled Structures, 156*, 106999.

23. Karakoti, A., Pandey, S., & Kar, V. R. (2021). Nonlinear transient analysis of porous functionally graded material plates under blast loading. *Materials Today: Proceedings, 46*, 8111–8113.

24. Phung-Van, P., Thai, C. H., Ferreira, A. J. M., & Rabczuk, T. (2020). Isogeometric nonlinear transient analysis of porous FGM plates subjected to hygro-thermo-mechanical loads. *Thin-Walled Structures, 148*, 106497.

25. Hancock, M. J. (2004). The 1-D heat equation: 18.303 linear partial differential equations. Available online at: https://ocw.mit.edu/courses/mathematics/18-303-linear-partial-differential-equations-fall-2006/lecture-notes/heateqni.pdf

26. El Ibrahimi, M., & Samaouali, A. (2021). Closed-form approximate solution for heat transfer analysis within functionally graded plate with temperature-dependent thermal conductivity. *Composite Structures, 271*, 114140.

27. Sladek, J., Sladek, V., & Zhang, C. (2003). Transient heat conduction analysis in functionally graded materials by the meshless local boundary integral equation method. *Computational Materials Science, 28*(3–4), 494–504.

28. Wang, H., Qin, Q. H., & Kang, Y. L. (2006b). A meshless model for transient heat conduction in functionally graded materials. *Computational Mechanics, 38*(1), 51–60.

29. Cao, L., Wang, H., & Qin, Q. H. (2012). Fundamental solution based graded element model for steady-state heat transfer in FGM. *Acta Mechanica Solida Sinica, 25*(4), 377–392.

30. Demirbas, M. D., & Apalak, M. K. (2019). Thermal stress analysis of one-and two-dimensional functionally graded plates subjected to in-plane heat fluxes. *Proceedings of the Institution of Mechanical Engineers, Part L: Journal of Materials: Design and Applications, 233*(4), 546–562.

4 Effect of Corrugation on the Deformation Behavior of Spatially Graded Composite Panels

Abhilash Karakoti and Vishesh Ranjan Kar

National Institute of Technology Jamshedpur,
Jamshedpur, India

K. Jayakrishna

School of Mechanical Engineering, VIT Vellore,
Vellore, India

Mohamed Thariq Hameed Sultan

University Putra Malaysia, UPM Serdang,
Selangor Darul Ehsan, Malaysia

CONTENTS

DOI: 10.1201/9781003158813-4

4.1 INTRODUCTION

Functionally graded materials (FGM) are advanced materials having location specific changes in the micro/macrostructure or chemical composition causing gradual changes in the properties of the final material obtained (Jha, Kant, and Singh 2013). Being an advanced composite material FGM has several advantages over the conventional laminated composites, such as improved load transfer mechanism, enhanced toughness, high corrosion, and thermal resistant (Li, Li, and Hu 2018). Defects such as delamination and crack formation can be prevented due to the smooth and gradual gradation. A large number of engineering components and tools require a material that will have performance varying with location (Dayyani et al. 2015). Corrugated panels have high bending stiffness with comparatively less use of material when compared to the flat panels (Biancolini 2005). Corrugated structures can be subjected to large deformation and have the advantage over the flat panels in that their local strains remain comparatively low (Winkler and Kress 2010). First-order shear deformation theory (FSDT) or higher-order shear deformation theories (HSDT) are widely used for FGM analysis.

Song et al. (2018) investigated bending and buckling analysis of FG multilayer graphene nanoplatelet polymer (GPL) composite using FSDT and concluded that GPL reinforcing pattern and geometry, among other aspects, influence the bending and buckling behavior of a composite plate. Zidi et al. (2014) used four variable refined plate theories to derive the bending behavior of FGM under hygro-thermo-mechanical loading without any shear correction factor. The transverse displacement can also be separated into the bending and shear parts, as concluded by Thai and Choi (2013), which reduces the number of unknowns and governing equations unlike conventional FSDT. Belabed et al. (2014) employed a new higher-order shear and normal deformation with only five unknowns for FGM plates considering "stretching effect" that was not considered earlier. Amirpour et al. (2017) concentrated on thin-plate FGM varying the in-plane stiffness and derived the analytical solution for displacement and stress that can be used for thin structures. Investigation of thermoelastic bending FG sandwich plates was performed by Tounsi et al. (2013) using the trigonometric shear deformation theory with only four unknown functions. Mahapatra, Kar, and Panda (2016) developed a nonlinear mathematical model based on HSDT to show the flexural behavior of laminated curved panels under uniform load. Instead of rectangular or square plates, Zhao et al. (2017) observed the bending behavior of graphene nanoplatelets-reinforced trapezoidal plates and concluded that with the decrease in the two base angles, deflection of the plates also decreases.

Kar and Panda (2016) also investigated the effect of various parameters on the bending behavior of different functionally graded curved panel (elliptical, hyperbolic, and cylindrical) in combination with thermomechanical loading. Mahdi and Babaghasabha (2016) investigated bending and buckling behavior of corrugated sandwich plates using ANSYS under uniaxial load. Kar and Panda (2015) investigated the free vibration behavior of a spherical FG shell panel. The mathematical model was HSDT based, and micromechanical approach for material property was applied. Hamilton's principle and direct iterative approach were employed to obtain the governing equations and final solution, respectively. Li et al. (2018) obtained the linear and nonlinear displacements by generalized differential quadrature method for two-dimensional FG beam and examined the nonlinear bending behavior using the Euler–Bernoulli beam kinematic theory. Thai et al. (2017) used the isogeometric analysis to show the post-buckling behavior of FG microplates under thermal and mechanical loads using Reddy's third-order shear deformation plate theory where material properties are considered as temperature dependent or temperature independent. Li et al. (2018) examined the bending behavior of simply supported FG sandwich plates with FG soft core under distributed load using Navier approach. Lv and Liu (2017) used minimum total potential energy principle to examine the impact of material uncertainties on nonlinear bending behavior of FG nano-beam.

Sid et al. (2013) developed a new higher-order shear and normal deformation theories where they distributed the transverse displacement into bending, shear, and thickness stretching part that doesn't require any shear correction factor and the number of unknowns reduced from six to four. Kar and Panda (2015) demonstrated the bending behavior of FG spherical shell and graded the material properties via power-law index. Shell model is HSDT based, and equilibrium equations are obtained using variational principle. Nejati et al. (2017) evaluated the static and free vibration responses of carbon nanotubes reinforced into FG conical shells under uniform loading; material properties were determined by four-parameter power-law, and generalized differential quadrature method (GDQ) was used for numerical solution. Viola et al. (2014) calculated tangential and normal stresses in FG conical shells and panels using unconstrained third-order shear deformation theory (UTSDT), and differential equations were solved using the GDQ technique. Viola et al. (2016) used UTSDT to signify the effect of material parameters on the static response of FG spherical shells and panels under uniform loadings, and GDQ method was implemented to solve the governing equations.

Over the years, researchers have mainly focused on the bending and vibration analysis of FGM panels. However, limited literature is available on geometrical consideration of corrugation. In this chapter, bending behavior of flat, noncorrugated and corrugated FGM panels is investigated using FSDT. The material properties of the FGM panels are varied continuously according to the Voigt's model via power-law distribution. A general curvature equation is used to introduce the sinusoidal corrugation, and governing equations are obtained using variational principle. A homemade computer code is used to obtain the deflection responses of the panels. The bending analysis for various geometric parameters is examined to show the adequacy of the present model.

4.2 MATHEMATICAL FORMULATION

In the present analysis, a metal-ceramic corrugated FG panel is considered. Here the FG panel with gradation has uniform thickness h, length a, breadth b, and wave

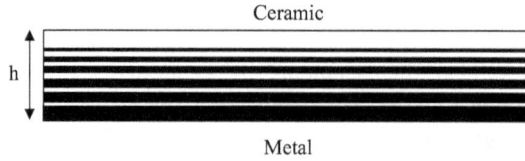

FIGURE 4.1 Schematic diagram of FGM.

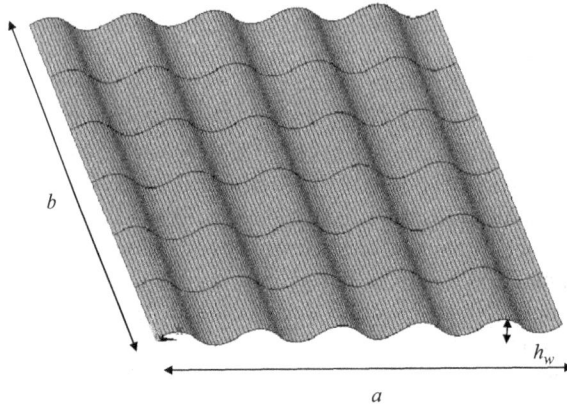

FIGURE 4.2 Geometrical representation of sinusoidally corrugated FGM panel.

height *hw* as shown in Figures 4.1 and 4.2. The sinusoidal corrugation is incorporated by using mathematical curvature equation (Karakoti and Kar 2019).

$$R_{\text{curvature}} = \frac{-a^2 \left(\dfrac{\phi^2 n_w^4}{a^4} \cos\left(\dfrac{n_w \pi x}{a} \right)^2 + 1 \right)^3}{2\phi n_w^3 \pi^2 \sin\left(\dfrac{n_w \pi x}{a} \right)} \tag{4.1}$$

where $\phi = \dfrac{h_w a}{n_w}$.

4.2.1 Effective Material Properties

Here, as we move from one surface to the next, properties are changing smoothly from top to bottom according to the power-law distribution (Shen 2009), such that the top surface consists of metal and the bottom surface consists of ceramics. Using the Voigt's model, Equation (4.2) is obtained, which gives the effective material properties of corrugated FG panel:

$$P = (P_m - P_c)V_{fm} + P_c \tag{4.2}$$

where P_m and P_c are the metal and ceramic material properties, respectively, and V_{fm} is the volume fraction of the metal that is the top surface and can be evaluated by power-law distribution (Shen 2001) as shown in Equation (4.3).

$$V_{fm} = \left(\frac{z}{h} + \frac{1}{2}\right)^{pi} \tag{4.3}$$

where power law index (pi) must be within the limits ($0 \le pi \le \infty$) and helps in describing the material property along the thickness of the panel. Material properties can vary through power-law index with respect to top or bottom layer to get different material profiles.

Now, with the help of Equations (4.2) and (4.3), effective material properties like Young's modulus (E) and Poisson's ratio (v) can be obtained.

$$E = (E_m - E_c)\left(\frac{z}{h} + \frac{1}{2}\right)^{pi} + E_c \tag{4.4}$$

$$v = (v_m - v_c)\left(\frac{z}{h} + \frac{1}{2}\right)^{pi} + v_c \tag{4.5}$$

where, E_m and v_m represents the properties of metal and E_c and v_c represents the properties of ceramics.

4.2.2 DISPLACEMENT FIELD

FSDT midplane kinetics is used in this study for the corrugated FG panel to derive the mathematical model (Reddy 1984), which is expressed as Equation (4.6).

$$\left. \begin{array}{l} u(x,y,z,t) = u_0(x,y,t) + zu_1(x,y,t) \\ v(x,y,z,t) = v_0(x,y,t) + zv_1(x,y,t) \\ w(x,y,z,t) = w_0(x,y,t) + zw_1(x,y,t) \end{array} \right\} \tag{4.6}$$

where z is the thickness coordinate varies from $-h/2$ to $+h/2$ and t is the time. u_0, v_0, and w_0 are the displacements in x, y, and z axis, respectively. u_1, v_1, and w_1 are the rotations about the axis.

4.2.3 STRAIN DISPLACEMENT RELATIONS

The strain displacement relation can be expressed by Equation (4.7).

$$\{\varepsilon\} = \begin{Bmatrix} \varepsilon_{xx} \\ \varepsilon_{yy} \\ \varepsilon_{zz} \\ \gamma_{xy} \\ \gamma_{xz} \\ \gamma_{yz} \end{Bmatrix} = \begin{Bmatrix} \left(\dfrac{\partial u}{\partial x}\right) \\ \left(\dfrac{\partial v}{\partial y}\right) \\ \left(\dfrac{\partial w}{\partial z}\right) \\ \left(\dfrac{\partial u}{\partial y}+\dfrac{\partial v}{\partial x}\right) \\ \left(\dfrac{\partial u}{\partial z}+\dfrac{\partial w}{\partial x}\right) \\ \left(\dfrac{\partial v}{\partial z}+\dfrac{\partial w}{\partial y}\right) \end{Bmatrix} \tag{4.7}$$

Strain displacement relation of Equation (4.7) can also be expressed by Equation (4.8).

$$\{\varepsilon\} = \begin{Bmatrix} \varepsilon_{xx}^o \\ \varepsilon_{yy}^o \\ \varepsilon_{zz}^o \\ \gamma_{xy}^o \\ \gamma_{xz}^o \\ \gamma_{yz}^o \end{Bmatrix} + z \begin{Bmatrix} k_x \\ k_y \\ k_z \\ k_{xy} \\ k_{xz} \\ k_{yz} \end{Bmatrix} \tag{4.8}$$

$$\{\varepsilon\} = [T]\{\bar{\varepsilon}\} \tag{4.9}$$

where $\{\bar{\varepsilon}\} = \{\varepsilon_{xx}^o\ \varepsilon_{yy}^o\ \varepsilon_{zz}^o\ \gamma_{xy}^o\ \gamma_{xz}^o\ \gamma_{yz}^o\ k_x\ k_y\ k_z\ k_{xy}\ k_{xz}\ k_{yz}\}^T$ is the mid-plane strain vector.

4.2.4 CONSTITUTIVE RELATION

The constitutive relation for the FG panel is expressed by Equation (4.10).

$$\begin{Bmatrix} \sigma_{xx} \\ \sigma_{yy} \\ \sigma_{zz} \\ \tau_{xy} \\ \tau_{xz} \\ \tau_{yz} \end{Bmatrix} = \begin{bmatrix} Q_{11} & Q_{12} & Q_{13} & 0 & 0 & 0 \\ Q_{12} & Q_{22} & Q_{23} & 0 & 0 & 0 \\ Q_{13} & Q_{23} & Q_{33} & 0 & 0 & 0 \\ 0 & 0 & 0 & Q_{66} & 0 & 0 \\ 0 & 0 & 0 & 0 & Q_{55} & 0 \\ 0 & 0 & 0 & 0 & 0 & Q_{44} \end{bmatrix} \begin{Bmatrix} \varepsilon_{xx} \\ \varepsilon_{yy} \\ \varepsilon_{zz} \\ \gamma_{xy} \\ \gamma_{xz} \\ \gamma_{yz} \end{Bmatrix} \tag{4.10}$$

Equation (4.10) can also be expressed by Equation (4.11)

$$\{\sigma\} = [Q]\{\varepsilon\} \tag{4.11}$$

where $[Q]$ is the stiffness matrix.

4.2.5 STRAIN ENERGY

The strain energy of the corrugated FG panel can be expressed by Equation (4.12).

$$U = \frac{1}{2}\int_{v}\{\varepsilon\}^{T}\{\sigma\}dV \tag{4.12}$$

Now, the above equations can be evaluated Equation (4.13).

$$U = \frac{1}{2}\int\left(\{\bar{\varepsilon}\}^{T}[D]\{\bar{\varepsilon}\}\right)dA \tag{4.13}$$

$$\text{where } [D] = \int_{-h/2}^{h/2}[T]^{T}[Q][T]dz$$

4.2.6 WORK DONE

When a uniform load of q is applied, the relation for work done can be obtained by Equation (4.14).

$$W = \int_{A}\{\delta\}^{T}\{q\}dA \tag{4.14}$$

4.2.7 FINITE ELEMENT FORMULATION

With six degrees of freedom per node, the displacement vector for each element can be expressed by Equation (4.15).

$$\{\delta\} = \sum_{i=1}^{6}N_{i}\{\delta_{i}\} \tag{4.15}$$

where $\{\delta_{i}\} = [u_{0i} \ v_{0i} \ w_{0i} u_{1i} \ v_{1i} \ w_{1i}]^{T}$ is the nodal displacement vector at node i. N_{i} is the shape function for the ith node (Cook, Malkus, and Plesha 1989).

The mid-plane strain vector having nodal displacement vector can be expressed by Equation (4.16).

$$\{\bar{\varepsilon}\} = [B]\{\delta_{i}\} \tag{4.16}$$

where [B] is the product of differential operators and the shape functions in the strain terms.

By substituting Equation (4.16) in (13), elemental strain energy can be expressed by Equation (4.17).

$$U_e = \frac{1}{2} \iint \left(\{\delta_i\}^T [B]^T [D][B]\{\delta_i\} \right) dxdy \tag{4.17}$$

$$U_e = \frac{1}{2} \{\delta_i\}^T [K]_e \{\delta_i\} \tag{4.18}$$

where $[K]_e$ is the elemental stiffness matrix.

Similarly, elemental work done can be expressed by Equation (4.19).

$$W_e = \{\delta_i\}^T \{F\}_e \tag{4.19}$$

where $\{F\}_e$ is elemental force vector.

4.2.8 GOVERNING EQUATIONS

The final governing equation of corrugated FG panel by using minimum total potential energy principle is given by Equation (4.20).

$$\delta\pi = \delta U - \delta W = 0 \tag{4.20}$$

where π is the total potential energy and δ is the variational symbol.

Equilibrium equation for static analysis under uniform load can be obtained from Equations (4.13), (4.14), (4.15), (4.16), and (4.17) and expressed by Equation (4.21).

$$[K]_e \{\delta\}_e = \{q\}_e \tag{4.21}$$

4.3 RESULTS AND DISCUSSION

In the study reported, a customized code is developed based on FSDT to address the deflection responses of flat, noncorrugated and corrugated FG composite panels. For the present analysis the following edge constraints are used:

a. All sides are simply supported (SSSS)

$$u = w = v_1 = 0 \text{ at } x = 0, a$$

$$v = w = u_1 = 0 \text{ at } y = 0, b$$

b. All sides are clamped (CCCC)

$$u = v = w = u_1 = v_1 = 0 \text{ at } x = 0, a \text{ and } y = 0, b$$

c. Simply supported and clamped (SCSC)

$$u = w = v_1 = 0 \text{ at } x = 0, a$$

$$u = v = w = u_1 = v_1 = 0 \text{ at } x = 0, a \text{ and } y = 0, b$$

4.3.1 CONVERGENCE AND VALIDATION

Deflection responses of corrugated FG panel are examined for various mesh sizes. Figure 4.3 is used for convergence study of simply supported sinusoidally corrugated FG panel ($a = b$, $h_w = a/50$, $a/h = 50$, $Q = 100$, $P_{ind} = 1$) under uniform load for varied number of waves of corrugation ($n_w = 1, 2, 5$ and 10). Results revealed that deflection responses after (5×5) mesh size are almost constant for all the cases. So, (6×6) mesh size will give reasonable results for numerical experimentations.

Now, to show the accuracy, the present model is validated with Zenkour et al. (2006). The bending responses for simply supported Al/ZrO$_2$ FGM flat panel for different power law indices ($P_{ind} = 0, 1, 2, 3, 4$ and 5) subjected to uniform load are shown in Table 4.1. The difference in the deflection responses of the proposed model

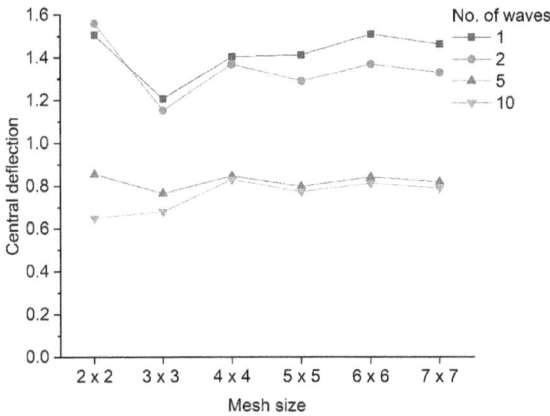

FIGURE 4.3 Deflection responses of sinusoidally corrugated functionally graded composite panel for various mesh densities.

TABLE 4.1
Central Deflection of Flat FGM Panel

P_{ind}	Zenkour et al. (26)	Present Results	% Error
0	0.4665	0.47956	−2.800%
1	0.9287	0.851352	8.329%
2	1.194	1.0355	13.275%
3	1.32	1.137454	13.829%
4	1.389	1.207336	13.079%
5	1.4356	1.262702	12.044%

as compared to the results of generalized shear deformation theory (Zenkour 2006) is nominal. Hence, the similarity among the present and published results confirms the accuracy of the FSDT based model.

4.3.2 NUMERICAL EXPERIMENTATION

For further investigations the effect of different parameters such as side-to-thickness ratio (a/h), aspect ratio ($ar = 1$), wave height (h_w), number of waves (n_w), power-law index (P_{ind}), and support conditions on the deflection responses of corrugated and noncorrugated FG composite panel under uniform loading is examined. The material properties considered for numerical experimentation are tabulated in Table 4.2. General considerations for numerical experimentation are simply supported (*SSSS*), aspect ratio ($ar = a/b = 1$), wave height ($h_w = a/50$), thickness ($h = a/50$), and power-law index ($P_{ind} = 1$), if not otherwise specified.

Figure 4.4 demonstrates the central deflection of flat and noncorrugated ($n_w = 1$) FG panel ($a/h = 50$, $h_w = a/50$, $P_{ind} = 1$, *SSSS*) with different aspect ratio ($ar = 1, 1.5, 2, 2.5$) under uniform load. The deflection responses are maximum for square panels and reducing with increasing aspect ratio. Noncorrugated panels exhibit lower

TABLE 4.2
Material Properties for FGM (Swaminathan and Sangeetha 2017; Thai et al. 2017)

Material	E (GPa)	v
Al	70	0.3
SiC	137	0.3

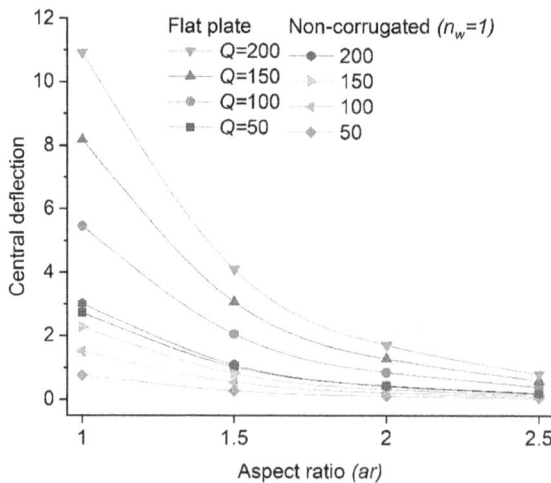

FIGURE 4.4 Central deflection of corrugated and noncorrugated FG panel for different aspect ratio.

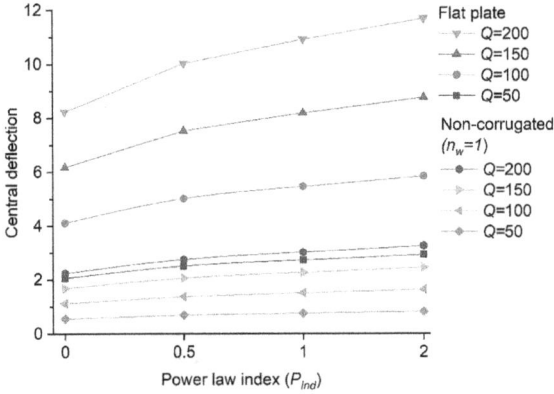

FIGURE 4.5 Central deflection of corrugated and noncorrugated FG panel for different power law indices.

deflections as compared to the flat panels because of the sinusoidal curvature, which is making them stiffer.

Figure 4.5 demonstrates the deflection parameters of flat and noncorrugated ($n_w = 1$) FG panel ($a/h = 50$, $h_w = a/50$, $ar = 1$, SSSS) with different power-law indices ($P_{ind} = 0, 0.5, 1, 2$). Increasing power-law indices increases the deflection parameters. This is because material properties are changing from ceramic rich to metal rich. Thus, the panel with the lowest power-law index ($P_{ind} = 0$) is stiffer.

Figure 4.6 demonstrates the deflection parameters of simply supported flat and noncorrugated ($n_w = 1$) FG panel ($h_w = a/50$, $ar = 1$, $P_{ind} = 1$, SSSS) under uniform load. Increasing side-to-thickness ratio ($a/h = 20, 30, 40, 50$) increases the deflection

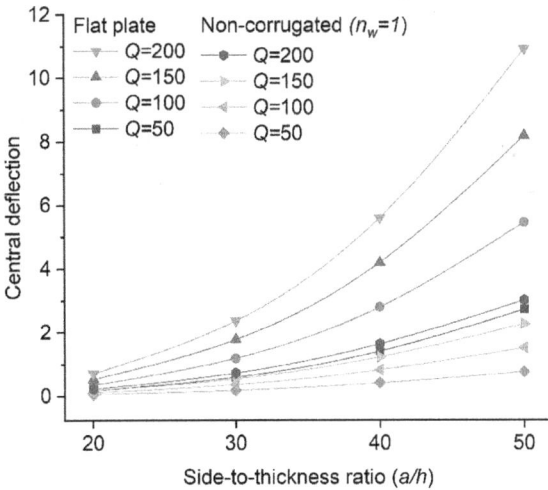

FIGURE 4.6 Central deflection of corrugated and noncorrugated FG panel for different side-to-thickness ratio.

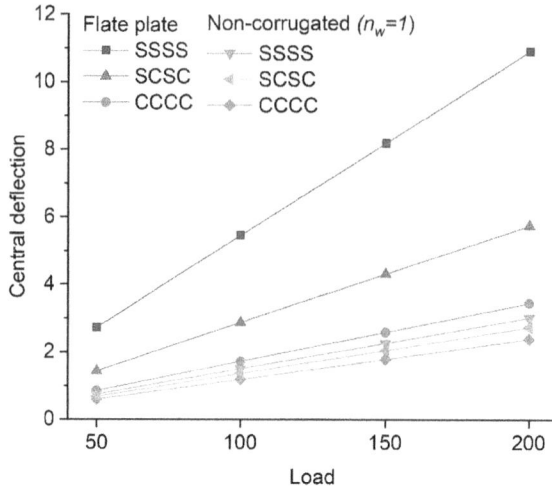

FIGURE 4.7 Central deflection of corrugated and noncorrugated FG panel for different support conditions.

responses. As the ratio is increasing, the thickness of the panel is reducing, thus the ones with lower side-to-thickness ratio will be stiffer.

Figure 4.7 exhibits the central deflection parameter of noncorrugated ($n_w = 1$) and flat FG panel ($a/h = a/50$, $h_w = a/50$, $ar = 1$, $P_{ind} = 1$) under uniform load with various support conditions (*SSSS, CCCC, SCSC*). Among all the different support conditions, simply supported panels are exhibiting maximum deflection and fully clamped panels are showing minimum deflection.

In Figures 4.4–4.7 effect of various geometric parameters on the deflection parameters is demonstrated. According to these figures, a noncorrugated ($n_w = 1$) FG panel is exhibiting lower deflection responses compared to the flat panel. This is due to introduction of sinusoidal curvature, which is providing stiffness to the panel.

Table 4.3 tabulates the central deflection of corrugated FG panel ($n_w = 2, 5, 7, 10$) with different wave height ($h_w = a/50$, $a/60$, $a/80$, $a/100$). The corrugated FG panel is subjected to uniform load and edges are SSSS. In all the cases considered, deflection parameters are enhancing significantly with decrement in the wave height.

TABLE 4.3

Central Deflection of Corrugated FG Panel with Varying Wave Height (h_w)

Wave Height (h_w)	No. of Waves (n_w)			
	2	5	7	10
$a/50$	0.684722	0.420611	0.418249	0.406202
$a/60$	0.731352	0.488936	0.483484	0.469256
$a/80$	0.785956	0.592176	0.583956	0.56891
$a/100$	0.814686	0.66265	0.654187	0.64058

Tables 4.4–4.7 exhibits the variation in deflection parameters of sinusoidally corrugated FG panel with different side-to-thickness ratio, aspect ratio, power-law index, and support conditions. In all the considered cases, deflection responses are reducing significantly as the number of waves (n_w) of corrugation is increasing. It reveals that the panels with higher wave height and number of corrugated waves will be stiffer.

TABLE 4.4

Central Deflection of Corrugated FG Panel with Varying Side-to-Thickness Ratio (a/h)

Side to Thickness Ratio (a/h)	No. of Waves (n_w)			
	2	5	7	10
20	0.055569	0.048556	0.048277	0.04748
30	0.173503	0.134053	0.13287	0.130009
40	0.380932	0.260809	0.258596	0.251935
50	0.684722	0.420611	0.418249	0.406202

TABLE 4.5

Central Deflection of Corrugated FG Panel with Varying Aspect Ratio (ar)

Aspect Ratio (Ar)	No. of Waves (n_w)			
	2	5	7	10
1	0.684722	0.420611	0.418249	0.406202
1.5	0.234439	0.123974	0.116345	0.108737
2	0.092328	0.048675	0.043782	0.039493
2.5	0.041978	0.023425	0.020516	0.01795

TABLE 4.6

Central Deflection of Corrugated FG Panel with Varying Power Law Index (P_{ind})

Power Law Index (P_{ind})	No. of Waves (n_w)			
	2	5	7	10
0	0.508967	0.314795	0.312941	0.303873
0.5	0.622227	0.381501	0.379334	0.36825
1	0.684722	0.420611	0.418249	0.406202
2	0.742483	0.460894	0.458274	0.44551

TABLE 4.7

Central Deflection of Corrugated FG Panel with Varying Support Conditions

Support Conditions	No. of Waves (n_w)			
	2	5	7	10
SSSS	0.684722	0.420611	0.418249	0.406202
CCCC	0.484756	0.256875	0.245102	0.241563
SCSC	0.571716	0.323223	0.316648	0.311422
SCSC	2.28674	1.29338	1.26882	1.24689

4.4 CONCLUSION

The deflection responses of flat noncorrugated ($n_w = 1$) and corrugated ($n_w = 2, 5, 7, 10$) FG panels are examined under uniform load with different geometrical parameters and support conditions. The sinusoidal geometry is introduced via mathematical curvature equation. FSDT kinematics is used to obtain the strain terms with six degrees-of-freedom. Material properties are evaluated through power law formulation. Final governing equations are obtained by minimum total potential energy principle. The significance of the proposed model is evaluated by comparing it with previously published results. For the first time, sinusoidal corrugation of FGM panels is taken into consideration and compared with the deflection responses of the flat FG panel. Numerical experimentation of flat noncorrugated and corrugated FG panels is under uniform load, and the following remarks are produced:

- The corrugated FG panels are highly stiff, followed by noncorrugated and flat FG panel.
- Corrugated FG panels with high aspect ratio, high wave height, and lower side-to-thickness ratio are stiff when subjected to uniform load.
- Number of waves (n_w) in a corrugated panel significantly affects the deflection responses. The higher the number of waves, the stiffer is the corrugated panel.
- In all the cases, with increase in the power-law index (P_{ind}), the stiffness of the FG panel is decreasing as the material properties change from ceramic rich ($P_{ind} = 0$) to metal rich.

Many different numerical and experimental analyses could be considered for future work. Limited literature is available related to the analyses of corrugated panels, and the presented study mainly has focused on the deflection behavior of corrugated FG panel. Hence, the presented model can modified to obtain the buckling and free vibration behavior of corrugated panels to examine its efficacy. Corrugated structures having different geometries can also be compared with the experimental results to show its feasibility and performance in the real world.

ACKNOWLEDGMENT

The authors would like to thank the Science and Engineering Research Board, Department of Science and Technology, Government of India (File No. ECR/2016/ 001829), for the financial support.

REFERENCES

Abhilash, K. and Kar, V. R. (2019), "Deformation Characteristics of Sinusoidally-Corrugated Laminated Composite Panel—A Higher-Order Finite Element Approach", *Composite Structures*, **216**, 151–158.

Amirpour, M., Das R., and Flores E.I.S. (2017), "Bending Analysis of Thin Functionally Graded Plate Under In-Plane Stiffness Variations", *Applied Mathematical Modelling*, **44**, 481–96.

Belabed, Z., Houari, M. S. A., Tounsi, A., Mahmoud, S. R. and Bég, O. A. (2014), "An Efficient and Simple Higher Order Shear and Normal Deformation Theory for Functionally Graded Material (FGM) Plates", *Composites Part B: Engineering*, **60**, 274–83.

Biancolini, M. E. (2005), "Evaluation of Equivalent Stiffness Properties of Corrugated Board", *Composite Structures*, **69**(3), 322–28.

Cook, R. D., D. S. Malkus, and Plesha M. E. (1989), *Concepts and Applications of Finite Element Analysis*, John Wiley and Sons, USA.

Dayyani, I., Shaw, A. D., Saavedra Flores, E. I. and Friswell, M. I. (2015), "The Mechanics of Composite Corrugated Structures: A Review with Applications in Morphing Aircraft", *Composite Structures*, **133**, 358–80.

Dongdong, L., Deng, Z., Xiao, H. and Jin, P. (2018), "Thin-Walled Structures Bending Analysis of Sandwich Plates with Di Ff Erent Face Sheet Materials and Functionally Graded Soft Core", *Thin Walled Structures*, **122**, 8–16.

Houari, M. S. A., Tounsi, A. and Bég, O. A. (2013), "Thermoelastic Bending Analysis of Functionally Graded Sandwich Plates Using a New Higher Order Shear and Normal Deformation Theory", *International Journal of Mechanical Sciences*, **76**, 102–11.

Shen, H. S. (2001), "Hygrothermal Effects on the Postbuckling of Shear Deformable Laminated Plates", *International Journal of Mechanical Sciences*, **43**(5), 1259–81.

Shen, H. S. (2009), "Nonlinear Bending of Functionally Graded Carbon Nanotube-Reinforced Composite Plates in Thermal Environments", *Composite Structures*, **91**(1), 9–19.

Jha, D. K., Kant, T. and Singh R. K. (2013), "A Critical Review of Recent Research on Functionally Graded Plates", *Composite Structures*, **96**, 833–49.

Kar, V. R. and Panda, S. K. (2015a), "Large Deformation Bending Analysis of Functionally Graded Spherical Shell Using FEM", *Structural Engineering and Mechanics*, **53**(4), 661–79.

Kar, V. R. and Panda, S.K. (2015b), "Nonlinear Flexural Vibration of Shear Deformable Functionally Graded Spherical Shell Panel", *Steel and Composite Structures*, **18**(3), 693–709.

Kar, V. R. and Panda, S.K. (2016), "Nonlinear Thermomechanical Behavior of Functionally Graded Material Cylindrical/Hyperbolic/Elliptical Shell Panel with Temperature-Dependent and Temperature-Independent Properties", *Journal of Pressure Vessel Technology*, **138**(6), 1–13.

Li, L., Li, X. and Hu, Y. (2018), "Nonlinear Bending of a Two-Dimensionally Functionally Graded Beam", *Composite Structures*, **184**, 1049–61.

Mahapatra, T. R., Kar, V. R. and Panda, S. K. (2016), "Large Amplitude Bending Behaviour of Laminated Composite Curved Panels", *Engineering Computations*, **33**(1),116–38.

Mahdi, K. M. and Babaghasabha, V. (2016), "Bending and Buckling Analysis of Corrugated Composite Sandwich Plates", *Journal of the Brazilian Society of Mechanical Sciences and Engineering,* **38**(8), 2571–88.

Nejati, M., Asanjarani, A., Dimitri, R. and Tornabene, F. (2017), "International Journal of Mechanical Sciences Static and Free Vibration Analysis of Functionally Graded Conical Shells Reinforced by Carbon Nanotubes", *International Journal of Mechanical Sciences,* **130**, 383–98.

Reddy, J. N. (1984), "A Simple Higher-Order Theory for Laminated Composite Plates", *Journal of Applied Mechanics,* **51**(4), 745–752.

Son, T., Thai, H. T., Vo, T. P. and Reddy, J. N. (2017), "Post-Buckling of Functionally Graded Microplates Under Mechanical and Thermal Loads Using Isogeomertic Analysis", *Engineering Structures,* **150**, 905–17.

Song, M., Yang, J. and Kitipornchai, S. (2018), "Bending and Buckling Analyses of Functionally Graded Polymer Composite Plates Reinforced with Graphene Nanoplatelets", *Composites Part B: Engineering,* **134**, 106–13.

Swaminathan, K. and Sangeetha, D. M. (2017), "Thermal Analysis of FGM Plates—A Critical Review of Various Modeling Techniques and Solution Methods", *Composite Structures,* **160**, 43–60.

Thai, H. T. and Choi, D. H. (2013), "A Simple First-Order Shear Deformation Theory for the Bending and Free Vibration Analysis of Functionally Graded Plates", *Composite Structures,* **101**, 332–40.

Tounsi, A., Houari, M. S. A., Benyoucef, S., Abbas, E. and Bedia, A. (2013), "A Refined Trigonometric Shear Deformation Theory for Thermoelastic Bending of Functionally Graded Sandwich Plates", *Aerospace Science and Technology,* **24**(1), 209–20.

Viola, E., Rossetti, L., Fantuzzi, N. and Tornabene, F. (2014), "Static Analysis of Functionally Graded Conical Shells and Panels Using the Generalized Unconstrained Third Order Theory Coupled with the Stress Recovery", *Composite Structures,* **112**, 44–65.

Viola, E., Rossetti, L., Fantuzzi, N. and Tornabene, F. (2016), "Generalized Stress—Strain Recovery Formulation Applied to Functionally Graded Spherical Shells and Panels under Static Loading", *Composite Structures,* **156**, 145–64.

Winkler, M. and Kress, G. (2010), "Deformation Limits for Corrugated Cross-Ply Laminates", *Composite Structures,* **92**(6), 1458–68.

Zenkour, A. M. (2006), "Generalized Shear Deformation Theory for Bending Analysis of Functionally Graded Plates", *Applied Mathematical Modelling,* **30**, 67–84.

Zhao, Z., Feng, C., Wang, Y. and Yang, J. (2017), "Bending and Vibration Analysis of Functionally Graded Trapezoidal Nanocomposite Plates Reinforced with Graphene Nanoplatelets (GPLs)", *Composite Structures,* **180**, 799–808.

Zheng, L. and Liu, H. (2017), "Nonlinear Bending Response of Functionally Graded Nanobeams with Material Uncertainties", *International Journal of Mechanical Sciences,* **134**, 123–35.

Zidi, M., Tounsi, A., Houari, M. S. A., Bedia, E. A. A. and Bég, O. A. (2014), "Bending Analysis of FGM Plates Under Hygro-Thermo-Mechanical Loading Using a Four Variable Refined Plate Theory", *Aerospace Science and Technology,* **34**(1), 24–34.

5 Graphene-Magnesium Core-Shell Nanocomposites
Physical, Mechanical, Thermal, and Electrical Properties

Dhiman K. Das

Gaighata Government ITI,
Ramchandrapur, India

Jit Sarkar

Boldink Technologies Private Limited,
Howrah, India

S. Sahoo

National Institute of Technology,
Durgapur, India

CONTENTS

DOI: 10.1201/9781003158813-5

5.1 INTRODUCTION

Magnesium finds a wide range of applications in automotive, aerospace, and electronics industry [1–7] because of its light weight and good strength and thermal and electrical conductivity. But the demand of enhanced properties, high performances, and specific applications at low cost by these industries has pushed several researchers to begin working on different composites of magnesium [8–12]. Graphene [13], the two-dimensional carbon-based nanomaterial, has recently drawn attention of researchers due to its unique properties. The superior mechanical, thermal, and electrical properties of graphene [14–19] can be attributed to its single-layered, two-dimensional hexagonal lattice structure. Carbon nanotube and graphene nanoplatelets have been used by several researchers as reinforcing agent in a magnesium matrix, thus forming a graphene-magnesium composite material with improved properties [20–26]. Previously, work was done to prepare graphene-magnesium nanocomposite by reinforcing graphene nanopallets in magnesium [27–29]. The mechanical properties of graphene-magnesium nanocomposite are evaluated using a MPa scale. To date, experimental synthesys of graphene-magnesium core shell nanocomposite with graphene in the core and magnesium shell has not been reported.

Recently, Das and Sarkar [30] analytically studied the properties of graphene-magnesium nanocomposite materials for its plausible advanced aerospace applications. They studied the mechanical and thermal properties of rectangular graphene-magnesium nanocomposite samples of varying sizes and graphene content. The mechanical and thermal properties were found to increase with the increase in graphene content of the graphene-magnesium nanocomposite samples, and they reported very high mechanical and thermal properties with Young's modulus and thermal conductivity of more than 65 GPa and 175 W/mK, respectively. Core-shell type nanostructures have drawn much attention among researchers and have been widely investigated both experimentally and theoretically [31–34] due to their excellent material properties, owing to their typical structure of solid core of one type and solid shell of another type. The excellent mechanical properties of core-shell structure applied to graphene-magnesium nanocomposite can greatly enhance their strength in comparison to other magnesium alloys currently used in aerospace applications.

In the present study, graphene-magnesium core-shell nanocomposites have been considered with a graphene core reinforced inside a magnesium shell. Cylindrical samples with varying core and shell thickness have been considered for convenience of calculations, and the evaluation of room temperature physical, mechanical, thermal, and electrical properties of the nanocomposites has been carried out using some mathematical and analytical models.

5.2 MATHEMATICAL MODELING OF THE PROPERTIES

Let us consider a cylindrical-shaped nanocomposite sample with a graphene core and a magnesium shell as shown in Figure 5.1. The different properties of the nanocomposite are being evaluated by varying both core and shell diameter. Later these same properties are recalculated by varying core the diameter only and keeping the shell diameter fixed.

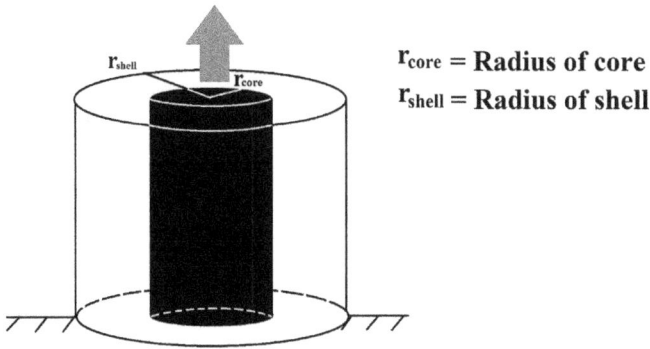

r_{core} = **Radius of core**

r_{shell} = **Radius of shell**

FIGURE 5.1 The designed magnesium-graphene composite with one end fixed and tensile force applied on the free end.

5.2.1 PHYSICAL PROPERTY

Density of a material is one of its physical properties. For composites, density can be calculated by [35]

$$\rho_c = V_m \rho_m + V_r \rho_r \tag{5.1}$$

where ρ_c, ρ_m, and ρ_r are the density of the composite, density of the matrix material, and density of the reinforced material, respectively.

5.2.2 MECHANICAL PROPERTY

Young's modulus of a composite along longitudinal direction can be calculated by [36]

$$E_{cl} = \left(\frac{A_m}{A_c}\right) E_m + \left(\frac{A_r}{A_c}\right) E_r \tag{5.2}$$

where E_{cl}, E_m, and E_r are the Young's modulus of the composite in longitudinal direction, Young's modulus of matrix material (magnesium here), and Young's modulus of the reinforced fiber (graphene), respectively, and A_c, A_m, and A_r are cross-sectional area of the composite in longitudinal direction, cross-sectional area of matrix material, and cross-sectional area of the reinforced material, respectively. This is known as rule of mixture.

For transverse loading, Young's modulus for composites can be calculated by inverse of rule of mixture. The equation can be written as [36]

$$E_{ct} = \frac{E_r E_m}{E_r V_m + E_m V_r} \tag{5.3}$$

where E_{ct} is the Young's modulus of the composite in transverse direction, $V_m = \left(\dfrac{A_m}{A_c}\right)$ and $V_r = \left(\dfrac{A_r}{A_c}\right)$. Ultimate tensile strength (UTS) for composite materials along longitudinal direction can be calculated using the equation [36]

$$\sigma_{cl} = V_m \sigma_m + V_r \sigma_r \qquad (5.4)$$

where σ_{cl}, σ_m, and σ_r are the UTS of the composite in longitudinal direction, UTS of the matrix material, and UTS of the reinforced material.

If some load is applied on the composite, then the part of load carried by reinforced material is governed by the equation [36]

$$\frac{P_r}{P_c} = V_m \frac{E_r}{E_c} \qquad (5.5)$$

where P_c and P_r are the load applied on the composite and load carried by the reinforced material. Load carried by the matrix is given by [35]

$$P_c = P_m + P_r \qquad (5.6)$$

where P_m is the load carried by the matrix material.

5.2.3 THERMAL PROPERTY

Coefficient of linear expansion for composite materials is given by [37, 38]

$$\alpha_c = \frac{E_r V_r \alpha_r + E_m V_m \alpha_m}{E_r V_r + E_m V_m} \qquad (5.7)$$

where α_c, α_m, and α_r are the coefficient of linear expansion of the composite, coefficient of linear expansion of the matrix material, and coefficient of linear expansion of the reinforced material, respectively. Again, thermal conductivity of composite material is governed by [39]

$$\frac{1}{k_c} = \frac{V_r}{k_r} + \frac{V_m}{k_m} \qquad (5.8)$$

where k_c, k_m, and k_r are the thermal conductivity of the composite, thermal conductivity of the matrix material, and thermal conductivity of the reinforced material, respectively.

5.2.4 ELECTRICAL PROPERTY

Electrical conductivity of a composite can be calculated using [40]

$$\sigma_{ec} = \sigma_{er} \left(\varphi - \varphi_c\right)^t \qquad (5.9)$$

where σ_{ec} and σ_{er} are the electrical conductivities of the composite and reinforced material, respectively, φ and φ_c are the volume fraction and percolation threshold, respectively, and t is the critical exponent, respectively. Electrical resistivity is the inverse of conductivity for any material and can be determined by the equation [41]

$$\rho'_{rc} = \frac{1}{\sigma_{ec}} \tag{5.10}$$

where ρ'_{rc} is the electrical resistivity of the composite. Lorenz number of a material can be calculated using the equation [42]

$$L_c = \frac{k_c}{\sigma_{ec}T} \tag{5.11}$$

where L_c is Lorenz number of the composite and T is the operation temperature.

5.2.5 WITH VARYING BOTH CORE AND SHELL DIAMETER

We have considered four cylindrical nanocomposite samples with graphene core of radius (r_{core}) 8, 10, 12, 14 nm and magnesium shell of radius (r_{shell}) 38, 40, 42, 44 nm, respectively.

5.2.6 WITH VARYING CORE DIAMETER AND FIXED SHELL DIAMETER

In this case, we have considered four cylindrical nanocomposite samples with graphene core of radius (r_{core}) 8, 10, 12, 14 nm and magnesium shell of radius (r_{shell}) 40 nm. For our samples with variable core and shell diameters the volume ratio of graphene and magnesium are 16:225, 25:225, 36:225, and 49:225 for samples with serial number 1, 2, 3, and 4 shown in Table 5.1. For samples with variable core and

TABLE 5.1
Calculated A_c, A_m, A_r, V_m, and V_r for the Composite Samples

Sl. No.	Sample Geometry	A_c (nm²)	A_m (nm²)	A_r (nm²)	V_m (nm²)	V_r (nm²)
Variable core and shell diameter (nm)						
1.	r_{core} = 8 nm; r_{shell} = 38 nm	4534.16	4333.2	200.96	0.9556	0.0443
2.	r_{core} = 10 nm; r_{shell} = 40 nm	5024	4710	314	0.9375	0.0625
3.	r_{core} = 12 nm; r_{shell} = 42 nm	5538.96	5086.8	452.16	0.9183	0.08163
4.	r_{core} = 14 nm; r_{shell} = 44 nm	5463.6	6079.04	615.44	0.8987	0.1011
Variable core diameter and fixed shell diameter (nm)						
5.	r_{core} = 8 nm; r_{shell} = 40 nm	5024	4823.04	200.96	0.96	0.04
6.	r_{core} = 10 nm; r_{shell} = 40 nm	5024	4710	314	0.9375	0.0625
7.	r_{core} = 12 nm; r_{shell} = 40 nm	5024	4571.84	452.16	0.91	0.09
8.	r_{core} = 14 nm; r_{shell} = 40 nm	5024	4408.56	615.44	0.8775	0.1225

fixed shell diameters the volume ratio of graphene and magnesium are 1:16, 1:25, 9:49, and 49:169 for samples with serial number 5, 6, 7, and 8 shown in Table 5.1.

The A_c, A_m, A_r, V_m, and V_r for the composite samples with above given specifications are calculated and presented in Table 5.1. Young's modulus of graphene and magnesium are 1,034 GPa and 45 GPa respectively [43, 44]. Mass density of graphene and magnesium are 2,300 kg/m³ [45] and 1,740 kg/m³ [44], coefficient of linear expansion for graphene and magnesium are -3.75×10^{-6}/K [46] and 26×10^{-6}/K [44] and thermal conductivities of graphene and magnesium are 2,500 W/mK [47] and 156 W/mK [44], respectively. Percolation threshold for honeycomb structures is $1 - 2\sin\left(\dfrac{\pi}{18}\right)$ [48], critical exponent is 0.333 [50], and electrical conductivity of graphene is 1.96×10^{10} S/m [42].

So sequentially arranging the above stated values we get the values of all the parameters of graphene and magnesium required to evaluate the mechanical, thermal, electrical, and physical properties of our desired magnesium-graphene nanocomposite. Matching them with previously given nomenclatures we get $E_r = 1034$ GPa, $E_m = 1034$ GPa, $\rho_r = 2300$ kg/m³, $\rho_m = 1740$ kg/m³, $\sigma_r = -3.75 \times 10^{-6}$/K, $\sigma_m = 26 \times 10^{-6}$/K, $\varphi = 1 - 2\sin\left(\dfrac{\pi}{18}\right)$, $t = 0.333$, and $\sigma_{er} = 1.96 \times 10^{10}$ S/m.

5.3   RESULTS AND DISCUSSION

Substituting these values in the above stated Equations (5.1)–(5.11), we have estimated physical property like density, mechanical properties like Young's modulus, UTS, load transmittance through matrix and fiber, thermal properties like coefficient of linear expansion, thermal conductivity and electrical properties like electrical conductivity, resistivity and Lorenz number for the graphene-magnesium nanocomposites. The presence of graphene in the core is expected to enhance the different material properties of the nanocomposites due to its own superior properties owing to the two-dimensional, single-layer, hexagonal structure.

5.3.1   ESTIMATION OF PHYSICAL AND MECHANICAL PROPERTIES

Different physical and mechanical properties of the graphene-magnesium nanocomposites under varying conditions of core and shell radius are listed in Table 5.2 and their variations are plotted in Figures 5.2–5.5.

Figure 5.2 shows the variation of density of both the nanocomposites with variable core and shell radius and with variable core and constant shell radius. The density of the nanocomposites is found to increase linearly with the increasing core radius in both cases. This can be directly attributed to the increase in both core and shell radius in the first case (Figure 5.2a) and only the core radius in the second case (Figure 5.2b). The variation of Young's modulus in the transverse and longitudinal directions for both the nanocomposites with variable core and shell radius and with variable core and constant shell radius is shown in Figure 5.3. It was observed that the Young's modulus along longitudinal loading direction is nearly double to that along transverse loading direction, with the rate of increase in

TABLE 5.2
Mechanical and Physical Properties of the Composite Samples

Sl. No.	Sample Geometry	Mechanical Property					Physical Property
		Young's Mmodulus (GPa)		UTS (GPa)	Load Carried (kN)		Density (kg/m³)
		Longitudinal	Transverse	Longitudinal	Fiber	Matrix	
Variable core and shell radius (nm)							
1.	$r_{core} = 8$; $r_{shell} = 38$	88.8302	46.99601	4.4096	5.1566	4.8433	1764.634
2.	$r_{core} = 10$; $r_{shell} = 40$	106.8125	47.8611	3.95875	6.0503	3.9496	1775
3.	$r_{core} = 12$; $r_{shell} = 42$	125.7346	48.8147	3.61051	6.7129	3.287	1785.591
4.	$r_{core} = 14$; $r_{shell} = 44$	145.0484	49.8283	3.2875	7.207	2.793	1796.268
Variable core diameter and fixed shell radius (nm)							
5.	$r_{core} = 8$; $r_{shell} = 40$	84.56	46.7901	3.9807	4.891	5.108	1762.4
6.	$r_{core} = 10$; $r_{shell} = 40$;	106.8125	47.8611	3.95875	6.0503	3.9496	1775
7.	$r_{core} = 12$; $r_{shell} = 40$	134.01	49.2386	3.9808	6.944	3.055	1790.4
8.	$r_{core} = 14$; $r_{shell} = 40$	166.1525	50.9723	3.98055	7.6234	2.3765	1808.6

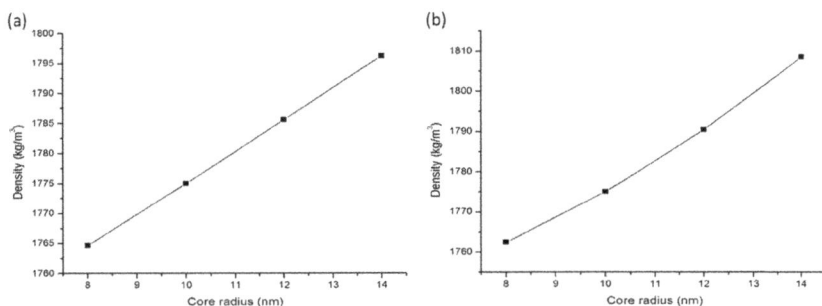

FIGURE 5.2 Variation of density with core radius for (a) variable core and variable shell radius and (b) variable core and constant shell radius.

Young's modulus along longitudinal loading direction to be much greater than that along transverse loading direction. The Young's modulus is found to increase linearly both for variable core and shell radius and for variable core and constant shell radius during both longitudinal and transverse loading due to increase in dimension of the nanocomposites.

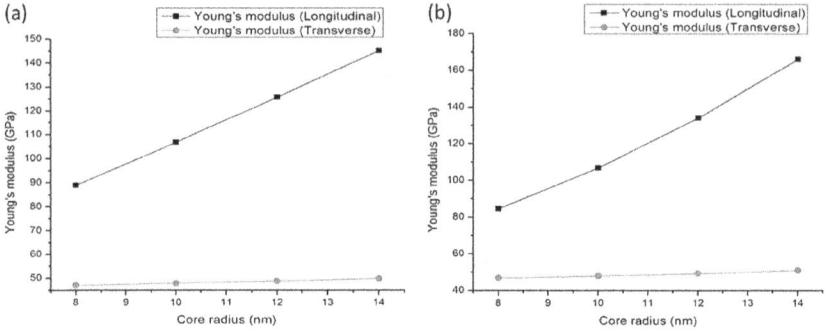

FIGURE 5.3   Variation of Young's modulus with core radius for (a) variable core and variable shell radius and (b) variable core and constant shell radius.

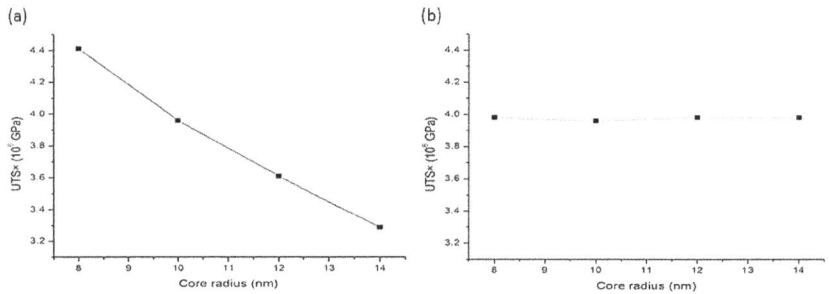

FIGURE 5.4   Variation of UTS with core radius for (a) variable core and variable shell radius and (b) variable core and constant shell radius.

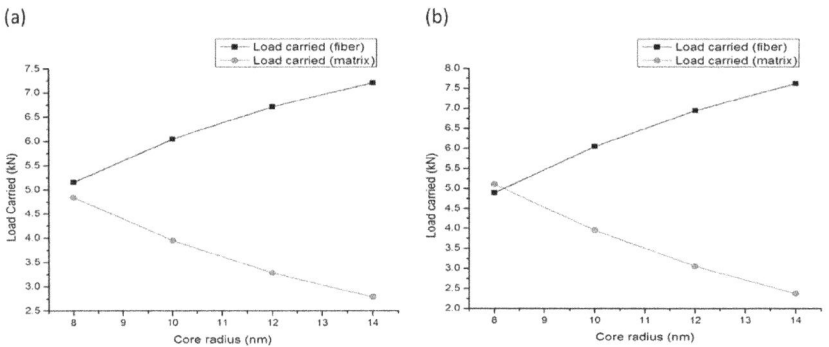

FIGURE 5.5   Variation of load carried by matrix and fiber with core radius for (a) variable core and variable shell radius and (b) variable core and constant shell radius.

Figure 5.4 shows the variation of ultimate tensile strength of both nanocomposites with variable core and shell radius and with variable core and constant shell radius. It was found that the ultimate tensile strength increases linearly with decreasing core radius (as well as shell radius), which can be attributed to the dominating effect of specific surface/interface area at the nanoscale. On the other hand, it slightly fluctuates with decreasing core radius for variable core and constant shell radius, which indicates that the variation of shell thickness mainly contributes to the change in ultimate tensile strength of the nanocomposites. Thus, for a constant shell radius, the change in core radius has no significant effect on the ultimate tensile strength of the nanocomposites.

The variation of load carrying capacity of the matrix and fiber with core radius for both the nanocomposites with variable core and shell radius and with variable core and constant shell radius is shown in Figure 5.5. It was found that the load-carrying capacity of the fiber in comparison to the matrix is enhanced with increase in core radius, due to the superior mechanical properties of graphene. It increases almost linearly for both variable core and shell radius and for variable core and constant shell radius, due to the increase in the amount of graphene. The combination of the high individual load carrying capacity of the fiber and matrix contributes to the higher load-carrying capacity of the nanocomposites.

5.3.2 ESTIMATION OF THERMAL PROPERTIES

Different thermal properties of the graphene-magnesium nanocomposites under varying conditions of core and shell radius are listed in Table 5.3 and their variations are plotted in Figures 5.6 and 5.7.

Figure 5.6 shows the variation of coefficient of linear thermal expansion of both the nanocomposites with variable core and shell radius and with variable core and constant shell radius. It was observed that the coefficient of linear thermal expansion decreases almost linearly with increasing core radius for both variable core and shell radius and for variable core and constant shell radius, which indicates lower thermal expansion of the nanocomposites with increasing size. This indicates that the variation of core thickness mainly contributes to the change in coefficient of linear

TABLE 5.3
Thermal Properties of the Composite Samples

Sl. No.	Sample Geometry	Thermal Property	
		Coefficient of Linear Expansion (/K)	Thermal Conductivity (W/Mk)
Variable core and shell radius (nm)			
1.	$r_{core} = 8$; $r_{shell} = 38$;	1.06553×10^{-5}	162.7773
2.	$r_{core} = 10$; $r_{shell} = 40$	8×10^{-6}	165.71
3.	$r_{core} = 12$; $r_{shell} = 42$	6.0279×10^{-6}	168.942
4.	$r_{core} = 14$; $r_{shell} = 44$	4.5486×10^{-6}	172.374
Variable core diameter and fixed shell radius (nm)			
5.	$r_{core} = 8$; $r_{shell} = 40$;	1.144×10^{-5}	162.0798
6.	$r_{core} = 10$; $r_{shell} = 40$	8×10^{-6}	165.71
7.	$r_{core} = 12$; $r_{shell} = 40$;	5.3408×10^{-6}	170.377
8.	$r_{core} = 14$; $r_{shell} = 40$	3.032×10^{-6}	176.2425

(a)

(b)

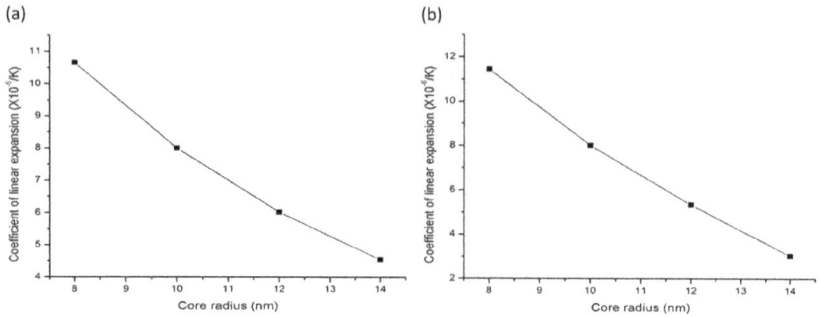

FIGURE 5.6 Variation of linear thermal expansion coefficient with core radius for (a) variable core and variable shell radius and (b) variable core and constant shell radius.

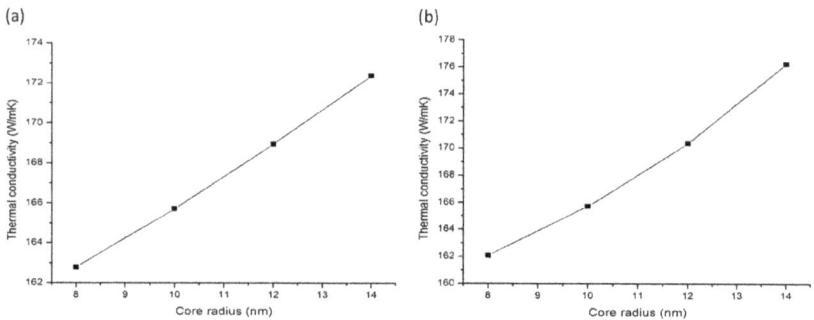

(a)

(b)

FIGURE 5.7 Variation of thermal conductivity with core radius for (a) variable core and variable shell radius and (b) variable core and constant shell radius.

thermal expansion of the nanocomposites. The variation of thermal conductivity with core radius for both the nanocomposites with variable core and shell radius and with variable core and constant shell radius is shown in Figure 5.7. It was found that the thermal conductivity linearly increases with increasing core radius for both variable core and shell radius and with variable core and constant shell radius, which can be attributed to the increase in the amount of graphene, which exhibits excellent thermal conductivity.

5.3.3 ESTIMATION OF ELECTRICAL PROPERTIES

Different electrical properties of the graphene-magnesium nanocomposites under varying conditions of core and shell radius are listed in Table 5.4 and their variations are plotted in Figures 5.8 and 5.9.

Figure 5.8 shows the variation of electrical conductivity of both the nanocomposites with variable core and shell radius and with variable core and constant shell radius. It was observed that the electrical conductivity increases linearly with the increase in core radius for variable core and shell radius and for variable core and constant shell radius, which can be attributed to the increase in the amount of

TABLE 5.4
Electrical Properties of the Composite Samples

Sl. No.	Sample Geometry	Electrical Property	
		Electrical Conductivity (S/m)	Electrical Resistivity (Ω/m)
Variable core and shell radius (nm)			
1.	$r_{core} = 8$; $r_{shell} = 38$	1.3547×10^{10}	7.3817×10^{-11}
2.	$r_{core} = 10$; $r_{shell} = 40$	1.37091×10^{10}	7.251×10^{-11}
3.	$r_{core} = 12$; $r_{shell} = 42$	1.40601×10^{10}	7.11232×10^{-11}
4.	$r_{core} = 14$; $r_{shell} = 44$	1.4345×10^{10}	6.97107×10^{-11}
Variable core diameter and fixed shell radius(nm)			
5.	$r_{core} = 8$; $r_{shell} = 40$	1.3489×10^{10}	7.4134×10^{-11}
6.	$r_{core} = 10$; $r_{shell} = 40$	1.37091×10^{10}	7.251×10^{-11}
7.	$r_{core} = 12$; $r_{shell} = 40$	1.4179×10^{10}	7.0526×10^{-11}
8.	$r_{core} = 14$; $r_{shell} = 40$	1.4667×10^{10}	6.81802×10^{-11}

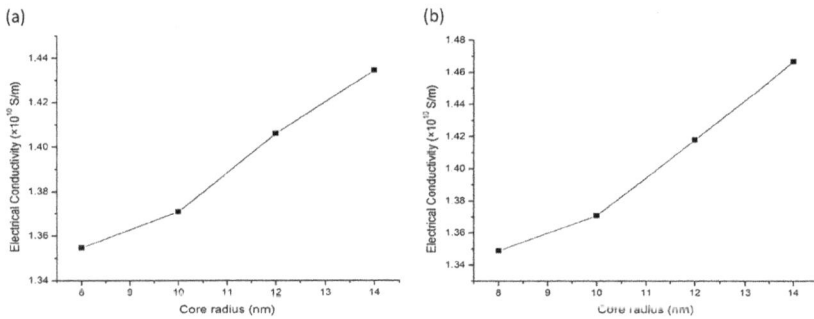

FIGURE 5.8 Variation of electrical conductivity with core radius for (a) variable core and variable shell radius and (b) variable core and constant shell radius.

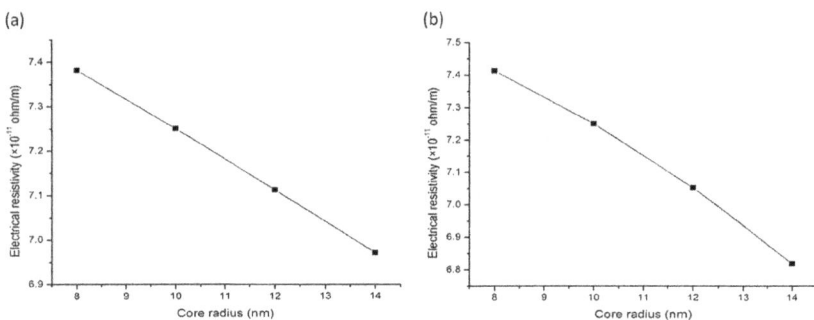

FIGURE 5.9 Variation of electrical resistivity with core radius for (a) variable core and variable shell radius and (b) variable core and constant shell radius.

graphene that exhibits excellent electrical conductivity. The variation of electrical resistivity for both the nanocomposites with variable core and shell radius and with variable core and constant shell radius is shown in Figure 5.9. On the other hand, it was observed that the electrical resistivity decreases linearly with increasing core radius, as the increase in amount of graphene allows higher electrical conductivity in the nanocomposites.

It is found that the Young's modulus, thermal conductivity, and electrical conductivity of the nanocomposites are enhanced up to 3.68, 0.12, and 643 times more than pure magnesium depending on the concentration of graphene reinforcement in it. Lorenz number of the nanocomposites is calculated as 4.0321×10^{-11} WΩ/K^2.

5.4 CONCLUSION

The physical, mechanical, thermal, and electrical properties of cylindrical (core-shell) graphene-reinforced magnesium nanocomposite are estimated using analytical models. The results show huge improvement in mechanical, thermal, and electrical properties due to the hexagonal, single-layer graphene reinforcement. The properties improve more with increasing core thickness for a particular shell thickness, as well as with increasing core and shell thickness. Again, the increasing core thickness negatively affects the strength of the nanocomposite. The graphene-magnesium nanocomposite shows enhanced Young's modulus, thermal conductivity, and electrical conductivity up to 3.68, 0.12, and 643 times more than normal magnesium. The Young's modulus for the nanocomposite is even greater than pure aluminum and copper. It also shows low coefficient of linear expansion, electrical resistivity, and higher electrical conductivity than copper and aluminum. Thus graphene-magnesium nanocomposite can be used as a potential candidate for aerospace, automotive, and electronic applications.

ACKNOWLEDGMENTS

S. Sahoo gratefully acknowledges the financial support of CSIR, New Delhi, India for this work through the project No. 03(1473)/19/EMR-II Dated 05/08/2019. Dhiman K. Das gratefully acknowledges Gaighata Government ITI, Ramchandrapur, India for initiation and continuation of this project.

REFERENCES

1. Blawert, C., Hort, N. and Kainer, K. U.: Automotive applications of magnesium and its alloys. *Trans. Indian Inst. Met.* **57**, 397–408 (2004).
2. Kulekci, M. K.: Magnesium and its alloys applications in automotive industry. *Inter. J. of Adv. Manuf. Technol.* **39**, 851–865 (2008).
3. Froes, F. H., Eliezer, D. and Aghion, E.: The science, technology, and applications of magnesium. *J. of the Miner. Met. & Mater. Soc. (TMS)* **50**, 30–34 (1998).
4. Landkof, B.: *Magnesium applications in aerospace and electronics industries, in magnesium alloys and their applications.* Wiley-VCH Verlag GmbH & Co., Germany (2000).
5. Ostrovsky, I. and Henn, Y.: Present state and future of magnesium application in aerospace industry. *International Conference "New challenges in aeronautics" ASTEC'07,* Moscow (2007).

6. Luo, A. A.: Magnesium casting technology for structural applications. *J. of Magnesium and Alloys* **1**, 2–22 (2013).

7. Easton, M.: Magnesium alloy applications in automotive structures. *J. of Metals* **60**, 57–62 (2008).

8. Tiwari, S., Balasubramaniam, R. and Gupta, M.: Corrosion behavior of SiC reinforced magnesium composites. *Corros. Sci.* **49**, 711–725 (2007).

9. Sankaranarayanan, S. et al.: Nano-AlN particle reinforced Mg composites: Microstructural and mechanical properties. *Mater. Sci. and Technol.* **31**, 1122–1131 (2015).

10. Seetharaman, S. et al.: Synthesis and characterization of nano boron nitride reinforced magnesium composites produced by the microwave sintering method. *Materials* **6**, 1940–1955 (2013).

11. Tun, K. S. et al.: Tensile and compressive responses of ceramic and metallic nanoparticle reinforced Mg composites. *Materials* **6**, 1826–1839 (2013).

12. Wang, H. Y. et al.: Fabrication of TiB_2 particulate reinforced magnesium matrix composites by powder metallurgy. *Mater. Lett.* **58**, 3509–3513 (2004).

13. Geim, A. K. and Novoselov, K. S.: The rise of graphene. *Nat. Mater.* **6**, 183–191 (2007).

14. Ovid'ko, I. A.: Mechanical properties of graphene. *Rev. Adv. Mater. Sci.* **34**, 1–11 (2013).

15. Scarpa, F., Adhikari, S. and Phani, A. S.: Effective elastic mechanical properties of single layer graphene sheets. *Nanotech.* **20**, 065709 (2009).

16. Xu, X. et al.: Length-dependent thermal conductivity in suspended single-layer graphene. *Nat. Commun.* **5**, 1–6 (2014).

17. Balandin, A. A. et al.: Superior thermal conductivity of single-layer graphene. *Nano Lett.* **8**, 902–907 (2008).

18. Wassei, J. K. and Kaner, R. B.: Graphene, a promising transparent conductor. *Mater. Today* **13**, 52–59 (2010).

19. Deka, M. J., Baruah, U. and Chowdhury, D.: Insight into electrical conductivity of graphene and functionalized graphene: Role of lateral dimension of graphene sheet. *Mater. Chem. and Phys.* **163**, 236–244 (2015).

20. Goh, C. S.: Development of novel carbon nanotube reinforced magnesium nanocomposites using the powder metallurgy technique. *Nanotechnol.* **17**, 7–12 (2006).

21. Morelli, E. C.: Carbon nanotube/magnesium composites. *Phys. Stat. Sol. (A)* **201**, 53–55 (2004).

22. Rashad, M.: Use of high energy ball milling to study the role of graphene nanoplatelets and carbon nanotubes reinforced magnesium alloy. *J. of Alloys and Compd.* **646**, 223–232 (2015).

23. Xiang, S. et al.: Graphene nanoplatelets induced heterogeneous bimodal structural magnesium matrix composites with enhanced mechanical properties. *Sci. Rep.* **6**, 38824 (2016).

24. Rashad, M. et al.: Enhanced tensile properties of magnesium composites reinforced with graphene nanoplatelets. *Mater. Sci. & Engg. A* **630**, 36–44 (2015).

25. Rashad, M. et al.: Improved strength and ductility of magnesium with addition of aluminum and graphene nanoplatelets (Al + GNPs) using semi powder metallurgy method. *J. of Ind. and Eng. Chem.* **23**, 243–250 (2015).

26. Rashad, M. et al.: Development of magnesium-graphene nanoplatelets composite. Journal of Composite Materials. **49**(3), 285–293 (2015).

27. Hu, Z. et al.: Graphene-reinforced metal matrix nanocomposites—A review. *Mater. Sci. Technol.* **32**(9), 930–953 (2016).

28. Xiang, S. et al.: Graphene nanoplatelets induced heterogeneous bimodal structural magnesium matrix composites with enhanced mechanical properties. *Sci. Rep.* **6**(38824), 1–12 (2016).

29. Das, D. K. and Sarkar, Jit: Graphene magnesium nanocomposite: An advanced material for aerospace application. *Mod. Phys. Lett. B* **32**, 1850075 (2018).
30. Sakar, J. et al.: Synthesis and characterizations of Cu–Ag core–shell nanoparticles. *Adv. Sci. Lett.* **22**, 193–196 (2016).
31. Sarkar, Jit and Das, D. K.: Study of the effect of varying core diameter, shell thickness and strain velocity on the tensile properties of single crystals of Cu–Ag core–shell nanowire using molecular dynamics simulations. *J. Nanopart. Res. B* **20**, 9 (2018).
32. Sarkar, Jit and Das, D. K.: Investigation of mechanical properties and deformation behavior of single-crystal Al-Cu core-shell nanowire generated using non-equilibrium molecular dynamics simulation. *J. Nanopart. Res. B* **20**, 153 (2018).
33. Sarkar, Jit and Das, D. K.: Molecular dynamics study of defect and dislocation behaviors during tensile deformation of copper-silver core-shell nanowires with varying core diameter and shell thickness *J. Nanopart. Res. B* **20**, 272 (2018).
34. Velmurugan, R.: Composite materials. http://nptel.ac.in/courses/101106038/mod03lec01.pdf (accessed 2 June 2020).
35. Soboyejo, W. O.: *Mechanical properties of engineered materials.* CRC Press, USA (2002).
36. Karch, C.: Micromechanical analysis of thermal expansion coefficients. *Model. Simul. Mater. Sci.* **4**, 104–118 (2014).
37. Zhiguo, R. et al.: Determination of thermal expansion coefficients for unidirectional fiber- reinforced composites. *Chin. J. of Aeronaut.* **27**, 1180–1187 (2014).
38. Chandana, E. and Hussian, S. A.: Thermal conductivity characterization of bamboo fiber reinforced in epoxy resin. *IOSR J. of Mech. and Civil Eng.* **9**, 07–14 (2013).
39. Clingerman, M. L.: Development and modelling of electrically conductive composite materials. PhD Thesis. Michigan Technological University, USA (2008).
40. Datta, D., Pal, B. and Chaudhuri, B.: *Elements of physics-2.* Publishing Syndicate, India (2003).
41. Das, D. K., Roy, S. and Sahoo, S.: Theoretical study of some electrical parameters of graphene. *Mod. Phys. Lett. B* **30**, 1650366 (2016).
42. Lee, C. et al.: Measurement of the elastic properties and intrinsic strength of monolayer graphene. *Science* **321**, 385–388 (2008).
43. https://en.wikipedia.org/wiki/Magnesium (accessed 02 June 2020)
44. Gupta, S. S. and Batra, R. C.: Elastic properties and frequencies of free vibrations of single-layer graphene sheets. *J. of Comput. and Theor. Nanosci.* **7**, 1–14 (2010).
45. Shaina, P. R.: Estimating the thermal expansion coefficient of graphene: The role of graphene–substrate interactions. *J. Phys.: Condens. Matter* **28**, 085301 (2016).
46. Cai, W. et al.: Thermal transport in suspended and supported monolayer graphene grown by chemical vapor deposition. *Nano Lett.* **10**, 1645–1651 (2010).
47. Stover, C. and Weisstein, E. W.: Percolation threshold. *MathWorld–A Wolfram Web Resource.* http://mathworld.wolfram.com/PercolationThreshold.html (accessed 2 June 2020).
48. Bastas, N. et al.: Method for estimating critical exponents in percolation processes with low sampling. *Phys. Rev. E* **90**, 062101 (2014).
49. Du, X. M.: Preparation and characterization of magnesium matrix composites reinforced with graphene nano-sheets. *J. of Non-Oxide Glasses* **9**, 25–32 (2017).

6 Free Vibration of Carbon Nanotube–Reinforced Composite Beams under the Various Boundary Conditions

Lazreg Hadji
University of Tiaret,
Tiaret, Algeria

Mehmet Avcar
Suleyman Demirel University,
Isparta, Turkey

Ömer Civalek
China Medical University,
Taichung, Taiwan

CONTENTS

6.1 INTRODUCTION

The desirable features of structural components in engineering applications are safety, functionality, aesthetics, and cost. The use of nonuniform, nonhomogeneous,

DOI: 10.1201/9781003158813-6

and reinforced components helps ensure the requirements, as well as enhanced strength and structural efficiency, while lowering total cost and weight. Composite material refers to any solid that consists of more than one component, in which they are in separate phases. The main advantages of composite materials are excellent strength-to-weight and stiffness-to-weight ratios. One of the composite materials consisting of fibers in a matrix that offer considerable benefit over traditional structural materials is a fiber-reinforced composite. In contemporary engineering applications, structural components made of composite materials have an extensive range of uses—for example, the fibrous composites, consisting of carbon, glass, aramid, and basalt fibers, have seen widespread use in modern engineering and industries, such as civil, automotive, mechanical, defence, marine, aviation, and aerospace [1–4].

The discovery of carbon nanotubes (CNTs) by Iijima [5] accelerated nanotechnology development. Because of their remarkable mechanical, chemical, thermal, physical, and electrical characteristics, CNTs have become utilized as reinforcing components instead of traditional composites to build high-performance structural and multifunctional composites for a variety of possible applications [6–12]. Therefore, most engineering structural components are now quickly replaced with CNTs, and they serve in a comparable state for longer periods. Consequently, CNTRC structures have gotten a lot of interest, and so examination of mechanical behaviors is becoming a hot topic for researchers. Based on Timoshenko beam theory and von Kármán geometric nonlinearity, Ke et al. [13] studied the nonlinear free vibration of FGCNT beams reinforced by SWCNTs. Free vibrations and buckling analysis of CNTRC Timoshenko beam resting on an elastic foundation are investigated by Yas and Samadi [14]. Based on Timoshenko beam theory, Ke et al. [15] presented a dynamic stability analysis for functionally graded nanocomposite beams reinforced by SWCNTs. Shen and Xiang [16] investigated the behaviors of CNTRC beams lying on an elastic foundation in thermal settings, including large-amplitude vibration, nonlinear bending, and thermal post-buckling. Wattanasakulpong and Ungbhakorn [17] investigated the bending, buckling, and vibration characteristics of CNTRC beams with the help of a few shear deformation theories. Taking into account two different types of CNT reinforced beams, Lin and Xiang [18] analyzed the linear free vibration of nanocomposite beams reinforced with SWCNTs. Based on the Timoshenko beam theory and von Kármán geometric nonlinearity, Ansari et al. [19] studied the forced vibration behavior of nanocomposite beams reinforced by SWCNTs. Within the context of Timoshenko beam theory, Wu et al. [20] examined the free vibration and elastic buckling of sandwich beams with a stiff core and FG-CNTRC face sheets. Taking into consideration the stretching effect, Hadji et al. [21, 22] dealt with the bending and free vibration behaviors of CNTRC beams resting on parameter elastic foundation using a quasi-3D shear deformation theory. Based on first-order shear deformation elasticity theory, Shi et al. [23] proposed an exact solution method for modeling the free vibration of FG-CNTRC beams with arbitrary boundary conditions. Salami [24] used extended high-order sandwich panel theory to investigate the free vibration analysis of a sandwich beam with softcore and CNTRC face sheets. Based on Bernoulli–Euler beam theory, Rosa et al. [25] investigated the dynamic characteristics of FG-CNTRC nanobeams made of multilayered polymer in hygrothermal conditions. Civalek et al. [26] performed a forced vibration analysis of a CNTRC beam subjected to a harmonic point load at the midpoint of the beam under simply supported boundary conditions.

Numerous researches devote efforts to investigating the mechanical characteristics of CNTRC beams, according to the findings of the open literature search. The majority of these researches use numerical solution methods; nevertheless, the number of studies that use exact solutions is relatively low, and the precise solutions obtained in these studies are usually for simply supported boundary conditions. Because free vibration of CNTRC beams is so important in current technology, finding a dependable precise solution for this behavior under various boundary circumstances has become a requirement. As a result, the current study attempts to solve this issue. A precise analytical solution for the free vibration of CNTRC beams under four distinct boundary conditions is presented using RBT in order to achieve this goal. The effects of carbon nanotube volume fraction, diverse types of CNT distribution patterns, span-to-depth ratio, and boundary conditions on the free vibration response of CNTRC beams are investigated in detail.

6.2   THEORETICAL FORMULATION

Consider a straight CNTRC beam comprised of a combination of SWCNTs and an isotropic polymer matrix with length L, width b, and thickness h in the x, y and z axes, as illustrated in Figure 6.1(a). In the thickness direction of the composite

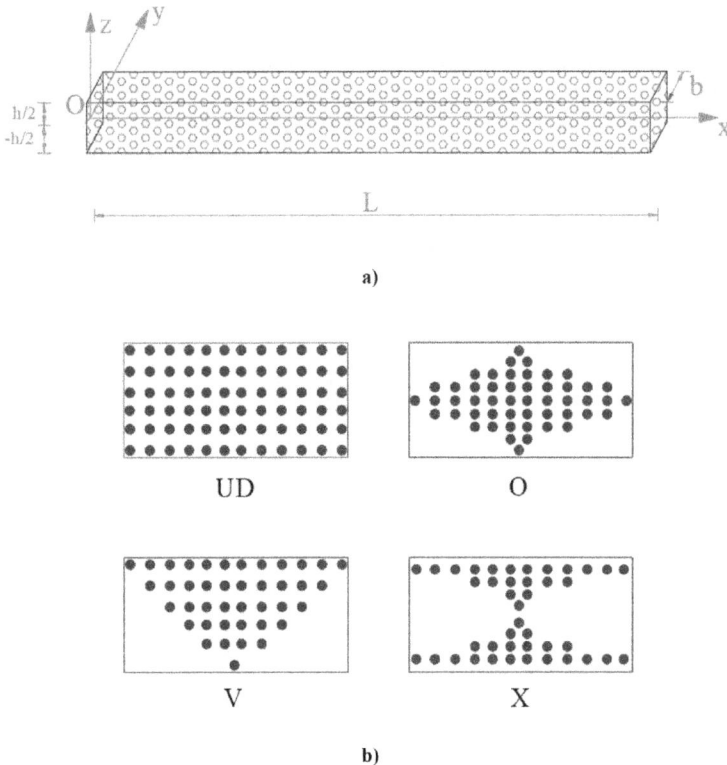

a)

b)

FIGURE 6.1   CNTRC beam (a) geometry and coordinate system and (b) cross-sections of UD, FG-O, FG-V, and FG-X patterns.

beams (z-axis direction), Figure 6.1(b) shows the uniform distribution (UD) and functionally graded (FG) distributions (FG-O, FG-X, and FG-V) of carbon nanotubes. Note that the density of CNTs within the area is constant in this illustration, but the volume fraction varies as the beam thickness increases. As a result, there is no sudden interface between the CNT and the polymer matrix throughout the beam.

The effective Young's modulus and shear modulus of CNTRC beams may be stated as according to the mixture model of rule [27]:

$$E_{11} = \eta_1 V_{cnt} E_{11}^{cnt} + V_p E_p \tag{6.1}$$

$$\frac{\eta_2}{E_{22}} = \frac{V_{cnt}}{E_{22}^{cnt}} + \frac{V_p}{E_p} \tag{6.2}$$

$$\frac{\eta_3}{G_{12}} = \frac{V_{cnt}}{G_{12}^{cnt}} + \frac{V_p}{G_p} \tag{6.3}$$

$$V_{cnt} + V_P = 1 \tag{6.4}$$

where E_p and G_p are the related material properties of the polymer matrix, V_{cnt} and V_P are the volume fractions for CNT and the polymer matrix, respectively, and $E_{11}^{cnt}, E_{22}^{cnt}$, and G_{12}^{cnt} are Young's modulus and shear modulus of SWCNT, respectively.

Similarly, the Poisson's ratio (ν) and mass density (ρ) of the CNTRC beam can be given as

$$\nu = V_{cnt} \nu^{cnt} + V_p \nu^p \tag{6.5}$$

$$\rho = V_{cnt} \rho^{cnt} + V_p \rho^p \tag{6.6}$$

where ν^{cnt}, ν^p and ρ^{cnt}, ρ^p are the Poisson's ratio and densities of the CNT and polymer matrix, respectively.

The definitions for different patterns of CNT reinforcements given in Figure 6.1(b) are described with the following mathematical functions [27]:

$$\text{UD: } V_{cnt} = V_{cnt}^* \tag{6.7}$$

$$\text{FG–O: } V_{cnt} = 2\left(1 - 2\frac{|z|}{h}\right)V_{cnt}^* \tag{6.8}$$

$$\text{FG–X: } V_{cnt} = 4\frac{|z|}{h}V_{cnt}^* \tag{6.9}$$

TABLE 6.1

The Efficiency Parameters of CNTs

V_{cnt}^*	η_1	$\eta_2 = \eta_3$
0.12	1.2833	1.0566
0.17	1.3414	1.7101
0.28	1.3238	1.7380

$$FG - V : V_{cnt} = \left(1 + 2\frac{z}{h}\right)V_{cnt}^* \tag{6.10}$$

where V_{cnt}^* is the specified volume of CNTs, which may be determined using the equation below:

$$V_{cnt}^* = \frac{W_{cnt}}{W_{cnt} + \left(\rho^{cnt}/\rho^m\right)\left(1 - W_{cnt}\right)} \tag{6.11}$$

where W_{cnt} is the mass fraction of CNTs.

In this chapter, size-dependent properties of the CNTRC beam are captured using the three efficiency parameters (η_i) given in Table 6.1, which are related to the volume fraction $\left(V_{cnt}^*\right)$ and offered by Yas and Samadi [14] depending on molecular dynamic simulations of Han and Elliott [28].

6.3 BASIC EQUATIONS

6.3.1 FUNDAMENTAL ASSUMPTIONS

The following are the current theory's assumptions:

- The FG beam's middle surface serves as the origin of the Cartesian coordinate system.
- Because the displacements are minor concerning the beam's height, the stresses involved are negligible.
- The transverse displacement is composed of the following: bending and shear, which are solely functions of the x and t as

$$w(x,z,t) = w_b(x,t) + w_s(x,t) \tag{6.12}$$

- The transverse normal stress σ_z is negligible in comparison with in-plane stresses σ_x.
- In comparison to in-plane stress, σ_x, the transverse normal stress is negligible, σ_z.

- Extension, bending, and shear components comprise the axial displacement, u in the x-direction.

$$u = u_0 + u_b + u_s \tag{6.13}$$

- The bending component, w_b, is found to be similar to the displacements predicted by classical beam theory, and so the expression for u_b is as follows:

$$u_b = -z\frac{\partial w_b}{\partial x} \tag{6.14}$$

- The shear components, u_s and w_s causes the hyperbolic variation of shear strain γ_{xz} and so shear stress, τ_{xz}, is zero at the top and bottom faces of the beam. As a result, u_s may be written as

$$u_s = -f(z)\frac{\partial w_s}{\partial x} \tag{6.15}$$

where $f(z)$ is the trigonometric shear deformation function and defined as

$$f(z) = z - \frac{h}{\pi}\sin\left(\frac{\pi z}{h}\right) \tag{6.16}$$

6.3.2 KINEMATICS

The displacement field of the n-order refined theory may be calculated using Equations (6.12)–(6.16) as a function of the assumptions established in the prior section.

$$u(x,z,t) = u_0(x,t) - z\frac{\partial w_b}{\partial x} - f(z)\frac{\partial w_s}{\partial x} \tag{6.17}$$

$$w(x,z,t) = w_b(x,t) + w_s(x,t) \tag{6.18}$$

The displacements are associated with the strains shown below:

$$\varepsilon_x = \varepsilon_x^0 + z k_x^b + f(z) k_x^s \tag{6.19}$$

$$\gamma_{xz} = g(z)\gamma_{xz}^s \tag{6.20}$$

where

$$\varepsilon_x^0 = \frac{\partial u_0}{\partial x}, \; k_x^b = -\frac{\partial^2 w_b}{\partial x^2}, \; k_x^s = -\frac{\partial^2 w_s}{\partial x^2}, \; \gamma_{xz}^s = \frac{\partial w_s}{\partial x} \tag{6.21}$$

$$g(z) = 1 - \frac{df(z)}{dz} \qquad (6.22)$$

Supposing that the CNTRC beam obeys Hooke's law, the stresses are expressed as

$$\sigma_x = Q_{11}(z)\varepsilon_x \qquad (6.23)$$

$$\tau_{xz} = Q_{55}(z)\gamma_{xz} \qquad (6.24)$$

where

$$Q_{11}(z) = \frac{E_{11}(z)}{1-v^2}; \quad Q_{55}(z) = G_{12}(z) \qquad (6.25)$$

6.3.3 EQUATIONS OF MOTION

The equations of motion are developed using Hamilton's principle [29]:

$$\int_{t_1}^{t_2} (\delta U - \delta T) dt = 0 \qquad (6.26)$$

where t is the time; t_1 and t_2 are the initial and end time, respectively; δU is the virtual change of the strain energy; and δT is the virtual change of the kinetic energy.

The change of the beam's strain energy may be expressed as

$$\begin{aligned}
\delta U &= \int_0^L \int_{-h/2}^{h/2} \left[\sigma_x \delta\varepsilon_x + \sigma_z \delta\varepsilon_z + \tau_{xz}\gamma_{xz} \right] dz dx \\
&= \int_0^L \left(N_x \frac{d\delta u_0}{dx} - M_b \frac{d^2 \delta w_b}{dx^2} - M_s \frac{d^2 \delta w_s}{dx^2} + Q_{xz} \frac{d\delta w_s}{dx} \right) dx
\end{aligned} \qquad (6.27)$$

where N_x, M_x^b, M_x^s and Q_{xz} are the stress resultants specified by

$$\left(N_x, M_x^b, M_x^s \right) = \int_{-\frac{h}{2}}^{\frac{h}{2}} (1, z, f)\sigma_x dz \text{ and } Q_{xz} = \int_{-\frac{h}{2}}^{\frac{h}{2}} g\tau_{xz} dz \qquad (6.28)$$

The change in kinetic energy may be stated in the following way:

$$\delta T = \int_0^L \int_{-\frac{h}{2}}^{\frac{h}{2}} \rho(z)\left[\dot{u}\delta\dot{u} + \dot{w}\delta\dot{w}\right]dzdx$$

$$= \int_0^L \left\{ I_0\left[\dot{u}_0\delta\dot{u}_0 + (\dot{w}_b + \dot{w}_s)(\delta\dot{w}_b + \delta\dot{w}_s)\right] - I_1\left(\dot{u}_0\frac{d\delta\dot{w}_b}{dx} + \frac{d\dot{w}_b}{dx}\delta\dot{u}_0\right)\right.$$

$$+ I_2\left(\frac{d\dot{w}_b}{dx}\frac{d\delta\dot{w}_b}{dx}\right) - J_1\left(\dot{u}_0\frac{d\delta\dot{w}_s}{dx} + \frac{d\dot{w}_s}{dx}\delta\dot{u}_0\right) + K_2\left(\frac{d\dot{w}_s}{dx}\frac{d\delta\dot{w}_s}{dx}\right)$$

$$\left. + J_2\left(\frac{d\dot{w}_b}{dx}\frac{d\delta\dot{w}_s}{dx} + \frac{d\dot{w}_s}{dx}\frac{d\delta\dot{w}_b}{dx}\right)\right\}dx$$

$$(6.29)$$

where the dot-superscript convention denotes differentiation concerning time, $\rho(z)$ is the mass density, and $(I_0, I_1, J_1, I_2, J_2, K_0)$ are the mass moment of inertias and defined as

$$(I_0, I_1, J_1, I_2, J_2, K_2) = \int_{-\frac{h}{2}}^{\frac{h}{2}} \left(1, z, f, z^2, zf, f^2\right)\rho(z)dz \qquad (6.30)$$

The subsequent equations of motion of the CNTRC beam are found by inserting Eqs. (6.27) and (6.29) into Equation (6.26), gathering the coefficients of δu_0, δw_b, and δw_s, and integrating by parts versus both spatial and time variables

$$\delta u_0: \quad \frac{dN_x}{dx} = I_0\ddot{u}_0 - I_1\frac{d\ddot{w}_b}{dx} - J_1\frac{d\ddot{w}_s}{dx} \qquad (6.31)$$

$$\delta w_b: \quad \frac{d^2M_b}{dx^2} = I_0(\ddot{w}_b + \ddot{w}_s) + I_1\frac{d\ddot{u}_0}{dx} - I_2\frac{d^2\ddot{w}_b}{dx^2} - J_2\frac{d^2\ddot{w}_s}{dx^2} \qquad (6.32)$$

$$\delta w_s: \quad \frac{d^2M_s}{dx^2} + \frac{dQ_{xz}}{dx} = I_0(\ddot{w}_b + \ddot{w}_s) + J_1\frac{d\ddot{u}_0}{dx} - J_2\frac{d^2\ddot{w}_b}{dx^2} - K_2\frac{d^2\ddot{w}_s}{dx^2} \qquad (6.33)$$

Equations (6.19)–(6.22), (6.23)–(6.25), and (6.28) can be used to formulate Equations (6.31)–(6.33) in terms of displacements (u_0, w_b, w_s) as follows:

$$A_{11}\frac{\partial^2 u_0}{\partial x^2} - B_{11}\frac{\partial^3 w_b}{\partial x^3} - B_{11}^s\frac{\partial^3 w_s}{\partial x^3} = I_0\ddot{u}_0 - I_1\frac{d\ddot{w}_b}{dx} - J_1\frac{d\ddot{w}_s}{dx} \qquad (6.34)$$

$$B_{11}\frac{\partial^3 u_0}{\partial x^3} - D_{11}\frac{\partial^4 w_b}{\partial x^4} - D_{11}^s\frac{\partial^4 w_s}{\partial x^4}$$
$$= I_0\left(\ddot{w}_b + \ddot{w}_s\right) + I_1\frac{d\ddot{u}_0}{dx} - I_2\frac{d^2\ddot{w}_b}{dx^2} - J_2\frac{d^2\ddot{w}_s}{dx^2} \tag{6.35}$$

$$B_{11}^s\frac{\partial^3 u_0}{\partial x^3} - D_{11}^s\frac{\partial^4 w_b}{\partial x^4} - H_{11}^s\frac{\partial^4 w_s}{\partial x^4} + A_{55}^s\frac{\partial^2 w_s}{\partial x^2}$$
$$= I_0\left(\ddot{w}_b + \ddot{w}_s\right) + J_1\frac{d\ddot{u}_0}{dx} - J_2\frac{d^2\ddot{w}_b}{dx^2} - K_2\frac{d^2\ddot{w}_s}{dx^2} \tag{6.36}$$

where $A_{11}, D_{11},\ldots$ etc., are the stiffnesses of the beam, and described as

$$\left(A_{11},B_{11},D_{11},B_{11}^s,D_{11}^s,H_{11}^s\right) = \int_{-\frac{h}{2}}^{\frac{h}{2}} Q_{11}\left(1,z,z^2,f(z),zf(z),f^2(z)\right)dz \tag{6.37}$$

$$A_{55}^s = \int_{-\frac{h}{2}}^{\frac{h}{2}} Q_{55}\left[g(z)\right]^2 dz \tag{6.38}$$

6.4 ANALYTICAL SOLUTION

It is possible to develop the exact solution for CNTRC beams under different boundary conditions. For the arbitrary ends with simply supported, clamped edge, and free conditions, the boundary conditions are:

Simply supported (S) end: $w_b = w_s = 0$ at $x = 0, L$ (6.39)

Clamped (C) end: $u_0 = w_b = \frac{\partial w_b}{\partial x} = w_s = \frac{\partial w_s}{\partial x} = 0$ at $x = 0, L$ (6.40)

Free (F) end: $N_x = \frac{\partial M_x}{\partial x} = Q_x = M_x = 0$ at $x = 0, L$ (6.41)

In the context of the present problem, the following representation for displacement quantities that meet the aforementioned boundary requirements is adequate:

$$\begin{Bmatrix} u_0 \\ w_b \\ w_s \end{Bmatrix} = \begin{Bmatrix} U_m X_m' e^{i\omega t} \\ W_{bm} X_m e^{i\omega t} \\ W_{sm} X_m e^{i\omega t} \end{Bmatrix} \tag{6.42}$$

where U_m, W_{bm} and W_{sm} are arbitrary parameters to be found, ω is the eigenfrequency associated with the mth eigenmode, and $\lambda = m\pi/L$. Table 6.2 lists the functions that fulfill at least the geometric boundary requirements given in Equations (6.39)–(6.41), and represents approximate deflected beam forms that were proposed by Reddy [30].

TABLE 6.2

The Admissible Functions versus Boundary Conditions

Boundary Condition	Admissible Function
S-S	$X_m(x) = \sin(\lambda_m x)$ $\lambda_m = m\pi/L$ "m = 1, 2, ..."
C-C	$X_m(x) = \sin(\lambda_m x) - \sinh(\lambda_m x) - \xi_m[\cos(\lambda_m x) - \cosh(\lambda_x x)]$ $\xi_m = [\sin(\lambda_m L) - \sinh(\lambda_m L)]/[\cos(\lambda_m L) - \cosh(\lambda_m L)]$ $\lambda_m = (m + 0.5)\pi/L$
C-S	$X_m(x) = \sin(\lambda x) - \sinh(\lambda x) - \xi_m[\cos(\lambda x) - \cosh(\lambda x)]$ $\xi_m = [\sin(\lambda_m L) + \sinh(\lambda_m L)]/[\cos(\lambda_m L) + \cosh(\lambda_m L)]$ $\lambda_m = (m + 0.25)\pi/L$
C-F	$X_m(x) = \sin(\lambda x) - \sinh(\lambda x) - \xi_m[\cos(\lambda x) - \cosh(\lambda x)]$ $\xi_m = [\sin(\lambda_m L) - \sinh(\lambda_m L)]/[\cos(\lambda_m L) - \cosh(\lambda_m L)]$ $\lambda_1 = 1.875/L, \lambda_2 = 4.694/L, \lambda_3 = 7.855/L, \lambda_4 = 10.966/L,$ $\lambda_m = (m - 0.25)\pi/L$ for $m \geq 5$

The analytical solutions may be found by inserting the formulas of u_0, w_b, w_s from Equation (6.42) into the equations of motion, Equations (6.34)–(6.36).

$$\left(\begin{bmatrix} a_{11} & a_{12} & a_{13} \\ a_{21} & a_{22} & a_{23} \\ a_{31} & a_{32} & a_{33} \end{bmatrix} - \omega^2 \begin{bmatrix} m_{11} & m_{12} & m_{13} \\ m_{21} & m_{22} & m_{23} \\ m_{31} & m_{32} & m_{33} \end{bmatrix} \right) \begin{Bmatrix} U_m \\ W_{bm} \\ W_{sm} \end{Bmatrix} = \begin{Bmatrix} 0 \\ 0 \\ 0 \end{Bmatrix} \qquad (6.42)$$

where

$$a_{11} = \int_0^L A_{11} X''' X' dx, \quad a_{12} = \int_0^L -B_{11} X''' X' dx, \quad a_{13} = \int_0^L -B_{11}^s X''' X' dx,$$

$$a_{21} = \int_0^L B_{11} X''' X dx, \quad a_{22} = \int_0^L -D_{11} X''' X dx, \quad a_{23} = \int_0^L -D_{11}^s X''' X dx \qquad (6.42)$$

$$a_{31} = \int_0^L B_{11}^s X''' X dx, \quad a_{32} = \int_0^L -D_{11}^s X''' X dx, \quad a_{33} = \int_0^L \left(-H_{11}^s X''' + A_{55}^s X'' \right) X dx$$

$$m_{11} = -I_0 \int_0^L X' X' dx, \quad m_{12} = I_1 \int_0^L X' X' dx, \qquad\qquad m_{13} = J_1 \int_0^L X' X' dx$$

$$m_{21} = -I_1 \int_0^L X'' X dx, \quad m_{22} = -\int_0^L \left(I_0 X - I_2 X'' \right) X dx, \quad m_{23} = -\int_0^L \left(I_0 X + J_2 X'' \right) X dx \quad (6.43)$$

$$m_{31} = -J_1 \int_0^L X'' X dx, \quad m_{32} = -\int_0^L \left(I_0 X - J_2 X'' \right) X dx, \quad m_{33} = -\int_0^L \left(I_0 X - K_2 X'' \right) X dx$$

6.5   NUMERICAL EXAMPLES AND DISCUSSION

Numerical findings of the free vibration behavior of CNTRC beams are provided and analyzed in this section. The following are the effective material properties of CNTRC beams at the room temperature used in studies by Yas and Samadi [14] and Tagrara et al. [31]:

- The matrix is made of polymethyl methacrylate (PMMA), which has the following material properties: $\nu^p = 0.3$; $\rho^p = 1190 kg/m^3$ and $E^p = 2.5 GPa$.
- The armchair (10, 10) SWCNTs are used as reinforcing material and have the following properties: $\nu^{cnt} = 0.19$; $\rho^{cnt} = 1400 kg/m^3$; $E_{11}^{cnt} = 600 GPa$; $E_{22}^{cnt} = 10 GPa$ and $G_{12}^{cnt} = 17.2 GPa$.

The following nondimensionalization is used for convenience:

$$\bar{\omega} = \omega L \sqrt{\frac{I_{00}}{A_{110}}} \tag{6.44}$$

where A_{110} and I_{00} are beam stiffness and mass moment of inertia of beam made of the pure matrix material, respectively.

Table 6.3 shows the first three nondimensional natural frequencies for CNTRC beams having different values of CNT volume fraction $\left(V_{cnt}^*\right)$ and several types of CNT distribution patterns under simply supported–simply supported boundary conditions for L/h = 15. Furthermore, the outcomes of the TrSDT are compared with the available results of Yas and Samadi [14], Lin and Xiang [18], Shi et al. [23], and Civalek et al. [32] employing first-order shear deformation theory FSDT, and Wattanasakulpong and Ungbhakorn [17] applying TrSDT.

Table 6.4 illustrates the first three nondimensional natural frequencies for CNTRC beams with various CNT volume fractions $\left(V_{cnt}^*\right)$ and several types of CNT distribution patterns under clamped–clamped boundary conditions for L/h = 15. Furthermore, the findings of the TrSDT are compared with the available results of Yas and Samadi [14] using FSDT.

Table 6.5 depicts the first three nondimensional natural frequencies for CNTRC beams with varying CNT volume fractions $\left(V_{cnt}^*\right)$ and diverse types of CNT distribution patterns under clamped–simply supported boundary conditions for L/h = 15. Furthermore, the results of the TrSDT are compared with the available results of Yas and Samadi [14] and Shi et al. [23] using FSDT.

Table 6.6 illustrates the first three nondimensional natural frequencies for CNTRC beams with different CNT volume fractions $\left(V_{cnt}^*\right)$ and distinct types of CNT distribution patterns under clamped–free boundary conditions for L/h = 15. Furthermore, the values of the TrSDT are compared with the available results of Yas and Samadi [14], Lin and Xiang [18], and Shi et al. [23] using FSDT.

Tables 6.3–6.6 demonstrate that the new findings are consistent with those previously reported. Besides, the CNTRC beams having an FG-X pattern appear to have the greatest natural frequency, while the FG-O pattern appears to have the lowest, as well as the natural frequencies of the UD pattern, are always higher than those for the

TABLE 6.3

The First Three Non-Dimensional Natural Frequencies for CNTRC Beams Under Simply Supported-Simply Supported Boundary Conditions (L/h= 15)

V^*_{CNT}	CNTR Type	Yas and Samadi [14]			Wattanasakulpong and Ungbhakorn [17]	Lin and Xiang [18]			Shi et al. [23]			Civalek et al. [32]	Present		
		ϖ_1	ϖ_2	ϖ_3	ϖ_1	ϖ_1	ϖ_2	ϖ_3	ϖ_1	ϖ_2	ϖ_3	ϖ_1	ϖ_1	ϖ_2	ϖ_3
0.12	UD	0.9753	2.8728	4.8704	0.9749	-	-	-	-	-	-	0.9905	0.9749	2.8851	4.9495
	FG-O	0.7527	2.4562	4.4320	0.7446	-	-	-	-	-	-	-	0.7446	2.3910	4.2807
	FG-X	1.1150	3.0814	5.0695	1.1163	-	-	-	-	-	-	1.1373	1.1163	3.1104	5.2002
	FG-V	0.9453	2.6424	4.6675	0.8443	-	-	-	-	-	-	0.8562	0.8443	2.6517	4.7037
0.17	UD	1.1999	3.6276	6.2363	-	-	-	-	-	-	-	-	1.1987	3.6344	6.3072
	FG-O	0.9155	3.0577	5.6139	-	-	-	-	-	-	-	-	0.9082	2.9940	5.4564
	FG-X	1.3830	3.9293	6.5447	-	-	-	-	-	-	-	-	1.3768	3.9178	6.6006
	FG-V	1.1609	3.3084	5.9498	-	-	-	-	-	-	-	-	1.030	3.3105	5.9545
0.28	UD	1.4401	4.1362	6.9245	-	1.4348	4.1050	6.8595	1.3985	4.0505	6.8086	-	1.4367	4.1426	7.0286
	FG-O	1.1202	3.6056	6.4434	-	-	-	-	-	-	-	-	1.1146	3.5645	6.3642
	FG-X	1.6493	4.4752	7.3068	-	1.6409	4.4333	7.2258	1.6086	4.3927	7.1907	-	1.6107	4.3301	7.1583
	FG-V	1.4027	3.8639	6.7618	-	1.3975	3.8370	6.6976	1.3639	3.7701	6.6301	-	1.2469	3.8386	6.7370

TABLE 6.4

Comparisons of the First Three Nondimensional Natural Frequencies for CNTRC Beams Under Clamped–Clamped Boundary Conditions (L/h = 15)

V^*_{CNT}	CNTR Type	Yas and Samadi [14]			Present		
		ϖ_1	ϖ_2	ϖ_3	ϖ_1	ϖ_2	ϖ_3
0.12	UD	1.5085	3.1353	4.9979	1.6134	3.5307	5.5562
	FG-O	1.3180	2.8762	4.6840	1.3400	3.0465	4.8822
	FG-X	1.6000	3.2629	5.1514	1.7372	3.7152	5.7853
	FG-V	1.4068	2.9997	4.8363	1.4855	3.3483	5.3654
0.17	UD	1.9144	4.0187	6.4348	2.0334	4.4961	7.1086
	FG-O	1.6500	3.6565	5.9970	1.6792	3.8784	6.2793
	FG-X	2.0498	4.2111	6.6753	2.1889	4.7138	7.3563
	FG-V	1.7721	3.8312	6.2139	1.8559	4.2351	6.8329
0.28	UD	2.1618	4.4556	7.0745	2.3154	5.0170	7.8646
	FG-O	1.9284	4.1740	6.7728	1.9974	4.5301	7.2479
	FG-X	2.3169	4.7051	7.4093	2.4170	5.1168	7.9514
	FG-V	2.0504	4.3414	6.9783	2.1492	4.7986	7.6548

TABLE 6.5

Comparisons of the First Three Nondimensional Natural Frequencies for CNTRC Beams Under Clamped–Simply Supported Boundary Conditions (L/h = 15)

V_{CNT}^*	CNTR Type	Yas and Samadi [14]			Shi et al. [23]			Present		
		ϖ_1	ϖ_2	ϖ_3	ϖ_1	ϖ_2	ϖ_3	ϖ_1	ϖ_2	ϖ_3
0.12	UD	1.2444	3.0159	4.9342	1.2154	2.9668	4.8734	1.3343	3.2310	5.2657
	FG-O	1.0331	2.6814	4.5619	1.0021	2.6224	4.4840	1.0620	2.7398	4.5979
	FG-X	1.3577	3.1817	5.1092	1.3315	3.1383	5.0557	1.4791	3.4344	5.5026
	FG-V	1.1529	2.8472	4.7474	1.1226	2.7932	4.6778	1.1909	3.0216	5.0487
0.17	UD	1.5602	3.8402	6.3370	1.5214	3.7701	6.2451	1.6614	4.0956	6.7264
	FG-O	1.2769	3.3772	5.8126	1.2374	3.2973	5.7032	1.3115	3.4614	5.8901
	FG-X	1.7188	4.0843	6.6094	1.6834	4.0219	6.5288	1.8456	4.3450	6.9930
	FG-V	1.4344	3.6064	6.0765	1.3949	3.5306	5.9733	1.4690	3.7997	6.4135
0.28	UD	1.8040	4.3112	6.9987	1.7622	4.2312	6.8867	1.9393	4.6108	7.4621
	FG-O	1.5229	3.9112	6.6127	1.4786	3.8195	6.4808	1.5866	4.0788	6.8305
	FG-X	1.9813	4.6030	7.3560	1.9416	4.5250	7.2448	2.0909	4.7490	7.5643
	FG-V	1.6933	4.1393	6.8633	-	-	-	1.7412	4.3491	7.2142

TABLE 6.6

Comparisons of the First Three Non-Dimensional Natural Frequencies for CNTRC Beams Under Clamped-Free Boundary Conditions (L/h= 15)

V_{CNT}^*	CNTR Type	Yas and Samadi [14] ϖ_1	ϖ_2	ϖ_3	Lin and Xiang [18] ϖ_1	ϖ_2	ϖ_3	Shi et al. [23] ϖ_1	ϖ_2	ϖ_3	Present ϖ_1	ϖ_2	ϖ_3
0.12	UD	0.3764	1.7006	3.6648	-	-	-	-	-	-	0.4020	1.6048	3.5313
	FG-O	0.2809	1.4266	3.2489	-	-	-	-	-	-	0.2904	1.3322	3.0470
	FG-X	0.4416	1.8497	3.8777	-	-	-	-	-	-	0.4852	1.7286	3.7158
	FG-V	0.3193	1.5473	3.4380	-	-	-	-	-	-	0.3340	1.4771	3.3489
0.17	UD	0.4587	2.1365	4.6614	-	-	-	-	-	-	0.4854	2.0224	4.4968
	FG-O	0.3394	1.7685	4.0913	-	-	-	-	-	-	0.3491	1.6691	3.8791
	FG-X	0.5413	2.3437	4.9706	-	-	-	-	-	-	0.5865	2.1779	4.7146
	FG-V	0.3866	1.9287	4.3500	-	-	-	-	-	-	0.4014	1.8451	4.2359
0.28	UD	0.5612	2.4614	5.2446	0.5600	2.4449	5.2005	0.5432	2.4061	5.1457	0.6054	2.3034	5.0178
	FG-O	0.4197	2.0993	4.7399	-	-	-	-	-	-	0.4357	1.9859	4.5309
	FG-X	0.6586	2.6987	5.6150	0.6566	2.6763	5.5589	0.6405	2.6446	5.5169	0.7281	2.4055	5.1176
	FG-V	0.4761	2.2685	5.0007	0.4753	2.2543	4.9590	0.4600	2.2106	4.8923	0.5000	2.1373	4.7994

FG-V pattern. Lastly, it has been discovered that CNTRC beams with greater CNT volume percentages exhibit higher frequency values.

Figures 6.2–6.5 report the first three nondimensional natural frequencies for CNTRC beams with several types of CNT distribution patterns versus varying span-to-depth ratio (L/h).

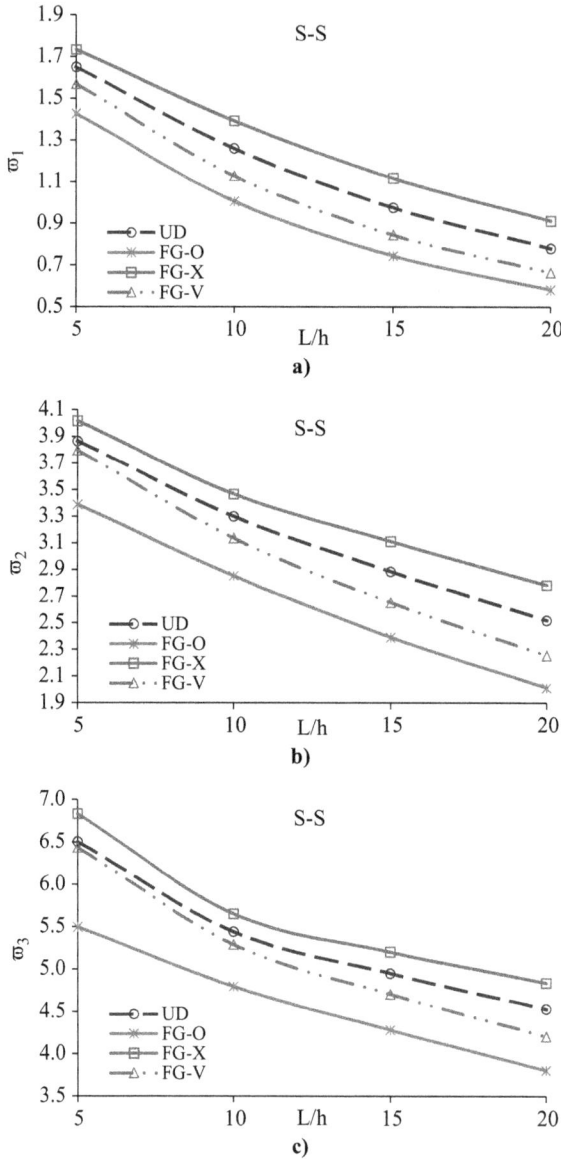

FIGURE 6.2 The first three nondimensional natural frequencies i.e., (a) first, (b) second and (c) third frequencies, of CNTRC beams under simply supported–simply supported boundary conditions versus span-to-depth ratio $\left(V_{CNT}^* = 0.12\right)$.

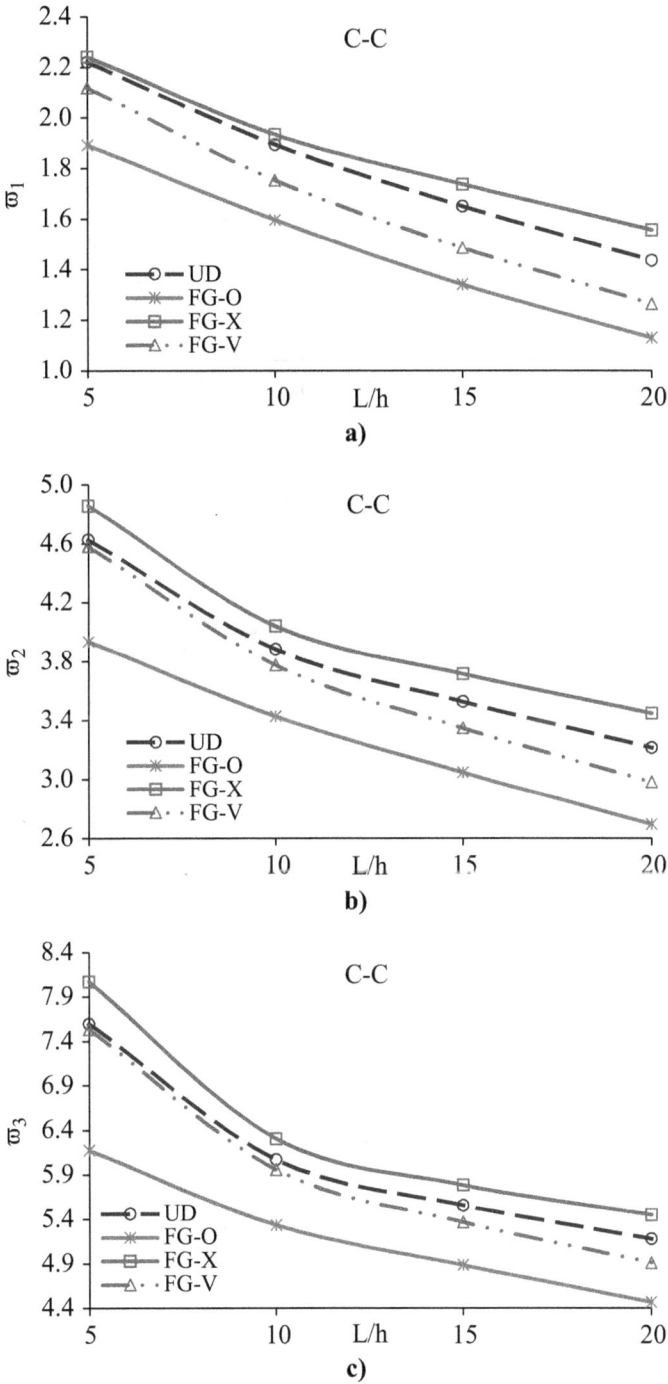

FIGURE 6.3 The first three nondimensional natural frequencies i.e., (a) first, (b) second and (c) third frequencies, of CNTRC beams under clamped–clamped boundary conditions versus span-to-depth ratio $\left(V_{CNT}^* = 0.12\right)$.

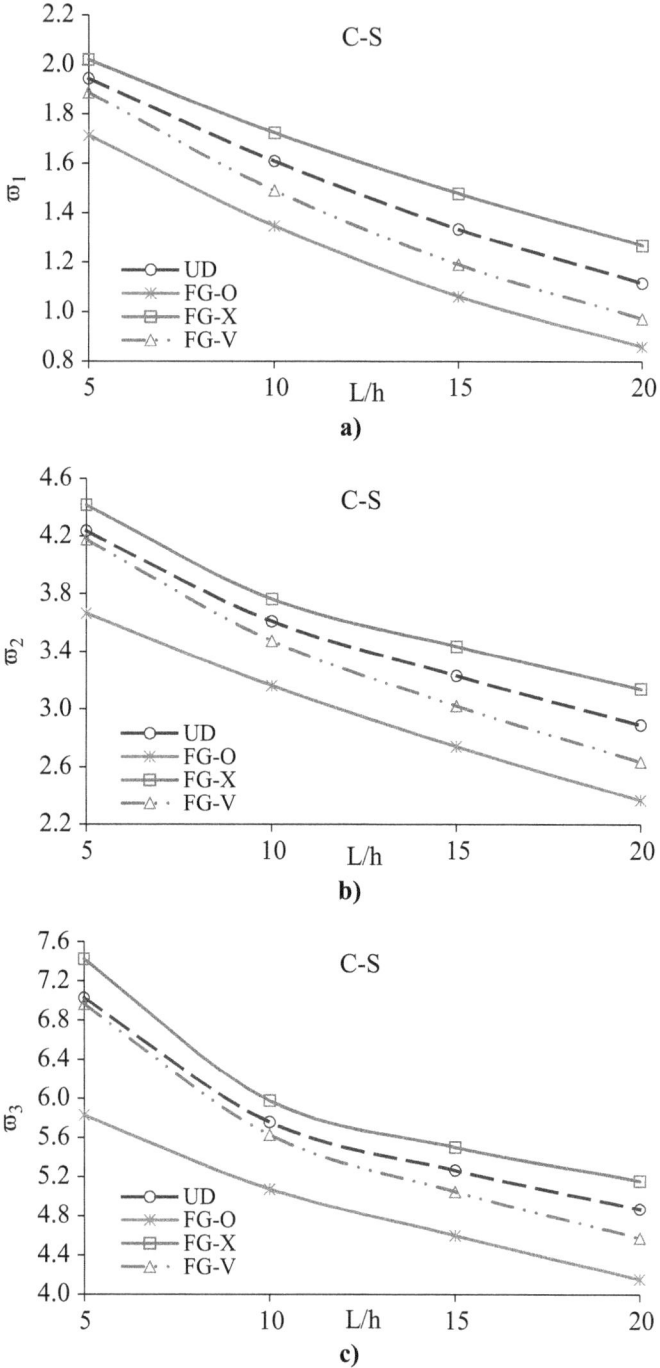

FIGURE 6.4 The first three nondimensional natural frequencies i.e., (a) first, (b) second and (c) third frequencies, of CNTRC beams under clamped–simply supported boundary conditions versus span-to-depth ratio $\left(V_{CNT}^* = 0.12\right)$.

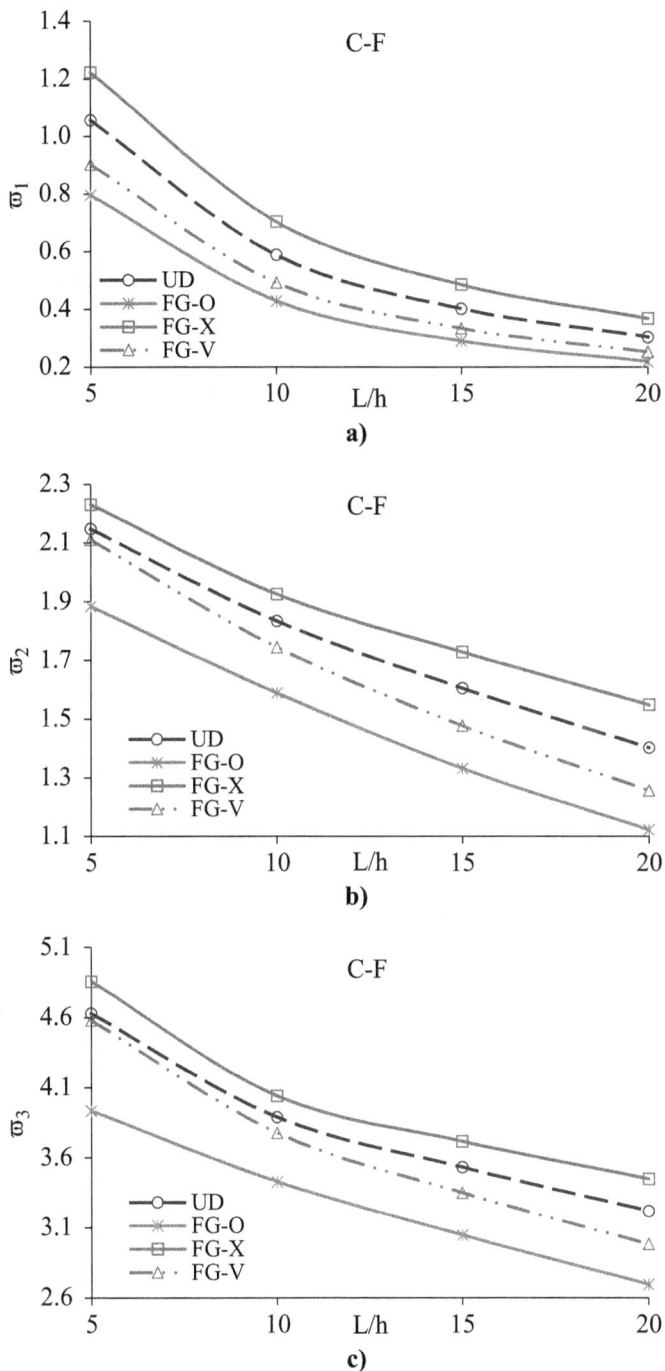

FIGURE 6.5 The first three nondimensional natural frequencies i.e., (a) first, (b) second and (c) third frequencies, of CNTRC beams under clamped-free boundary conditions versus span-to-depth ratio $\left(V_{CNT}^* = 0.12\right)$.

It has been noticed that as the span-to-depth ratios increase, nondimensional natural frequencies of CNTRC beams decrease. Furthermore, the effect of the variation of span-to-depth ratio is more pronounced on the fundamental natural frequency of CNTRC beams, for all boundary conditions. Lastly, as the span-to-depth ratios increase, the effect of the CNT distribution patterns becomes more dominant.

6.6 CONCLUSIONS

The goal of this chapter is to present an RBT for the free vibration analysis of CNTRC beams under four different boundary conditions. The developed RBT is notable for not only addressing the shear deformation effect but also dealing with only three unknowns as FSDT of Timoshenko without including a shear correction factor. The material characteristics of the CNTRC beams are figured out using the rule of mixture. The equations of motion of the CNTRC beam are derived based on TrSDT employing Hamilton's principle. The resulting equations are solved analytically for four different boundary conditions. In the numerical examples, the effects of carbon nanotube volume fraction, diverse types of CNT distribution patterns, span-to-depth ratio, and boundary conditions on the free vibration response of CNTRC beams are investigated in detail.

In conclusion, the following findings are reached:

a. As compared to uniform and asymmetric distributions, FG-CNTs with symmetric distributions through the beam thickness have more ability to lower or raise the natural frequency.
b. Larger frequency values are seen in CNTRC beams with higher CNT volume fractions.
c. The resulting frequency changes are due to boundary conditions so that the C-C CNTRC beams have the highest natural frequencies under C-C boundary conditions, and it is followed by the C-S and S-S boundary conditions, while the CNTRC beams have the lowest natural frequencies under C-F boundary conditions for every volume fraction.
d. The natural frequencies of CNTRC beams decrease as the span-to-depth ratio ratios increases for all types of CNT distribution patterns.
e. For all boundary conditions, the influence of change in the span-to-depth ratio is more evident on the fundamental natural frequency of CNTRC beams.
f. The effect of CNT distribution patterns becomes increasingly significant as the span-to-depth ratios increase.

Finally, it is concluded that the CNT distribution patterns, CNT volume fraction, span-to-depth ratio, and boundary conditions have an obvious impact on the free vibration of CNTRC beams. Furthermore, the suggested analytical solution technique not only handled the current problem adequately and produced good results but also made free vibration analysis of CNTRC beams under various boundary conditions easier.

REFERENCES

1. Reddy JN. *Mechanics of Laminated Composite Plates and Shells Theory and Analysis.* 2003.
2. Panda SK, Mahapatra TR. Nonlinear finite element analysis of laminated composite spherical shell vibration under uniform thermal loading. *Mecc* 2013; 49: 191–213. doi:10.1007/S11012-013-9785-9.
3. Hajianmaleki M, Qatu MS. Vibrations of straight and curved composite beams: A review. *Compos Struct* 2013; 100: 218–32. doi:10.1016/J.COMPSTRUCT.2013.01.001.
4. Singh VK, Panda SK. Nonlinear free vibration analysis of single/doubly curved composite shallow shell panels. *Thin-Walled Struct* 2014; 85: 341–9. doi:10.1016/J.TWS.2014.09.003.
5. Iijima S. Helical microtubules of graphitic carbon. *Nature* 1991. doi:10.1038/354056a0.
6. Shen HS. Nonlinear bending of functionally graded carbon nanotube-reinforced composite plates in thermal environments. *Compos Struct* 2009; 91: 9–19. doi:10.1016/j.compstruct.2009.04.026.
7. Shen HS. Postbuckling of nanotube-reinforced composite cylindrical shells in thermal environments, Part I: Axially-loaded shells. *Compos Struct* 2011; 93: 2096–108. doi:10.1016/j.compstruct.2011.02.011.
8. Shen HS. Postbuckling of nanotube-reinforced composite cylindrical shells in thermal environments, Part II: Pressure-loaded shells. *Compos Struct* 2011; 93: 2496–503. doi:10.1016/j.compstruct.2011.04.005.
9. Lau KT, Gu C, Gao GH, Ling HY, Reid SR. Stretching process of single- and multi-walled carbon nanotubes for nanocomposite applications. *Carbon N Y* 2014; 42: 426–8. doi:10.1016/j.carbon.2003.10.040.
10. Mehar K, Panda SK, Dehengia A, Kar VR. Vibration analysis of functionally graded carbon nanotube reinforced composite plate in thermal environment. *J Sandw Struct Mater* 2016; 18: 151–73. doi:10.1177/1099636215613324.
11. Mehar K, Panda SK, Patle BK. Stress, deflection, and frequency analysis of CNT reinforced graded sandwich plate under uniform and linear thermal environment: A finite element approach. *Polym Compos* 2018; 39: 3792–809. doi:10.1002/pc.24409.
12. Civalek Ö, Avcar M. Free vibration and buckling analyses of CNT reinforced laminated non-rectangular plates by discrete singular convolution method. *Eng Comput* 2022; 38: 489–521. doi:10.1007/s00366-020-01168-8.
13. Ke LL, Yang J, Kitipornchai S. Nonlinear free vibration of functionally graded carbon nanotube-reinforced composite beams. *Compos Struct* 2010; 92: 676–83. doi:10.1016/j.compstruct.2009.09.024.
14. Yas MH, Samadi N. Free vibrations and buckling analysis of carbon nanotube-reinforced composite Timoshenko beams on elastic foundation. *Int J Press Vessel Pip* 2012; 98: 119–28. doi:10.1016/J.IJPVP.2012.07.012.
15. Ke LL, Yang J, Kitipornchai S. Dynamic stability of functionally graded carbon nanotube-reinforced composite beams. *Mech Adv Mater Struct* 2013; 20: 28–37. doi:10.1080/15376494.2011.581412.
16. Shen HS, Xiang Y. Nonlinear analysis of nanotube-reinforced composite beams resting on elastic foundations in thermal environments. *Eng Struct* 2013; 56: 698–708. doi:10.1016/j.engstruct.2013.06.002.
17. Wattanasakulpong N, Ungbhakorn V. Analytical solutions for bending, buckling and vibration responses of carbon nanotube-reinforced composite beams resting on elastic foundation. *Comput Mater Sci* 2013; 71: 201–8. doi:10.1016/j.commatsci.2013.01.028.

18. Lin F, Xiang Y. Vibration of carbon nanotube reinforced composite beams based on the first and third order beam theories. *Appl Math Model* 2014; 38: 3741–54. doi:10.1016/j.apm.2014.02.008.

19. Ansari R, Faghih Shojaei M, Mohammadi V, Gholami R, Sadeghi F. Nonlinear forced vibration analysis of functionally graded carbon nanotube-reinforced composite Timoshenko beams. *Compos Struct* 2014; 113: 316–27. doi:10.1016/j.compstruct.2014.03.015.

20. Wu H, Kitipornchai S, Yang J. Free vibration and buckling analysis of sandwich beams with functionally graded carbon nanotube-reinforced composite face sheets. *Int J Struct Stab Dyn* 2015; 15: 1540011. doi:10.1142/S0219455415400118.

21. Hadji L, Zouatnia N, Meziane MA, Kassoul A. A simple quasi-3D sinusoidal shear deformation theory with stretching effect for carbon nanotube-reinforced composite beams resting on elastic foundation. *Earthquakes Struct* 2017; 13: 509–18. doi:10.12989/eas.2017.13.5.509.

22. Hadji L, Meziane MAA, Safa A. A new quasi-3D higher shear deformation theory for vibration of functionally graded carbon nanotube-reinforced composite beams resting on elastic foundation. *Struct Eng Mech An Int J* 2018; 66: 771–81.

23. Shi Z, Yao X, Pang F, Wang Q. An exact solution for the free-vibration analysis of functionally graded carbon-nanotube-reinforced composite beams with arbitrary boundary conditions. *Sci Rep* 2017; 7: 1–18. doi:10.1038/s41598-017-12596-w.

24. Jedari Salami S. Free vibration analysis of sandwich beams with carbon nanotube reinforced face sheets based on extended high-order sandwich panel theory. *J Sandw Struct Mater* 2018; 20: 219–48. doi:10.1177/1099636216649788.

25. Penna R, Lovisi G, Feo L. Dynamic response of multilayered polymer functionally graded carbon nanotube reinforced composite (FG-CNTRC) nano-beams in hygro-thermal environment. *Polymers (Basel)* 2021; 13: 2340. doi:10.3390/polym13142340.

26. Civalek Ö, Dastjerdi S, Akbaş ŞD, Akgöz B. Vibration analysis of carbon nanotube-reinforced composite microbeams. *Math Methods Appl Sci* 2021:mma.7069. doi:10.1002/mma.7069.

27. Shen HS. Nonlinear bending of functionally graded carbon nanotube-reinforced composite plates in thermal environments. *Compos Struct* 2009; 91: 9–19. doi:10.1016/j.compstruct.2009.04.026.

28. Han Y, Elliott J. Molecular dynamics simulations of the elastic properties of polymer/carbon nanotube composites. *Comput Mater Sci* 2007; 39: 315–23. doi:10.1016/j.commatsci.2006.06.011.

29. Reddy JN. Theory and analysis of elastic plates and shells. CRC; 2006.

30. Reddy JN. Mechanics of laminated composite plates and shells : theory and analysis. CRC Press; 2004.

31. Tagrara SH, Benachour A, Bouiadjra MB, Tounsi A. On bending, buckling and vibration responses of functionally graded carbon nanotube-reinforced composite beams. *Steel Compos Struct* 2015; 19: 1259–77. doi:10.12989/scs.2015.19.5.1259.

32. Civalek Ö, Akbaş ŞD, Akgöz B, Dastjerdi S. Forced vibration analysis of composite beams reinforced by carbon nanotubes. *Nanomaterials* 2021; 11: 1–17. doi:10.3390/nano11030571.

7 Transient Characteristics of Carbon Nanotube–Reinforced Composite Plates under Blast Load

Sunkesula Mohammad Bilal, Samarjeet Kumar, and Vishesh Ranjan Kar

National Institute of Technology Jamshedpur, Jamshedpur, India

CONTENTS

DOI: 10.1201/9781003158813-7

7.1   INTRODUCTION

On a microscopic scale, the mixture of two or more components is called *composite material*. They are widely used because of their superior strength, stiffness, corrosion resistance, smaller weight, thermal characteristics, wear resistance, and fatigue life [1]. Fiber and matrix are the two parts of a composite illustrated in Figure 7.1. Fibers are utilized in contemporary composites because of their superior specific mechanical qualities compared to typical bulk materials. Carbon and graphite have been popular fiber materials in many weight-sensitive sectors for decades [2]. The external breakage or damage is protected by a matrix that works as a bonding element. Polymer, ceramic, and metals are the most frequent matrix phase materials [3]. The bonding contact between the reinforcement and the matrix determines the load transformation. Bonding is affected by the type of reinforcement and matrix used and the production procedure [4].

Functionally graded material (FGM) is a novel type of advanced composite material in which the constituents are gradually modified regarding the spatial coordinate across the volume, resulting in a consistent change in the material's characteristics [5]. The overall qualities of functionally graded material are distinct from the various materials that comprise it. Nowadays, many FGM applications are present in engineering, and their use is expected to grow as the cost of material manufacturing procedures decreases as these processes improve [6,7]. The research is going on fabrication methods and their areas of use. As a result, material qualities are affected by their spatial location inside the structure. Materials can be tailored to specific functions and applications. Chemical, mechanical, thermal, and electrical characteristics are among those that may be designed/controlled to provide desired functionality. They provide the capacity to manage deformation, dynamic response, wear, corrosion, and other factors and design for various complicated environments and stress concentrations.

Carbon nanotubes (CNT) play a very important role in the engineering field. It is a cylindrical macromolecule consisting of carbon atoms arranged in a periodic hexagonal structure [8] and was invented by Sumio Iijima in 1991. CNT is constantly being applied in new fields of study for the thorough analysis of nanostructures. Because of its superior mechanical, thermal, and electrical characteristics, CNT is widely employed as a nanoscale reinforcing material in developing novel nanocomposites [9]. CNTs in polymer matrices can dramatically improve the stiffness and

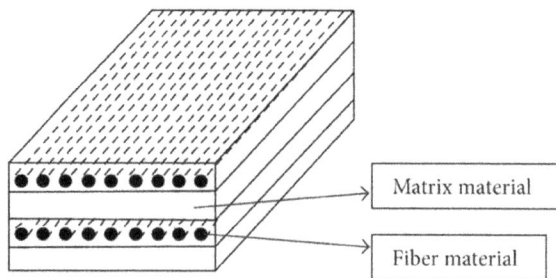

FIGURE 7.1   Composite plate.

strength of composites compared to those reinforced with traditional carbon fibres [10]. However, maintaining these exceptional qualities at the macro scale is a significant task. CNTs are widely recognized for their high-yield strength, Young's modulus, conductivity, and flexibility [11–16]. They also have 20 times the strength of high-strength steel alloys, are half the density of aluminum, and have a current-carrying capacity 10,000 times that of copper.

7.1.1 CLASSIFYING CARBON NANOTUBES

Single-walled carbon nanotubes and multi-walled carbon nanotubes are the two categories of CNT and are abbreviated as SWCNT and MWCNT, respectively. Rolling a single graphene sheet into a nano-meter diameter cylinder that makes a tube is called SWCNT, whereas rolling many graphene sheets into a cylinder that makes a tube is called MWCNT [17] (see Figures 7.2–7.4).

7.1.2 CARBON NANOTUBE STRUCTURE

Carbon nanotubes feature three distinct geometrical configurations of carbon atoms. Based on the rolling of graphene sheets into CNT, they are classified as chiral, armchair, and zig-zag [18] as depicted in Figure 7.5 due to their physical and mechanical characteristics dependent on their atomic arrangement.

7.1.3 APPLICATIONS OF CARBON NANOTUBES

CNT has many applications in all the fields like agriculture, mechanical, civil, aeronautical, etc. They act as an additive for different structural materials in nanotechnology since 21st century. They are used as ventilators for tractor engines, oil filters, air filters, radiator grills, etc. in the mechanical field. In civil engineering, these are utilized like embossed or perforated plates for floors, cable trays, etc. In aeronautical, they are used in spark arrestors, embossed floors, etc. [19].

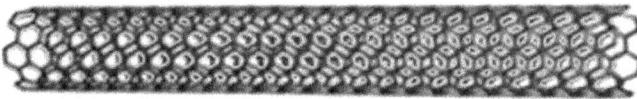

FIGURE 7.2   Graphene sheet rolled into SWCNTs [18].

FIGURE 7.3   Graphene sheet rolled into MWCNTs [18].

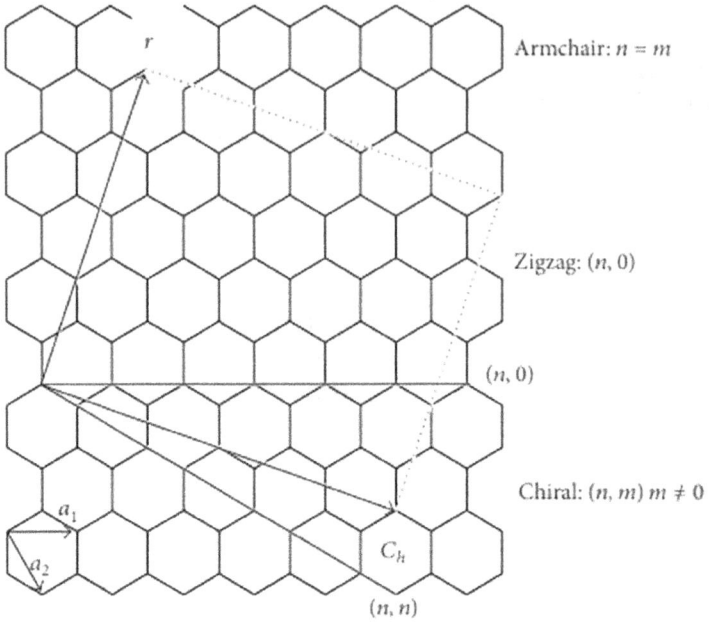

FIGURE 7.4   Formation of single-walled carbon nanotubes by rolling of a graphene sheet along lattice vectors [18].

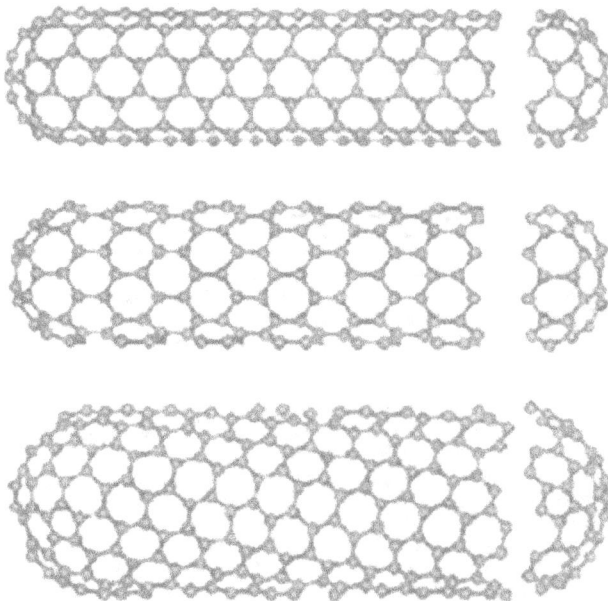

FIGURE 7.5   Classification of carbon nanotubes [18].

Due to its excellent thermal, physical, and mechanical properties, the CNT-RC plates attract the engineering field [20–22]. CNT provides structure, fatigue, better wear resistance, stiffness-to-weight ratio, strength-to-weight ratio, efficient size, shape, and best-elevated temperature properties and dynamic response of the system, ability to control the deformation, wear, corrosion of parts, etc. In the last couple of decades, there has been a drastic increase in the utilization of composite structures, particularly in aeronautical engineering, which attracted the engineers for its study. These composite structures are elevated to various thermal conditions and subjected to different types of loading combinations while they are functioning, resulting in deflections in geometry and shape. The present work analyzes the FG-CNT for different time-dependent loading conditions with different boundary conditions and structural instability.

7.2   MICROMECHANICAL PROPERTY OF FG CNT

In the present study, four types of CNT plates are considered, namely (i) uniformly distributed CNTRC plate, (ii) functionally graded – V type, (iii) functionally graded – O type (iv) functionally graded – X type CNTRC plate with length a, breadth b, and thickness h were considered and denoted by (i) UD, (ii) FG-V, (iii) FG-O, and (iv) FG-X respectively as shown in Figure 7.6. The effective material properties of CNT,

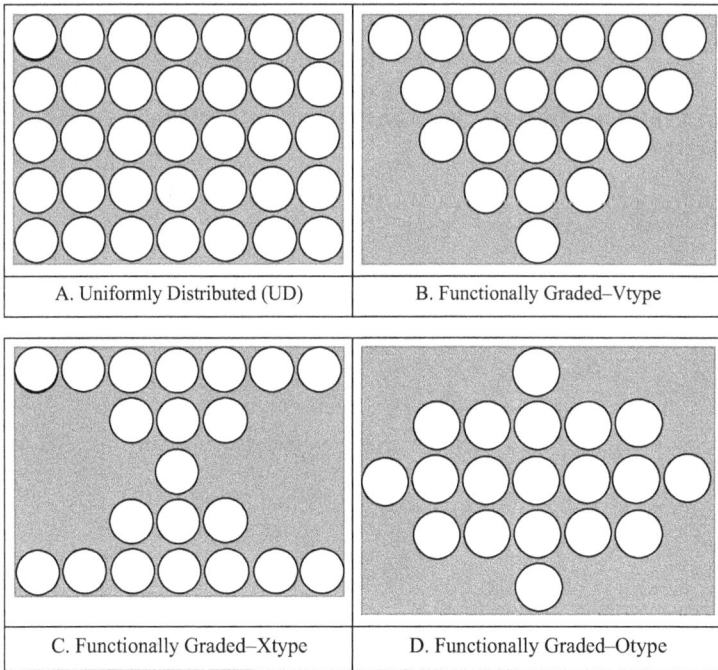

| A. Uniformly Distributed (UD) | B. Functionally Graded–Vtype |
| C. Functionally Graded–Xtype | D. Functionally Graded–Otype |

FIGURE 7.6   Model of the FG-CNTRCs plates. (A) UD CNTRC plate, (B) FG-V CNTRC plate, (C) FG-X CNTRC plate, and (D) FG-O CNTRC plate [27].

an isotropic polymer, and two-phase nanocomposites are evaluated by the Mori–Tanaka rule [23] or by using the Extended Rule of Mixture [24,25]. To evaluate effective material properties, the Rule of Mixture is easy and convenient compared to the Mori–Tanaka rule. Hence, in the present study Rule of Mixture is selected. Using this rule the effective material properties and efficiency parameters of CNTRC plates can be expressed as [24].

$$E_{11} = \eta_1 V_{CNT} E_{11}^{CNT} + V_m E^m$$

$$\frac{\eta_2}{E_{22}} = \frac{V_{CNT}}{E_{22}^{CNT}} + \frac{V_m}{E^m}$$

$$\frac{\eta_3}{G_{12}} = \frac{V_{CNT}}{G_{12}^{CNT}} + \frac{V_m}{G^m}$$

$$\rho = V_{CNT} \rho^{CNT} + V_m \rho^m$$

(7.1)

Here E_{11}^{CNT}, E_{22}^{CNT} are the young's modulus of the SWCNTs, and G_{12}^{CNT} is the Shear modulus of the SWCNT, E^m is Young's modulus of the matrix and G^m is the shear modulus of the matrix. By comparing the effective properties of CNTRC obtained from MD simulation with those from the mixture rule, the effective material properties and efficiency parameters of CNT were calculated. V_{CNT} and V_m are the volume fractions of CNT and matrix, respectively. Here the noticeable note is "Unity is the answer by adding volume fractions of CNT and matrix" ($V_{CNT} + V_m = 1$).

The four types of CNTRC plates, as shown in Figure 7.6, the carbon nanotubes distribution along the thickness direction of the plate [26] is assumed to be as shown in the Equation (7.2) below:

$$V_{CNT}(Z) = V_{CNT}^* \text{ For UD CNTRC}$$

$$V_{CNT}(Z) = 2\left(1 - \frac{2Z}{h}\right) V_{CNT}^* \text{ For FG-V}$$

$$V_{CNT}(Z) = 2\left(1 + \frac{2Z}{h}\right) V_{CNT}^* \text{ For FG-O}$$

$$V_{CNT}(Z) = 2\left(\frac{2Z}{h}\right) V_{CNT}^* \text{ For FG-X}$$

(7.2)

where, $V_{CNT}^* = \dfrac{w_{CNT}}{w_{CNT} + \left(\dfrac{\rho^{CNT}}{\rho^m}\right) - \left(\dfrac{\rho^{CNT}}{\rho^m}\right) w_{CNT}}$

Here, we present CNT total volume fraction $\left(V_{CNT}^*\right)$, the CNT mass fraction (w_{CNT}), and the matrix density (ρ^m). The CNT density (ρ^{CNT}).

For all four types of CNT, $V_{CNT} = w_{CNT}$

TABLE 7.1

Properties of 10 × 10 SWCNTs, with R (= 0.68 nm), h (= 0.067 nm), L (= 9.26 nm), v_{12}^{CNT} (= 0.175) That Depend on Temperature [28]

Temperature (K)	E_{11}^{CNT} (GPa)	E_{22}^{CNT} (GPa)	G_{12}^{CNT} (GPa)
300	5646.6	7080.0	1944.5
500	5530.8	6934.8	1964.3
700	5474.4	6864.1	1964.4

TABLE 7.2

Carbon Nanotube Efficiency Parameters at Different Percentages of V_{CNT}^{*} [28]

V_{CNT}^{*}	η_1	η_2
11%	0.149	0.934
14%	0.150	0.941
17%	0.149	1.381

Similarly, the density (ρ) and Poisson's ratio (v_{12}) are calculated by Equations 7.3 and 7.4

$$v_{12} = V_{CNT}^{*} v_{12}^{CNT} + V_m v^m \tag{7.3}$$

$$\rho = V_{CNT} \rho^{CNT} + V_m \rho^m \tag{7.4}$$

Here v_{12}^{CNT} and v^m are Poisson's ratios.

The transient behavior of functionally graded carbon nanotube–reinforced composite plate is investigated in this section. At 300K, the material characteristics for matrix, v^m is equal to 0.34, ρ^m is equal to $1.15 \frac{g}{cm^3}$ and E^m is equal to 2.1 GPa. The armchair (10 × 10) type SWCNT is explored in this work. The SWCNT materials characteristics are provided in Table 7.1. CNT efficiency parameters η_j are supplied in Table 7.2 to derive the effective material attributes of the CNTRC plate.

7.3 FINITE ELEMENT FORMULATION

The element SHELL 281 is selected for transient analysis. It consists of eight identical linear shell components, each with 6 degrees of freedom. The element library shown in Figure 7.7 was chosen from the Mechanical APDL 15.0 element library. Three translations are available along the x, y, and z axes, and three rotations around the x, y, and z axes. It is well suited for linear, big-rotation, and big-strain nonlinear applications,

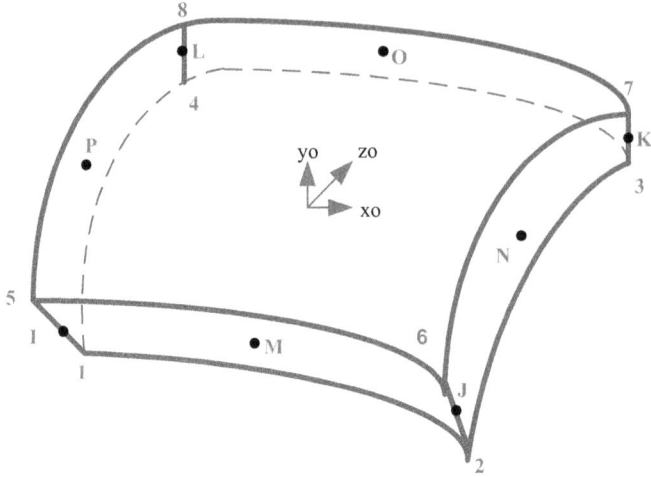

FIGURE 7.7 Element description of Shell 281.

with x, y, and z axes per node. It is working on First-Order Shear Deformation Theory (FSDT). True stress and logarithmic strain metrics are used in the element formulation. Figure 7.7 depicts the understanding of the SHELL 281 element.

If element orientation is provided, then x is element x-axis. If the element orientation is not provided, X_0 represents element x-axis.

It is generally known that utilizing the inbuilt steps in ANSYS, the midplane kinematics of carbon nanotubes–based composites has been taken as the FSDT [29] and acknowledged as follows:

$$
\left.
\begin{aligned}
u\left(x,y,z\right) &= u_0\left(x,y\right)+z\varphi_x\left(x,y\right) \\
v\left(x,y,z\right) &= v_0\left(x,y\right)+z\varphi_y\left(x,y\right) \\
w\left(x,y,z\right) &= w_0\left(x,y\right)+z\varphi_z\left(x,y\right)
\end{aligned}
\right\}
\tag{7.5}
$$

Here:

u, v, and w are the x, y, and z axes displacements, respectively.

u_0, v_0, and w_0 represent mid-plane displacement in x, y, and z axes, respectively.

φ_x and φ_y represent the rotations perpendicular to the mid-plane around the x and y axes, respectively.

7.3.1 STRAIN DISPLACEMENT RELATIONSHIP

Strains are derived from displacements as

$$
\{\varepsilon\} = \left\{u_x, v_y, w_z, u_y+v_x, v_z+w_y, w_x+u_z\right\}^T
\tag{7.6}
$$

Here $\{\varepsilon\} = \{\varepsilon_x, \varepsilon_y, \varepsilon_z, \gamma_{xy}, \gamma_{yz}, \gamma_{zx}\}^T$ and is called strain matrix.

The strain components are again arranged in the plane and out of the plane by following these steps.

Strain vector in the plane:

$$
\begin{Bmatrix} \varepsilon_x \\ \varepsilon_y \\ \gamma_{xy} \end{Bmatrix} = \begin{Bmatrix} \varepsilon_{x0} \\ \varepsilon_{y0} \\ \gamma_{xy0} \end{Bmatrix} + Z \begin{Bmatrix} K_x \\ K_y \\ K_{xy} \end{Bmatrix}
\tag{7.7}
$$

Strain vector in transverse direction:

$$
\begin{Bmatrix} \varepsilon_z \\ \gamma_{yz} \\ \gamma_{xz} \end{Bmatrix} = \begin{Bmatrix} \varepsilon_{z0} \\ \gamma_{yz0} \\ \gamma_{xz0} \end{Bmatrix} + Z \begin{Bmatrix} K_x \\ K_{yz} \\ K_{xz} \end{Bmatrix}
\tag{7.8}
$$

Where the deformation components are described as

$$
\begin{Bmatrix} \varepsilon_{x0} \\ \varepsilon_{y0} \\ \gamma_{xy0} \end{Bmatrix} = \begin{Bmatrix} \dfrac{\partial u_0}{\partial x} \\[2mm] \dfrac{\partial v_0}{\partial y} \\[2mm] \dfrac{\partial u_0}{\partial y} + \dfrac{\partial v_0}{\partial x} \end{Bmatrix}, \quad \begin{Bmatrix} K_x \\ K_y \\ K_{xy} \end{Bmatrix} = \begin{Bmatrix} \dfrac{\partial \varphi_x}{\partial x} \\[2mm] \dfrac{\partial \varphi_y}{\partial y} \\[2mm] \dfrac{\partial \varphi_x}{\partial y} + \dfrac{\partial \varphi_y}{\partial x} \end{Bmatrix}
\tag{7.9}
$$

$$
\begin{Bmatrix} \varepsilon_{z0} \\ \gamma_{yz0} \\ \gamma_{xz0} \end{Bmatrix} = \begin{Bmatrix} \varphi_z \\[2mm] \dfrac{\partial w_0}{\partial y} + \varphi_y \\[2mm] \dfrac{\partial w_0}{\partial x} + \varphi_x \end{Bmatrix}, \quad \begin{Bmatrix} K_x \\ K_{yz} \\ K_{xz} \end{Bmatrix} = \begin{Bmatrix} 0 \\[2mm] \dfrac{\partial \varphi_z}{\partial y} \\[2mm] \dfrac{\partial \varphi_z}{\partial x} \end{Bmatrix}
\tag{7.10}
$$

In terms of the nodal displacement vector, the strain vector expression is as follows:

$$
\{\varepsilon\} = [B]\{\delta\}
\tag{7.11}
$$

where $[B]$ = strain displacement matrix and $\{\delta\}$ = nodal displacement vector.

7.3.2 Constitutive Relation

The relation between stress and strain is

$$
\{\sigma\} = [D]\{\varepsilon\}
\tag{7.12}
$$

where $\{\sigma\}$ denotes linear-stress, $\{\varepsilon\}$ denotes linear-strain vector, and $[D]$ = rigidity matrix.

As seen below, the nodal displacement may be expressed as

$$
\left.\begin{aligned}
u^0 &= \sum_{j=1}^{8} N_j u_j^0 \\
v^0 &= \sum_{j=1}^{8} N_j v_j^0 \\
w^0 &= \sum_{j=1}^{8} N_j w_j^0 \\
\varphi_x &= \sum_{j=1}^{8} N_j \varphi_{xj}^0 \\
\varphi_y &= \sum_{j=1}^{8} N_j \varphi_{yj}^0 \\
\varphi_z &= \sum_{j=1}^{8} N_j \varphi_{zj}^0
\end{aligned}\right\}
\tag{7.13}
$$

For j^{th} nodal displacement, the equation is as follows:

$$
\left\{ \delta_j^* \right\} = \left\{ u_j^0 \ v_j^0 \ w_j^0 \ \varphi_{xj}^0 \ \varphi_{yj}^0 \ \varphi_{wj}^0 \right\}
\tag{7.14}
$$

$$
\left\{ \delta_j^* \right\} = \left[N_j \right] \left\{ \delta_j \right\}
\tag{7.15}
$$

Here

$$
\left[N_j \right] \text{ is equal to }
\begin{bmatrix}
N_j & 0 & 0 & 0 & 0 & 0 \\
0 & N_j & 0 & 0 & 0 & 0 \\
0 & 0 & N_j & 0 & 0 & 0 \\
0 & 0 & 0 & N_j & 0 & 0 \\
0 & 0 & 0 & 0 & N_j & 0 \\
0 & 0 & 0 & 0 & 0 & N_j
\end{bmatrix}
\tag{7.16}
$$

The strain energy is derived from the following equation. Here the nodal displacement values have to substitute:

$$
\left\{ \varepsilon_j \right\} = \left[B_j \right] \left\{ \delta_j \right\}
\tag{7.17}
$$

$$
U = \frac{1}{2} \left[B \right]^T \left\{ \delta_j \right\} \left[D \right] \left[B_j \right] \left\{ \delta_j \right\} dA - \left\{ F \right\}_{mj}
\tag{7.18}
$$

where $[B_j]$ represents the matrix of strain displacement relation and $\{F\}_{mj}$ represents the mechanical force.

The higher-order terms are represented by φ_z in Taylor's series shape functions (N_i) equation as

$$\delta = \sum_{i=1}^{j} N_i \delta_i \tag{7.19}$$

where $\delta_i = (u_{oi}\, v_{oi} w_{oi} \varphi_{xi} \varphi_{yi} \varphi_{zi})^T$

Shape functions for eight-nodded shell elements, $j = 8$, are expressed in natural coordinates η, ξ in Equation (7.20), and the elements' explanations are provided as follows [30]:

$$
\begin{aligned}
N^1 &= \frac{1}{4}(1-\xi)(1-\eta)(-\xi-\eta-1),\\[4pt]
N^2 &= \frac{1}{4}(1+\xi)(1-\eta)(\xi-\eta-1),\\[4pt]
N^3 &= \frac{1}{4}(1+\xi)(1+\eta)(1+\xi-\eta),\\[4pt]
N^4 &= \frac{1}{4}(1-\xi)(1+\eta)(-\xi+\eta-1),\\[4pt]
N^5 &= \frac{1}{2}(1-\xi^2)(1-\eta),\\[4pt]
N^6 &= \frac{1}{2}(1+\xi)(1-\eta^2),\\[4pt]
N^7 &= \frac{1}{2}(1-\xi^2)(1+\eta),\\[4pt]
N^8 &= \frac{1}{4}(1-\xi)(1-\eta^2),
\end{aligned}
\tag{7.20}
$$

Equation 7.21 is the final equation and is obtained by solving the total potential energy (TPE) equation

$$\partial\Pi = 0 \tag{7.21}$$

Here TPE is denoted by Π.

$$[K]\{\delta\} = \{F\}_m \tag{7.22}$$

Here $[K]$ = The stiffness matrix.

K matrix is solved by Gauss quadrature integration method for universal stiffness matrix

$$[K] = \int\limits_{-1-1}^{+1+1} [B]^T [D][B] |J| d\zeta \, dx d\eta \qquad (7.23)$$

Here made J represents Jacobian matrix.

7.3.3 DESCRIPTION OF STRUCTURAL AND OTHER SECOND-ORDER SYSTEMS

For almost all types of structurally dynamic problems in a mechanical system, finite element semidiscrete equation of motion is obtained from the principle of virtual work, which is specially discretized with the help of finite element method as follows:

$$[M]\{\ddot{\Delta}(t)\} + [C]\{\dot{\Delta}(t)\} + \{F^i(t)\} = \{F^a(t)\} \qquad (7.24)$$

where:

$\{F^a(t)\}$ is applied load factor

$\{F^i(t)\}$ is internal load vector

$\Delta(t)$ is the nodal displacement vector

$\{\dot{\Delta}(t)\}$ is the nodal velocity vector

$\{\ddot{\Delta}(t)\}$ is the nodal acceleration vector

$[C]$ is structural damping matrix

$[M]$ is the structural mass matrix

7.3.4 TIME INTEGRATION SCHEME FOR LINEAR SYSTEMS

The initial load is linearly proportional to the nodal displacement, and the structural stiffness matrix remains constant in a linear structural dynamics system. Hence, Equation 7.24 is written as

$$[M]\{\ddot{\Delta}(t)\} + [C]\{\dot{\Delta}(t)\} + [K]\{u(t)\} = \{F^a(t)\} \qquad (7.25)$$

where $[K]$ is the structural stiffness matrix.

In Equation 7.25, a finite element semidiscrete equation of motion can be numerically solved by the number of methods incorporated in the program, namely generalized – $\propto$ method and Newmark method. A Newmark method is used in this work because of its popularity among all other methods.

From Equation 7.25, we get

$$[M]\{\ddot{\Delta}_{n+1}\} + [C]\{\dot{\Delta}_{n+1}\} + [K]\{\Delta_{n+1}\} = \{F^a_{n+1}\} \qquad (7.26)$$

where

$\{F^a_{n+1}\}$ = The applied load $\{F^a_{n+1}(t_{n+1})\}$ at time t_{n+1}

$\{\Delta_{n+1}\}$ = The nodal displacement vector $\{\Delta(t_{n+1})\}$ at time t_{n+1}

$\left\{\dot{\Delta}_{n+1}\right\}$ =The nodal velocity vector $\left\{\dot{\Delta}\left(t_{n+1}\right)\right\}$ at time t_{n+1}

$\left\{\ddot{\Delta}_{n+1}\right\}$ = The nodal acceleration vector $\left\{\ddot{\Delta}\left(t_{n+1}\right)\right\}$ at time t_{n+1}

In Equation 7.26, the velocity and displacement are updated as

$$\left\{\dot{\Delta}_{n+1}\right\}=\left\{\dot{\Delta}_{n}\right\}+\left[(1-\alpha)\left\{\ddot{\Delta}_{n}\right\}+\alpha\left\{\ddot{\Delta}_{n+1}\right\}\right]\delta t \tag{7.27}$$

$$\left\{\Delta_{n+1}\right\}=\left\{\Delta_{n}\right\}+\left\{\dot{\Delta}_{n}\right\}\delta t+\left[\left(\frac{1}{2}-\beta\right)\left\{\ddot{\Delta}_{n}\right\}+\beta\ddot{\Delta}_{n+1}\right]\delta t^{2} \tag{7.28}$$

where

α, β is the Newmark's integration parameters

$\left\{\ddot{\Delta}_{n}\right\}$ is the nodal acceleration vector $\left\{\ddot{\Delta}\left(t_{n}\right)\right\}$ at time t_{n}

$\left\{\dot{\Delta}_{n}\right\}$ is the nodal velocity vector $\left\{\dot{\Delta}\left(t_{n}\right)\right\}$ at time t_{n}

$\left\{\Delta_{n}\right\}$ is the displacement vector $\left\{\Delta(t_{n})\right\}$ at time t_{n}

Finally, the three unknowns $\left\{\ddot{\Delta}_{n+1}\right\}$, $\left\{\dot{\Delta}_{n+1}\right\}$, and $\left\{\Delta_{n+1}\right\}$ can be calculated by solving the above three equations (Equation 7.26–7.28) with the three known quantities $\left\{\ddot{\Delta}_{n}\right\}, \left\{\dot{\Delta}_{n}\right\}$, and $\left\{\Delta_{n}\right\}$.

By using the above three equations (Equations 7.26–7.28), in terms of unknown and known quantities, a single-step time integrator is written as follows:

$$\left(a_{0}[M]+a_{1}[C]+[K]\right)\left\{\Delta_{n+1}\right\}\left\{F_{n+1}^{a}\right\}+[M]\left(a_{0}\left\{\Delta_{n}\right\}+a_{2}\left\{\dot{\Delta}_{n}\right\}\right.$$
$$=+a_{3}\left\{\ddot{\Delta}_{n}\right\})+[C]\left(a_{1}\left\{\Delta_{n}\right\}+a_{4}\left\{\dot{\Delta}_{n}\right\}+a_{5}\left\{\ddot{\Delta}_{n}\right\}\right. \tag{7.29}$$

where

$a_{0} = \dfrac{1}{\alpha\delta t^{2}}$

$a_{1} = \dfrac{\beta}{\alpha\delta t}$

$a_{2} = \dfrac{1}{\alpha\delta t}$

$a_{3} = \dfrac{1}{2\alpha} - 1$

$a_{4} = \dfrac{\beta}{\alpha} - 1$

$a_{5} = \dfrac{\delta t}{2}\left(\dfrac{\beta}{\alpha} - 2\right)$

$a_{6} = \delta t(1 - \beta)$

$a_{7} = \beta\delta t$

First, the unknown $\{\Delta_{n+1}\}$ is calculated using Equation 7.29. Then, the program computes the two unknowns $\{\dot{\Delta}_{n+1}\}$ and $\{\ddot{\Delta}_{n+1}\}$ by using the following equations:

$$\{\dot{\Delta}_{n+1}\} = a_1\left(\{\Delta_{n+1}\} - \{\Delta_n\}\right) - a_4\{\dot{\Delta}_n\} - a_5\{\ddot{\Delta}_n\} \tag{7.30}$$

$$\{\ddot{\Delta}_{n+1}\} = a_0\left(\{\Delta_{n+1}\} - \{\Delta_n\}\right) - a_2\{\dot{\Delta}_n\} - a_3\{\ddot{\Delta}_n\} \tag{7.31}$$

Accuracy, dissipation, and stability are the factors that play a vital role in selecting a proper time integration scheme for the equation of motion given in Equation 7.24. Time step size affects the stability in conditionally stable time integration algorithms. In contrast, time step size doesn't affect the stability in unconditionally stable time integration algorithms.

With the help of Newmark's parameters, β, the amount of numerical algorithms dissipation can be controlled, as follows:

$$\beta \geq \frac{1}{4}, \quad \alpha \geq \frac{1}{4}\left(\frac{1}{2}+\beta\right)^2 \tag{7.32}$$

The Newmark family of methods may be unconditionally stable with the Newmark parameters satisfying the above conditions. By introducing the amplitude decay factor $\gamma \geq 0$, the above conditions can be written as

$$\beta = \frac{1}{2}+\gamma, \quad \alpha = \frac{1}{4}(1+\gamma)^2, \gamma \geq 0 \tag{7.33}$$

Consequently, an unconditionally stable Newmark integration procedure is done via γ provided by the user on the TINTP command. The direct input parameters are α, β using the TINTP command.

7.3.5 Dynamic Loading

Transient means the applied loads are a function of time. Here the sinusoidal load is considered, which is time dependent. The sinusoidal load is classified into four types, namely exponential blast load, sine load, triangular load, and step load, which act on the surface of the plate.

In the present analysis, the plate is subjected to a transverse load that is sinusoidally distributed in the spatial domain and varies with time as follows:

$$q(x,y,t) = q_0 \sin\left(\frac{\pi x}{a}\right)\sin\left(\frac{\pi y}{b}\right)F(t) \tag{7.34}$$

where q_0 is transverse load intensity load (3.448 MPa), $F(t)$ is a time function that varies for different loading conditions according to one of the expressions given below, a is the length of the plate, and b is the width of the plate.

7.3.5.1   Exponential Blast Load

The time-dependent expression for explosive blast load is taken as

$$F(t) = \left\{ e^{-\gamma t} \right\} \tag{7.35}$$

where $\gamma = 330\ S^{-1}$, t is the time in sec, and Δt is taken as 0.0001 sec.

7.3.5.2   Sine Load

The expression for sine load, which depends on time, is

$$F(t) = \left\{ \begin{array}{ll} \sin\left(\dfrac{\pi t}{t_1}\right) & 0 \le t \le t_1 \\ 0 & t > t_1 \end{array} \right\} \tag{7.36}$$

where t = positive phase duration and its values are taken as 0.006 sec, t is the time in sec, and Δt is taken as 0.0001 sec.

7.3.5.3   Triangular Load

The time-dependent expression for triangular load is taken as

$$F(t) = \left\{ \begin{array}{ll} 1 - \dfrac{t}{t_1} & 0 \le t \le t_1 \\ 0 & t > t_1 \end{array} \right\} \tag{7.37}$$

7.3.5.4   Step Load

The expression for step load, which depends on time, is

$$F(t) = \left\{ \begin{array}{ll} 1 & 0 \le t \le t_1 \\ 0 & t > t_1 \end{array} \right\} \tag{7.38}$$

7.4   RESULTS AND DISCUSSION

To solve physical and engineering problems, a numerical technic is used called *finite element analysis* (FEA). It is beneficial for complex geometry problems, material quality, and different loadings without analytical solutions. Structural, fluid flow analysis biomechanics and electromagnetic fields are some of the key uses of FEA.

Mechanical APDL 15.0, a commercial FEA programme, was used to create the finite element simulation. The FG-CNTRC transient response was derived utilizing the effective material characteristics using the APDL algorithm.

7.4.1   SUPPORT CONDITIONS

A rectangular plate with dimensions of length a, thickness h, and width b is used throughout the study. To eliminate stiff motion and decrease unknown field variables, three alternative support conditions are simply supported, denoted by S, clamped support denoted by C, and evaluated singly and in combination. The following are the terms of assistance:

SSSS

All four edges are simply supported.

$$At\ x = 0, a : v_0 = w_0 = \theta_y = v_0^* = \theta_x^* = \theta_y^* = 0$$

$$At\ y = 0, b : u_0 = w_0 = \theta_x = u_0^* = \theta_x^* = 0$$

CCCC
All the edges are clamped.

$$At\ x = 0, a\ and\ y = 0, b : u_0 = v_0 = w_0 = \theta_x = \theta_y = u_0^* = v_0^* = \theta_x^* = \theta_y^* = 0$$

SCSC
Two opposite edges are clamped, and another two are simply supported.

$$At\ x = 0, a : v_0 = w_0 = \theta_y = v_0^* = \theta_x^* = \theta_y^* = 0$$

$$At\ y = 0, b : u_0 = v_0 = w_0 = \theta_x = \theta_y = u_0^* = v_0^* = \theta_x^* = \theta_y^* = 0$$

7.4.2   CONVERGENCE TEST

On a rectangular plate of $1 \times 1\ m^2$, pressure of $1.0 \times 10^5\ \frac{N}{m^2}$ is applied on the surface of the plate for different supporting conditions. The nondimensional central deflection of a simply supported FG-CNT reinforced composite plate is depicted in Figure 7.8.

7.4.3   VALIDATION TEST

Table 7.3. shows the effect of simply supported carbon nanotubes on applying a uniformly distributed load of $-1.0 \times 10^5\ N/m^2$

A simply supported square laminated plate with three layers is considered. The orientation is arranged as $0°$ for the first plate, $90°$ for the middle plate, and $0°$ for the last plate taken by [32]. The same is chosen to compare the dynamic response of plates under different transient loads, including triangular, explosive blast load, sine load, step load. The material properties are:

$E_1 = 172.369$ GPa
$E_2 = 6.895$ GPa
$G_{12} = 3.448$ GPa

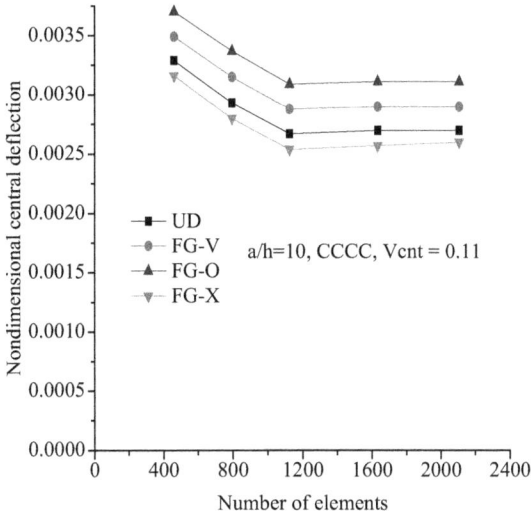

FIGURE 7.8 The nondimensional central deflection performed for FG-CNTRC plate at different mesh densities.

TABLE 7.3

Nondimensional Central Deflection of a Simply Supported FGCNT Reinforced Composite Plate Subjected to Uniformly Distributed Load

V_{CNT}	b/h	Present	Reference [31]	% Difference
0.11	10	0.003740	0.003739	0.027
	20	0.03630	0.03628	0.055
	50	1.16000	1.15500	0.130
0.14	10	0.00330	0.00331	0.182
	20	0.03000	0.03001	0.200
	50	0.91600	0.91750	0.164
0.17	10	0.00240	0.00239	0.042
	20	0.02350	0.02348	0.085
	50	0.75400	0.75150	0.265

$\vartheta_{12} = 0.25$
$\rho = 1603.03 \text{ Kg/m}^3$

The dimensions of the plates are taken as $a = b = 20h$. Here $h = 0.0381$ m which is the total thickness. On the plate, a transverse load is applied, which is sinusoidally distributed in the spatial domain and changes with time as follows:

$$q(x,y,t) = q_0 \sin\left(\frac{\pi x}{a}\right) \sin\left(\frac{\pi y}{a}\right) F(t) \tag{7.39}$$

$F(t)$ is taken as

$$
\left\{
\begin{array}{ll}
\sin\left(\dfrac{\pi t}{t_1}\right) & 0 \le t \le t_1 \\[2mm]
0 & t > t_1
\end{array}
\right\} \quad \text{for sine load} \tag{7.40}
$$

$$
\left\{
\begin{array}{ll}
1 - \dfrac{t}{t_1} & 0 \le t \le t_1 \\[2mm]
0 & t > t_1
\end{array}
\right\} \quad \text{for triangular load} \tag{7.41}
$$

$$
\left\{
\begin{array}{ll}
1 & 0 \le t \le t_1 \\
0 & t > t_1
\end{array}
\right\} \quad \text{for step load} \tag{7.42}
$$

$$
\left\{ e^{-\gamma t} \right\} \quad \text{for explosive blast load} \tag{7.43}
$$

where $t_1 = 0.006$ s, $\gamma = 330\ S^{-1}$, and $q_0 = 3.448$ MPa.

Figure 7.9 shows the transverse deflection at the middle of the plate at different times for various loads. The comparisons are made between present and reference results by [32]. From Figure 7.9, it is noticed that the present solutions are in good agreement with the strip element method.

7.4.4   NUMERICAL ILLUSTRATION

7.4.4.1   Effect of Different Types of Loading

This section investigates the transient response FG-CNTRC plate under different sinusoidal loads, i.e., exponential blast load, sine load, triangular load, and step load. Figure 7.10 presents the transient response of four types of FG carbon nanotube RC plate having 11% of V_{CNT} and $b/h = 10$, subjected to exponential blast, triangular, step, and sine loads. Here it is noted the all the three loads, i.e., exponential blast, triangular, and step load, have the same trend, whereas the trend for sine load is different from remain loads in all cases of carbon nanotubes.

Figure 7.11 presents the transient response of four types of FG carbon nanotube RC plate having 14% of V_{CNT} and b/h = 10, subjected to exponential blast, triangular, step, and sine loads. Here it is noted the all the three loads, i.e., exponential blast, triangular, and step load, have the same trend, whereas the trend for sine is different from remain loads in all cases of carbon nanotubes.

Figure 7.12 shows the transient response of different types of FG-CNTRC plates with $V_{CNT} = 0.14$, $b/h = 10$, subjected to exponential blast, triangular, step load, and sine loads.

7.4.4.2   Effect of Boundary Condition

In this section, the effect of boundary conditions for 11% of volume fractions of CNTs along with the plate thickness ratio of (b/h) is 10. FG-CNTRC plates under

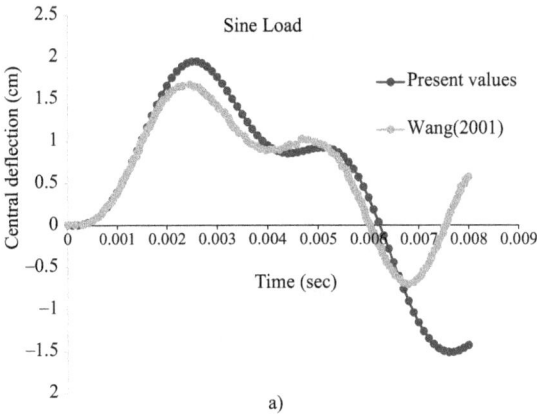

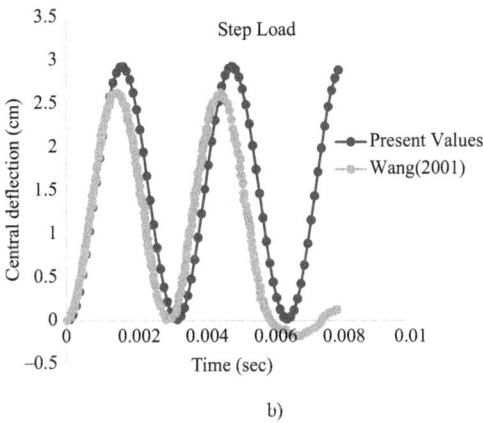

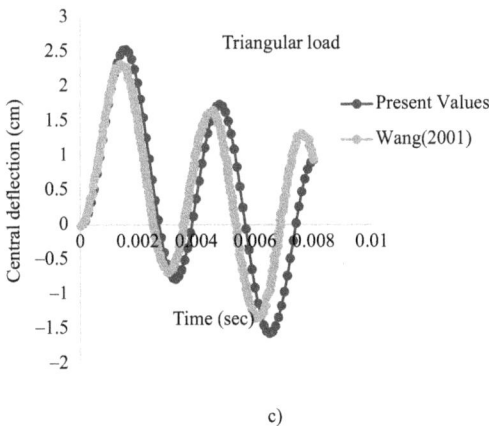

FIGURE 7.9   Comparison of the transient response of the rectangular laminated plate. (a) Comparison of present results with Wang et al. (2001) for sine load. (b) Comparison of present results with Wang et al. (2001) for the step load. (c) Comparison of present results with Wang et al. (2001) for Triangular load.

(Continued)

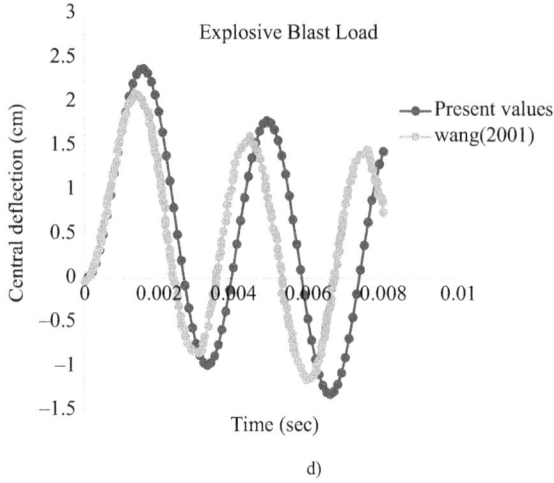

d)

FIGURE 7.9 (Continued) Comparison of the transient response of the rectangular laminated plate. (d) Comparison of present results with Wang et al. (2001) for explosive blast load.

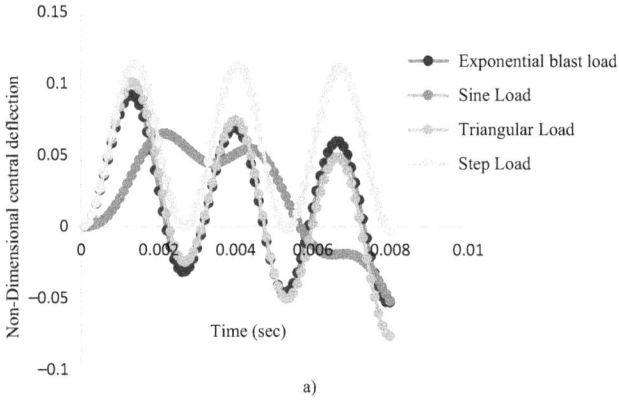

a)

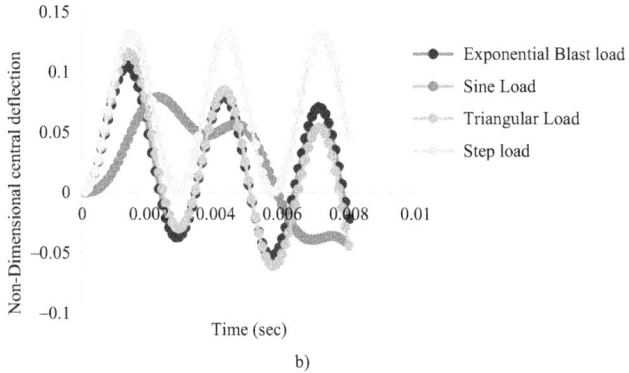

b)

FIGURE 7.10 Transient response of different types of FG-CNTRC plate on different blast loads for $V_{CNT} = 0.11$, $b/h = 10$. (a) UD FG-CNTRC plate. (b) FG-V type CNTRC plate.

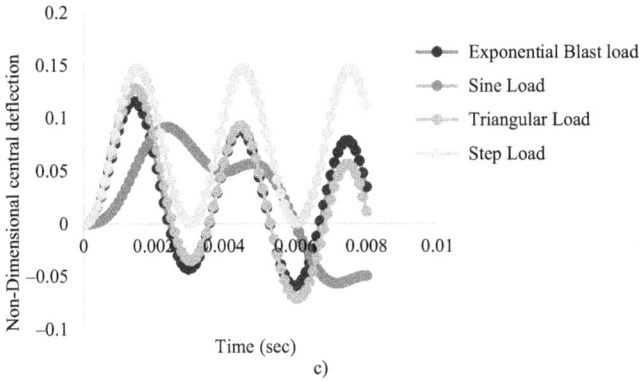

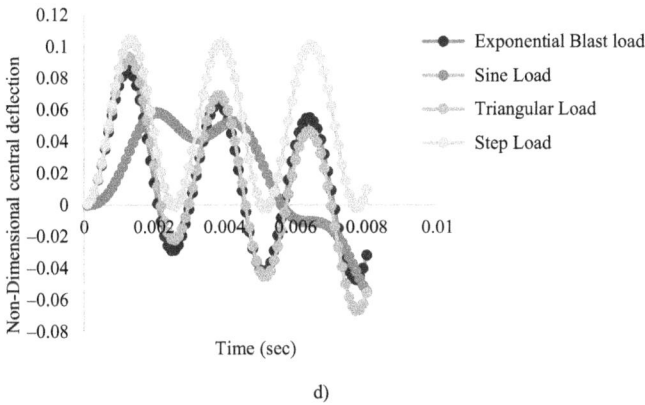

FIGURE 7.10 (Continued) Transient response of different types of FG-CNTRC plate on different blast loads for $V_{CNT} = 0.11$, $b/h = 10$. (c) FG-O type CNTRC plate. (d) FG-X type CNTRC plate.

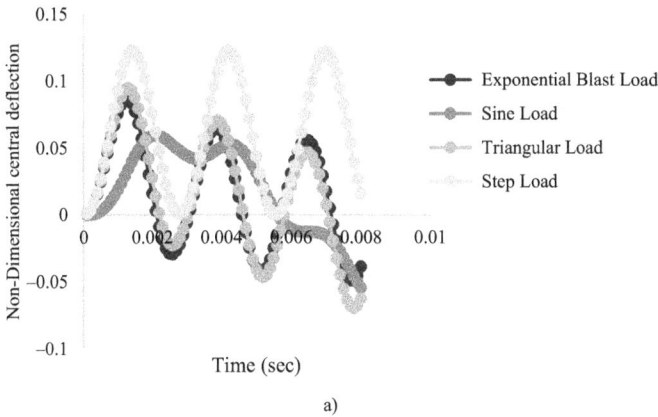

FIGURE 7.11 Transient response of different types of FG-CNTRC plate on different blast loads for $V_{CNT} = 0.14$, b/h = 10. (a) UD-FG CNTRC plate.

(Continued)

b)

c)

d)

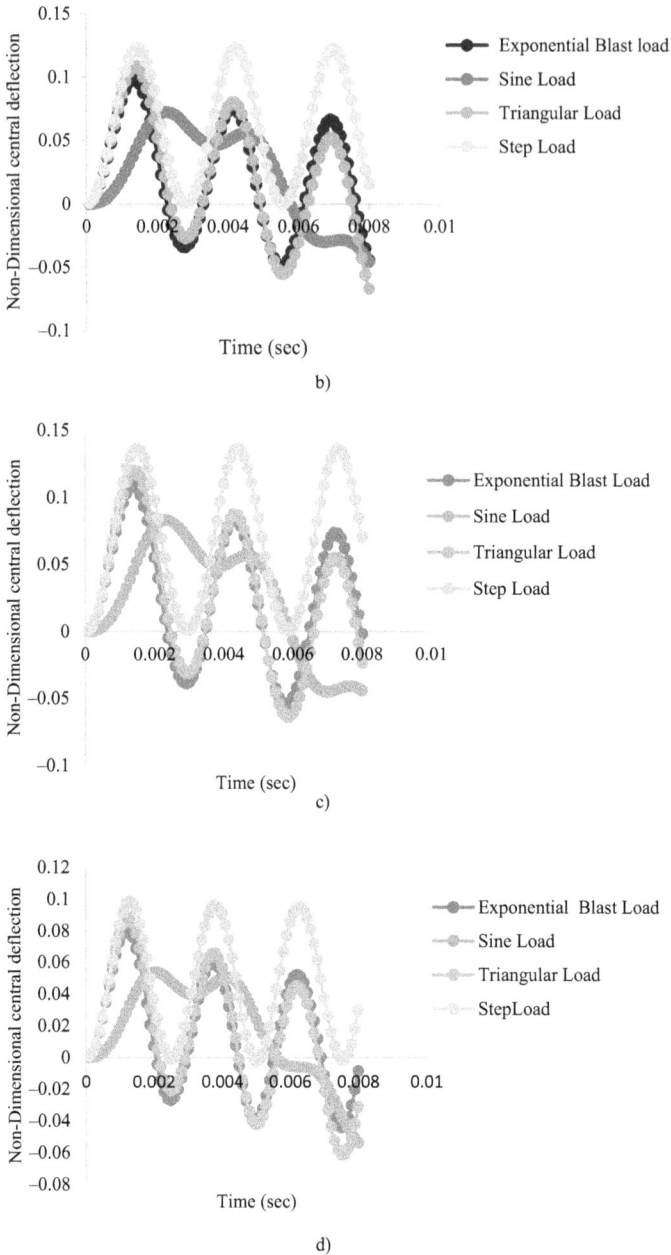

FIGURE 7.11 (Continued) Transient response of different types of FG-CNTRC plate on different blast loads for V_{CNT} = 0.14, b/h = 10. (b) FG-V type CNTRC plate. (c) FG-O type CNTRC plate. (d) FG-X type CNTRC plate.

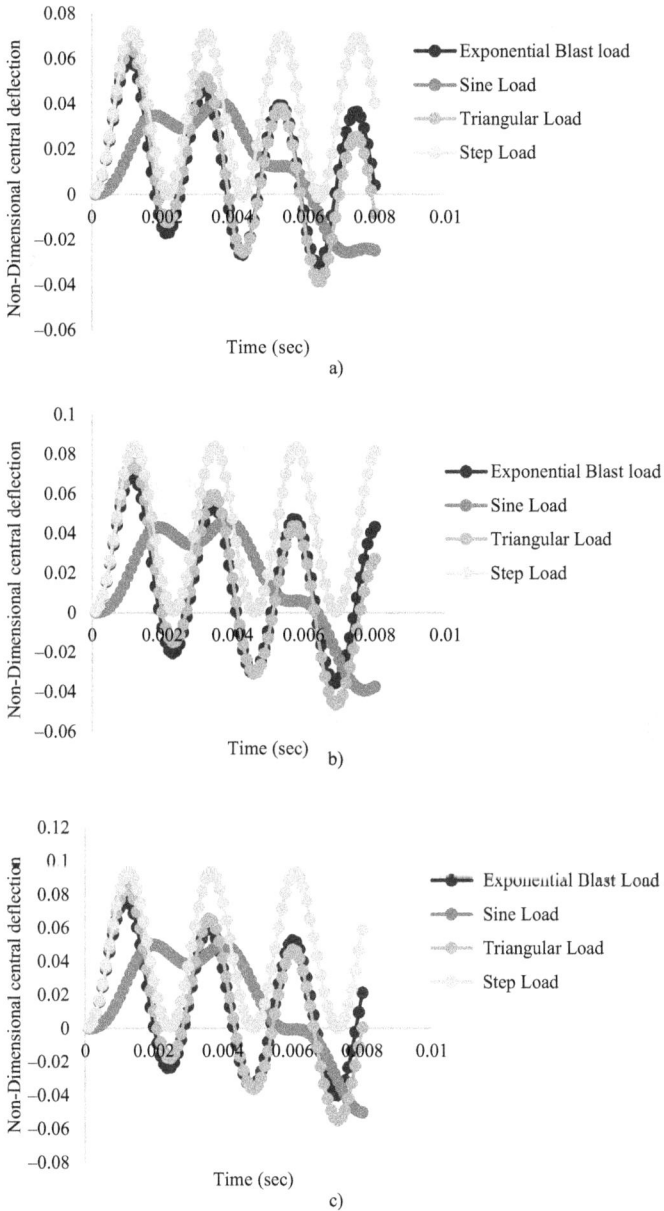

FIGURE 7.12 Transient response of different types of FG-CNTRC plate on different blast loads for V_{CNT} = 0.17, b/h = 10. (a) UD-FG CNTRC plate. (b) FG-V type CNTRC plate. (c) FG-O type CNTRC plate.

(*Continued*)

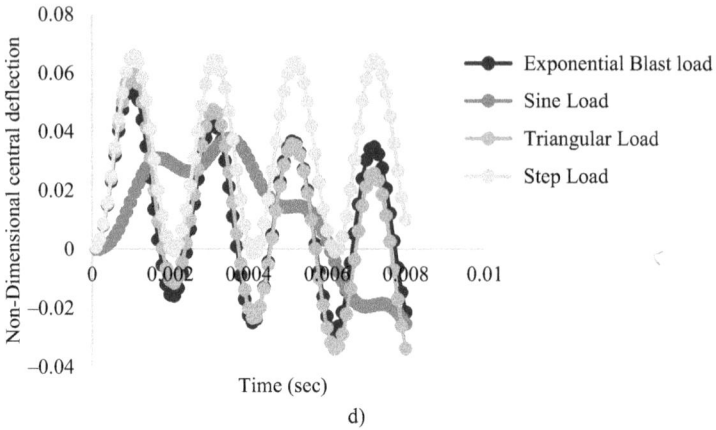

FIGURE 7.12 (Continued)   Transient response of different types of FG-CNTRC plate on different blast loads for $V_{CNT} = 0.17$, $b/h = 10$. (d) FG-X type CNTRC plate.

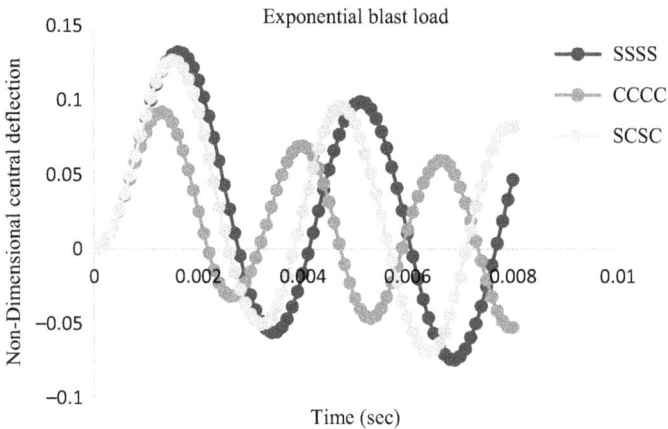

FIGURE 7.13   Transient response of FG-CNT plate with $V_{CNT} = 11\%$ and $b/h = 10$ for various boundary conditions under exponential blast load.

different sinusoidal loads, i.e., exponential blast load, sine load, triangular load, and step load, are evaluated (see Figure 7.13–7.16). Here, three types of boundary conditions are considered, namely SSSS, CCCC, and SCSC.

The FG CNT is affected by boundary conditions like SSSS type of boundary has higher amplitude, followed by SCSC and CCCC, respectively, as observed from Figures 7.13–7.16.

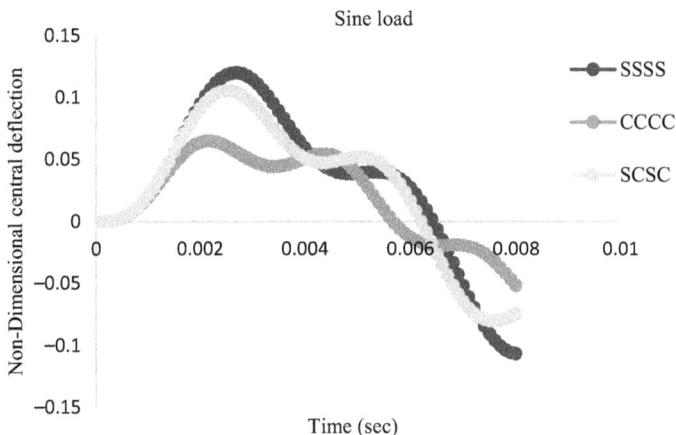

FIGURE 7.14   Transient response of FG-CNT plate with $V_{CNT} = 11\%$ and $b/h = 10$ for various boundary conditions under sine load.

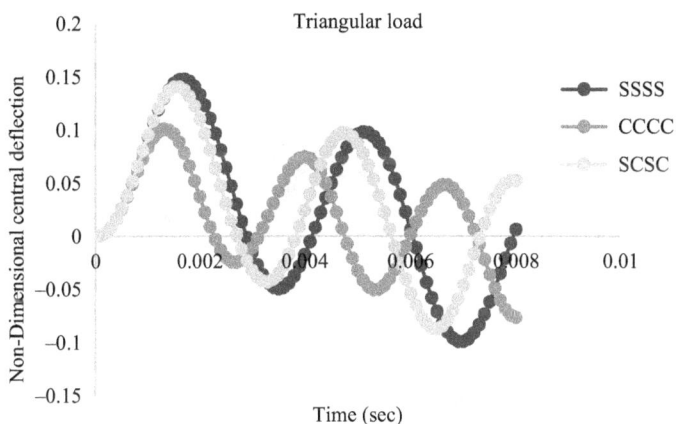

FIGURE 7.15   Transient response of FG-CNT with $V_{CNT} = 11\%$ and $b/h = 10$ for various boundary conditions under triangular load.

7.4.4.3   Effect of Geometrical Parameter

This section evaluates the effect of ratios (b/h) for FG-carbon nanotube RC plate under different sinusoidal loads, i.e., exponential blast load, sine load, triangular load, and step load. The thickness of the plate has been decreased with increasing the thickness ratio, and the values are 10, 12, and 15.

As observed from Figures 7.17–7.20, the response of deflection increases as the b/h ratio increases. When we increase the b/h ratio, the plate thickness decreases, which increases deflection.

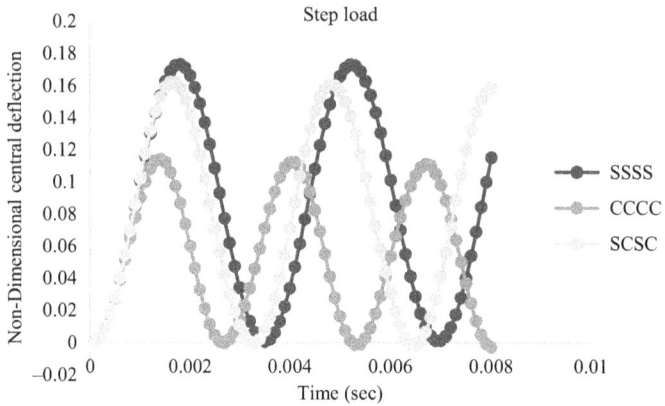

FIGURE 7.16   Transient response of FG-CNT plate with V_{CNT} = 11% and b/h = 10 for various boundary conditions under step load.

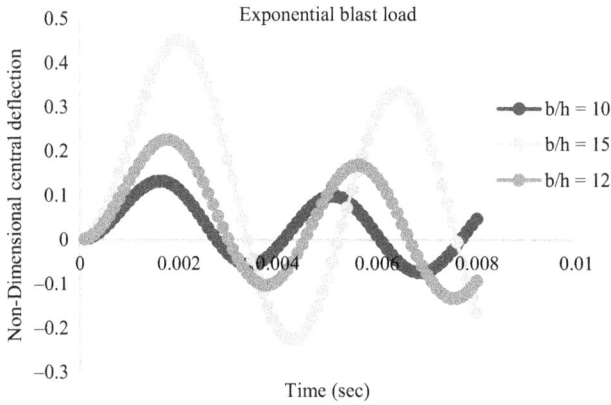

FIGURE 7.17   Effect of thickness ratio on FG-CNTRC plate with V_{CNT} = 0.11 under exponential blast load.

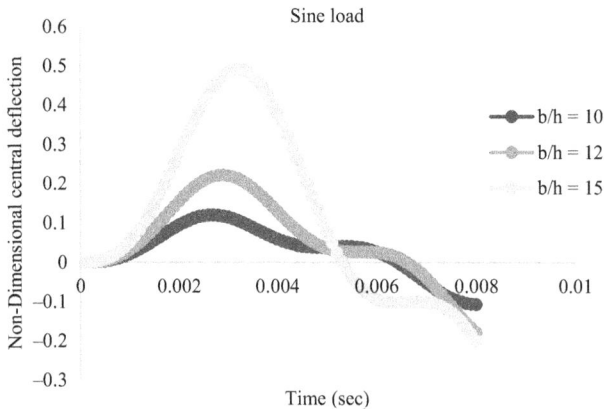

FIGURE 7.18   Effect of thickness on FG-CNTRC plate with V_{CNT} = 0.11 under sine load.

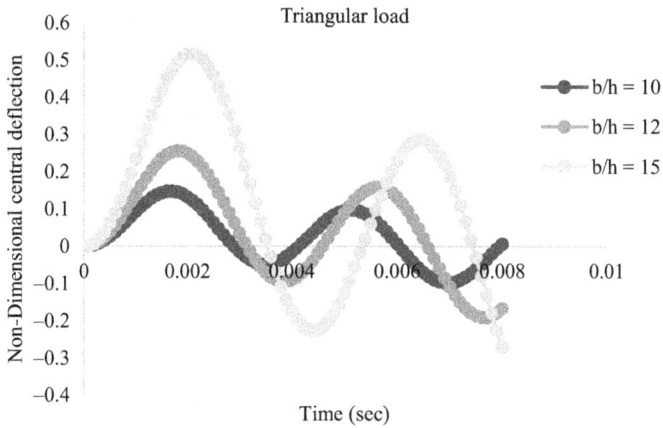

FIGURE 7.19   Effect of thickness ratio on FG-CNTRC plate with V_{CNT} = 0.11 under triangular load.

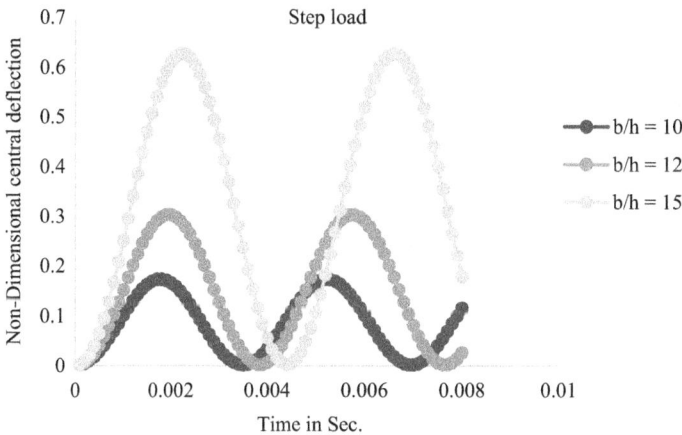

FIGURE 7.20   Effect of thickness ratio on FG-CNTRC plate with V_{CNT} = 0.11 under step load.

7.5   CONCLUDING REMARKS

This chapter analyzed the transient behaviors of carbon nanotube-reinforced composite plates under blast load. The blast load is distributed sinusoidally on the whole surface of the rectangular plate and varies with time. Here four types (explosive blast load, sine load, triangular load, and step load) are considered. Four types (UD, FG-V, FG-O, and FG-X CNT) are considered, and the effective material properties of CNTRC are calculated based on the extended rule of mixture, in which the uncertainty of CNT distribution is also considered. Static analysis of CNT plate under the application of pressure is determined first. Then the transient response for all the above loading cases is found. The accuracy of the results is compared with the previously published literature. Finally, the transient behavior of thin square FG-CNTRC plate on sinusoidal loads is calculated in mechanical APDL 15.0 platform for the first time. The following conclusions can be drawn:

➢ With an increase in width-to-thickness ratio (b/h), the transient response of all types of FG-CNT reinforced composites increases for all load cases.
➢ The SSSS type of boundary condition has a higher deflection response than the SCSC type of boundary condition. For the CCCC type of boundary condition, the response is very low in all types of CNTs.
➢ It is also noticed that exponential blast load, triangular load, and step load have the same trend, whereas the sine load shows a different trend for all types of CNTs and all supporting conditions.
➢ It is also found that for all types of loads, FG-O type CNTs have maximum deflection, while FG-X type has minimum deflection for all percentages of CNT volume fractions.
➢ As an increase in the percentage of CNT volume fraction, the transient response decreases for all types of loading cases.

REFERENCES

1. Daniel, I. M., Ishai, O., Daniel, I. M., & Daniel, I. (2006). *Engineering mechanics of composite materials* (Vol. 1994). New York: Oxford University Press.
2. Kausar, A., Rafique, I., & Muhammad, B. (2016). Review of applications of polymer/carbon nanotubes and epoxy/CNT composites. *Polymer-Plastics Technology and Engineering*, *55*(11), 1167–1191.
3. Evans, A., San Marchi, C., & Mortensen, A. (2003). Metal matrix composites. In *Metal Matrix Composites in Industry* (pp. 9–38). Boston, MA: Springer.
4. Erden, S., & Ho, K. (2017). Fiber reinforced composites. In Edited by: M. Özgür Seydibeyoğlu, Amar K. Mohanty and Manjusri Misra *Fiber Technology for Fiber-Reinforced Composites* (pp. 51–79). Sawston, United Kingdom, Woodhead Publishing.
5. Sheokand, S. K., Kalkal, K. K., & Deswal, S. (2021). Thermoelastic interactions in a functionally graded material with gravity and rotation under dual-phase-lag heat conduction. *Mechanics Based Design of Structures and Machines*, 1–21.
6. Bhavar, V., Kattire, P., Thakare, S., & Singh, R. K. P. (2017, September). A review on functionally gradient materials (FGMs) and their applications. In *IOP Conference Series: Materials Science and Engineering* (Vol. 229, No. 1, p. 012021). IOP Publishing.
7. El-Wazery, M. S., & El-Desouky, A. R. (2015). A review on functionally graded ceramic-metal materials. *Journal of Materials and Environmental Science*, *6*(5), 1369–1376.
8. Hussain, M. A. (2014). *Buckling analysis of functionally graded carbon nanotubes reinforced composite (FG-CNTRC) plate* (Doctoral dissertation).
9. Dabbagh, A., Rastgoo, A., & Ebrahimi, F. (2021). Thermal buckling analysis of agglomerated multiscale hybrid nanocomposites via a refined beam theory. *Mechanics Based Design of Structures and Machines*, *49*(3), 403–429.
10. Gopi, S., Balakrishnan, P., Sreekala, M. S., Pius, A., & Thomas, S. (2017). Green materials for aerospace industries. In Edited by: Dipa Ray *Biocomposites for High-Performance Applications* (pp. 307–318). Woodhead Publishing.
11. Ruoff, R. S., & Lorents, D. C. (1995). Mechanical and thermal properties of carbon nanotubes. *Carbon*, *33*(7), 925–930.

12. Sun, X., Qin, Z., Ye, L., Zhang, H., Yu, Q., Wu, X., ... & Yao, F. (2020). Carbon nanotubes reinforced hydrogel as flexible strain sensor with high stretchability and mechanically toughness. *Chemical Engineering Journal*, *382*, 122832.
13. Wang, X., Bradford, P. D., Liu, W., Zhao, H., Inoue, Y., Maria, J. P., ... & Zhu, Y. (2011). Mechanical and electrical property improvement in CNT/Nylon composites through drawing and stretching. *Composites Science and Technology*, *71*(14), 1677–1683.
14. Zhang, L., Ding, S., Li, L., Dong, S., Wang, D., Yu, X., & Han, B. (2018). Effect of characteristics of assembly unit of CNT/NCB composite fillers on properties of smart cement-based materials. *Composites Part A: Applied Science and Manufacturing*, *109*, 303–320.
15. Li, M., Chen, M., & Wu, Z. (2014). Enhancement in thermal property and mechanical property of phase change microcapsule with modified carbon nanotube. *Applied energy*, *127*, 166–171.
16. Codan, B. (2003). Space and Nanotechnology: the Versatility of Nanotubes Based Materials. In *54th International Astronautical Congress of the International Astronautical Federation, the International Academy of Astronautics, and the International Institute of Space Law* (pp. I–3).
17. Das, R., Bee Abd Hamid, S., Eaqub Ali, M., Ramakrishna, S., & Yongzhi, W. (2015). Carbon nanotubes characterization by X-ray powder diffraction–a review. *Current Nanoscience*, *11*(1), 23–35.
18. Saifuddin, N., Raziah, A. Z., & Junizah, A. R. (2013). Carbon nanotubes: a review on structure and their interaction with proteins. *Journal of Chemistry*, Volume 2013, Article ID 676815, 18 pages.
19. Bellucci, S., Balasubramanian, C., Micciulla, F., & Rinaldi, G. (2007). CNT composites for aerospace applications. *Journal of Experimental Nanoscience*, *2*(3), 193–206.
20. Kausar, A., Rafique, I., & Muhammad, B. (2016). Review of applications of polymer/carbon nanotubes and epoxy/CNT composites. *Polymer-Plastics Technology and Engineering*, *55*(11), 1167–1191.
21. Mahapatra, T. R., Mehar, K., & Panda, S. K. (2016). Flexural behaviour of functionally graded nanotube reinforced sandwich spherical panel.
22. Mahapatra, T. R., Mehar, K., Panda, S. K., Dewangan, S., & Dash, S. (2017, February). Flexural strength of functionally graded nanotube reinforced sandwich spherical panel. In *IOP Conference Series: Materials Science and Engineering* (Vol. 178, No. 1, p. 012031). IOP Publishing.
23. Nejati, M., Eslampanah, A., & Najafizadeh, M. (2016). Buckling and vibration analysis of functionally graded carbon nanotube-reinforced beam under axial load. *International Journal of Applied Mechanics*, *8*(01), 1650008.
24. Mehar, K., & Panda, S. K. (2016). Geometrical nonlinear free vibration analysis of FG-CNT reinforced composite flat panel under uniform thermal field. *Composite Structures*, *143*, 336–346.
25. Mehar, K., Panda, S. K., Dehengia, A., & Kar, V. R. (2016). Vibration analysis of functionally graded carbon nanotube reinforced composite plate in thermal environment. *Journal of Sandwich Structures & Materials*, *18*(2), 151–173.
26. Mehar, K., Panda, S. K., Bui, T. Q., & Mahapatra, T. R. (2017). Nonlinear thermoelastic frequency analysis of functionally graded CNT-reinforced single/doubly curved shallow shell panels by FEM. *Journal of Thermal Stresses*, *40*(7), 899–916.
27. Lei, Z. X., Liew, K. M., & Yu, J. L. (2013). Buckling analysis of functionally graded carbon nanotube-reinforced composite plates using the element-free kp-Ritz method. *Composite Structures*, *98*, 160–168.
28. Shen, H. S. (2009). Nonlinear bending of functionally graded carbon nanotube-reinforced composite plates in thermal environments. *Composite Structures*, *91*(1), 9–19.

29. Hosseini-Hashemi, S., Taher, H. R. D., Akhavan, H., & Omidi, M. (2010). Free vibration of functionally graded rectangular plates using first-order shear deformation plate theory. *Applied Mathematical Modelling*, *34*(5), 1276–1291.
30. Cook, R. D. (2007). *Concepts and applications of finite element analysis*. Hoboken, NJ: John Wiley & Sons.
31. Zhu, P., Lei, Z. X., & Liew, K. M. (2012). Static and free vibration analyses of carbon nanotube-reinforced composite plates using finite element method with first order shear deformation plate theory. *Composite Structures*, *94*(4), 1450–1460.
32. Wang, Y. Y., Lam, K. Y., & Liu, G. R. (2001). A strip element method for the transient analysis of symmetric laminated plates. *International journal of solids and structures*, *38*(2), 241–259.

8 Micromechanics-Based Finite Element Analysis of *HAp-Ti* Biocomposite Sinusoid Structure Using Homogenization Schemes

Abhilash Karakoti and Vishesh Ranjan Kar

National Institute of Technology Jamshedpur,
Jamshedpur, India

Karunesh Kumar Shukla

National Institute of Technology Jamshedpur,
Jamshedpur, India
MN National Institute of Technology,
Allahabad, India

CONTENTS

8.1   INTRODUCTION

Biocomposites are the prominent replacements of conventional biomaterials in many biomedical applications due to their enhanced material and mechanical properties.

DOI: 10.1201/9781003158813-8

For a material to be used for medical implant, it must be biocompatible, corrosion resistant, and should have appropriate strength depending on its use [1]. Out of many nature-inspired materials, functionally graded materials (FGMs) are the advanced composite, which have smooth gradation of material phase from one surface to another [2]. Many structural issues have been raised in biomedical areas in the past, such as high aspect ratio and vulnerability to impact that makes some of the lower limbs prone to fracture. Various types of external and internal fixation devices are used to treat the fracture, such as boneplate/screw, intermedullary rod, external fixator, and screw wires. Some of the major complications of using conventional metals are corrosion, fatigue failure, implant loosening, allergy, and mismatch of stiffness between bone and prosthesis, which can be avoided by using customized FGMs [3–4]. In addition, FGM properties can vary accordingly to match the biomechanical properties of bones at different locations [5]. In recent years, functionally graded biocomposite (FGBC) materials have gained enough attention because of their excellent tailor-made properties, which can enhance the prosthesis performance [6].

In view of the above, many researchers have examined the mechanical behavior of graded composite structures used in biomedical and in other engineering applications to demonstrate their strength and capabilities [2, 6]. In this regard, static and dynamic behavior of graded composite panels have been a subject of interest among many researchers [7–11]. Tornabene et al. [12] used generalized differential quadrature to obtain the governing equations and utilized Carrera unified formulation to examine the static responses of doubly curved laminated composites. Kiani et al. [13] investigated the static and dynamic behavior of doubly curved FGM analytically using Laplace domain using first-order shear deformation theory (FSDT). Zhao et al. [14] analyzed the static and free vibration behavior of FGM cylindrical shells by employing kp-Ritz method using Sander's FSDT kinematics. Wang et al. [15] explored the thermomechanical behavior of hydroxyapatite/titanium (*HAp/Ti*) FGM dental implants experiencing occlusal force. Here, material properties were evaluated by exponential distribution with the effect of temperature variation on the implant. Viola et al. [16] emphasized static analysis of FGM conical shells using constrained third-order shear deformation theory (TSDT) using generalized differential quadrature method. Ovesy et al. [17] examined the post-buckling responses for FGM plates by semianalytical finite strip method by employing minimum potential energy principle and Newton-Raphson method. Sherafat et al. [18] examined buckling behavior of rectangular FGM plates subjected to compression and tension load using higher-order plate theory and principle of minimum total potential energy.

Mojtaba et al. [19] used FGM in dental implants to reduce the stress concentration to avoid mismatch in the stiffness of crowns and dental tissue. Cheng et al. [20] optimized the shape of the dental implants by creating a 3D finite element model. In the proposed model, volume was reduced by 17.9%, which would allow ingrowth of new bones and consequently save the material. Oshkour et al. [21] used FGM to develop a 3D model of femoral prosthesis for hip replacement where they concluded that strain energy increased and interface stresses decreased with increasing volume fraction exponent. Guild and Bonfield [22] employed finite element method (FEM) to illustrate the mechanical and failure processes in hydroxyapatite-polyethylene composites. Jalali et al. [23] employed FEM to analyze the stress distribution in the dental implants where modeling was done in commercially available modeling software SolidWorks and properties were evaluated using power-law formulation. Marcian et al. [24] employed Mechanostat strain intervals and tensile/compression yield strain to analyze the micro-strain and displacement in dental implants. Rego and Sacks [25] modeled the aortic valve with FGM using high-resolution morphological measurement. Barao et al. [26] focused on various designs used for implant retained overdentures and full arch implant prosthesis. Here, they implemented 3D-FEM to compare and analyze the stress distribution.

Based on the available and noted literature, studies of the effect of different geometrical and material parameters on the flexural responses of FGM structure for biomedical applications are very limited. Therefore, at first, an effort has been made to give the insight on the flexural responses of FGBC sinusoid structure under uniform and sinusoidal pressure. Here, effective elastic properties are evaluated using Voigt's and Mori–Tanaka micromechanical model via power-law function. The equilibrium equations are obtained using minimum total potential energy principle and 2D finite element approximations. Moreover, computational results on the deflection responses of proposed biocomposite model are executed for different sets of parametric conditions including power-law indices, aspect ratios, side-to-thickness ratios, amplitude ratios, and support conditions.

8.2   MATHEMATICAL FORMULATION

In the present work, shallow FGBC sinusoid structure in rectangular $(a \times b)$ platform is considered with uniform thickness h and amplitude h_w (see Figure 8.1). The sinusoidal form of the curved panel is introduced by employing general curvature equation [27], as

$$R_x = \frac{-a^2 \left(\dfrac{\gamma^2}{a^4} \cos\left(\dfrac{\pi x}{a} \right)^2 + 1 \right)^3}{2\gamma\pi^2 \sin\left(\dfrac{\pi x}{a} \right)} \tag{8.1}$$

where $\gamma = h_w a$ indicates the curvature ratio.

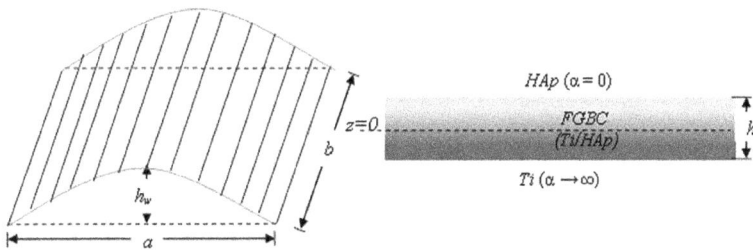

FIGURE 8.1   Description of FGBC sinusoid panel structure.

8.2.1 EFFECTIVE MATERIAL PROPERTIES

In the present analysis, hydroxyapatite and titanium are considered at the top and bottom surfaces of the structure, whereas smooth gradation is assumed in-between the top and bottom surfaces. Therefore, to evaluate the volume fractions of these biocompatible materials, power-law function is adopted as [28, 29]:

$$
\left.
\begin{aligned}
V_{HAp} &= \left(\frac{2z+h}{2h} \right)^{\alpha} \\
V_{Ti} &= 1 - \left(\frac{2z+h}{2h} \right)^{\alpha}
\end{aligned}
\right\} \quad 0 \le \alpha < \infty
\tag{8.2}
$$

where V_{HAp} and V_{Ti} are the volume fractions of hydroxyapatite and titanium, respectively. α denotes the power-law index, which describes and controls the material distribution from top to bottom surfaces of FGBC structure.

Here, Voigt's rule-of-mixture and Mori–Tanaka micromechanical material models are utilized to evaluate the in-homogeneous material properties of FGBC structure. Voigt's rule-of-mixture scheme can predict the overall material properties in average sense, whereas Mori–Tanaka scheme accounts for phase interaction effects [28].

8.2.1.1 Voigt's Rule-of-Mixture

The Voigt's micromechanical homogenization scheme is utilized to compute the average and overall material property of FGBC sinusoid structure (P_{FGBC}), as

$$
P_{FGBC} = \left(P_{HAp} - P_{Ti} \right) V_{HAp} + P_{Ti}
\tag{8.3}
$$

$$
P_{FGBC} = \left(P_{HAp} - P_{Ti} \right) \left(\frac{2z+h}{2h} \right)^{\alpha} + P_{Ti}
\tag{8.4}
$$

where P_{HAp} and P_{Ti} are the material properties of hydroxyapatite and titanium, respectively.

Here Equation (8.4) is utilized to compute the elastic properties of FGBC structure, such as Young's modulus (E_{FGBC}) and Poisson's ratio (ν_{FGBC}), as

$$
E_{FGBC} = \left(E_{HAp} - E_{Ti} \right) \left(\frac{2z+h}{2h} \right)^{\alpha} + E_{Ti}
\tag{8.4}
$$

$$
\nu_{FGBC} = \left(\nu_{HAp} - \nu_{Ti} \right) \left(\frac{2z+h}{2h} \right)^{\alpha} + \nu_{Ti}
\tag{8.5}
$$

where subscripts Ti and HAp are used to present the corresponding properties of titanium and hydroxyapatite, respectively.

8.2.1.2 Mori–Tanaka Scheme

In Mori–Tanaka scheme, bulk modulus (K_{FGBC}) and shear modulus (G_{FGBC}) for FGBC sinusoid structure are expressed as

$$\frac{K_{FGBC} - K_m}{K_c - K_m} = \frac{V_{HAp}}{1 + \left(\dfrac{\left(1 - V_{HAp}\right)\left(K_c - K_m\right)}{3K_m + 4G_m} \right)} \tag{8.6}$$

$$\frac{G_{FGBC} - G_m}{G_c - G_m} = \frac{V_{HAp}}{1 + \dfrac{\left(1 - V_{HAp}\right)\left(G_c - G_m\right)}{G_m + f_1}} \tag{8.7}$$

where $f_1 = G_m \left(\dfrac{9K_m + 8G_m}{6\left(K_m + 2G_m\right)} \right)$

The effective properties such as Young's Modulus (E_{FGBC}) and Poisson's ratio (ν_{FGBC}) can be written as

$$E_{FGBC} = \left(\frac{9K_{FGBC}G_{FGBC}}{3K_{FGBC} + G_{FGBC}} \right) \tag{8.8}$$

$$\nu_{FGBC} = \left(\frac{3K_{FGBC} - 2G_{FGBC}}{2\left(3K_{FGBC} + G_{FGBC}\right)} \right) \tag{8.9}$$

Figure 8.2 illustrates the Young's moduli of (HAp/Ti) FGBC structure using Voigt's rule-of-mixture and Mori–Tanaka schemes for various power-law indices ($\alpha = 0.2, 0.5, 1, 2, 5, 10, \infty$) along the thickness coordinate (z/h).

8.2.2 KINEMATIC FIELD

The present FGBC sinusoid panel structure utilizes TSDT mid-plane kinematics with 9 degrees-of-freedom [30–31]. Here the displacement field vector $\{\Phi\} = \lfloor u \ v \ w \rfloor^T$ is expressed using mid-plane displacement (u_0, v_0, w_0), rotation (u_1, v_1), and higher-order (u_2, v_2, u_3, v_3) terms as

$$\left. \begin{aligned} u\left(x,y,z\right) &= u_0\left(x,y\right) + zu_1\left(x,y\right) + z^2 u_2\left(x,y\right) + z^3 u_3\left(x,y\right) \\ v\left(x,y,z\right) &= v_0\left(x,y\right) + zv_1\left(x,y\right) + z^2 v_2\left(x,y\right) + z^3 v_3\left(x,y\right) \\ w\left(x,y,z\right) &= w_0\left(x,y\right) \end{aligned} \right\} \left(-\frac{h}{2} \le z \le \frac{h}{2} \right) \tag{8.10}$$

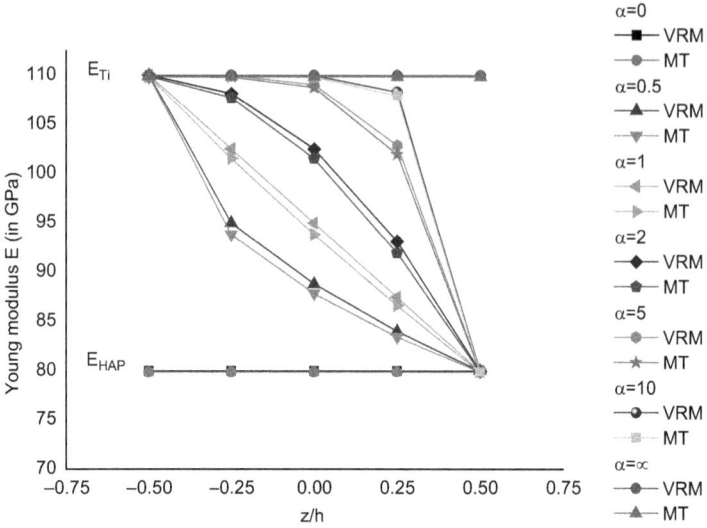

FIGURE 8.2 Variation of Young's modulus along with *HAp* volume fraction (V_{HAp}) and thickness coordinate (z/h) for different values of power-law indices.

The strain tensor $\{\bar{\varepsilon}\} = \left\lfloor \bar{\varepsilon}_1 \ \bar{\varepsilon}_2 \ \bar{\varepsilon}_4 \ \bar{\varepsilon}_5 \ \bar{\varepsilon}_6{}^T \right\rfloor$ of the FGBC sinusoid panel structure is expressed in terms of mid-plane strain terms as

$$
\begin{Bmatrix} \bar{\varepsilon}_1 \\ \bar{\varepsilon}_2 \\ \bar{\varepsilon}_4 \\ \bar{\varepsilon}_5 \\ \bar{\varepsilon}_6 \end{Bmatrix} = \begin{Bmatrix} \varepsilon_1 + zk_1 + z^2 l_1 + z^3 m_1 \\ \varepsilon_2 + zk_2 + z^2 l_2 + z^3 m_2 \\ \dfrac{1}{2}\left(\varepsilon_4 + zk_4 + z^2 l_4 + z^3 m_4 \right) \\ \dfrac{1}{2}\left(\varepsilon_5 + zk_5 + z^2 l_5 + z^3 m_5 \right) \\ \dfrac{1}{2}\left(\varepsilon_6 + zk_6 + z^2 l_6 + z^3 m_6 \right) \end{Bmatrix} \tag{8.11}
$$

where

$\varepsilon_1 = u_{0,x},\ k_1 = u_{1,x},\ l_1 = u_{2,x},\ m_1 = u_{3,x},$

$\varepsilon_2 = v_{0,y},\ k_2 = v_{1,y},\ l_2 = v_{2,y},\ m_2 = v_{3,y},$

$\varepsilon_4 = w_{0,y} + v_1,\ k_4 = 2v_2,\ l_4 = 3v_3,$

$\varepsilon_5 = w_{0,x} + u_1,\ k_5 = 2u_2 - \dfrac{u_1}{R_x},\ l_5 = 3u_3 - \dfrac{u_2}{R_x},\ m_5 = -\dfrac{u_3}{R_x},$

$\varepsilon_6 = u_{0,y} + v_{0,x},\ k_6 = u_{1,y} + v_{1,x}, l_6 = u_{2,y} + v_{2,x},\ m_6 = u_{3,y} + v_{3,x}.$

Equation (8.7) can also be expressed as

$$
\{\bar{\varepsilon}\} = [Z]\{\varepsilon\} \tag{8.12}
$$

where $\{\varepsilon\}$ is the mid-plane strain vector and $[Z]$ is the thickness-coordinate matrix.

8.2.3 CONSTITUTIVE AND ENERGY EQUATIONS

The stress tensor of isotropic and heterogeneous FGBC structure is expressed as

$$
\begin{Bmatrix} \sigma_1 \\ \sigma_2 \\ \sigma_4 \\ \sigma_5 \\ \sigma_6 \end{Bmatrix} = \begin{bmatrix} Q_{11} & Q_{12} & 0 & 0 & 0 \\ Q_{21} & Q_{22} & 0 & 0 & 0 \\ 0 & 0 & Q_{44} & 0 & 0 \\ 0 & 0 & 0 & Q_{55} & 0 \\ 0 & 0 & 0 & 0 & Q_{66} \end{bmatrix} \begin{Bmatrix} \bar{\varepsilon}_1 \\ \bar{\varepsilon}_2 \\ \bar{\varepsilon}_4 \\ \bar{\varepsilon}_5 \\ \bar{\varepsilon}_6 \end{Bmatrix} \tag{8.13}
$$

where

$$
Q_{11} = Q_{22} = \frac{E_{FGBC}}{\left(1 - \upsilon_{FGBC}^2\right)}, Q_{12} = Q_{21} = \frac{\upsilon_{FGBC} E_{FGBC}}{\left(1 - \upsilon_{FGBC}^2\right)}, Q_{44} = Q_{55} = Q_{66} = \frac{E_{FGBC}}{2\left(1 + \upsilon_{FGBC}\right)}.
$$

Equation (8.8) can also be rewritten as

$$
\{\sigma\} = \left[Q_{FGBC}\right]\{\varepsilon\} \tag{8.14}
$$

where $[Q_{FGBC}]$ is the elastic constant matrix for a typical FGBC material.

The strain energy of the FGBC sinusoid panel structure is expressed as

$$
S_\varepsilon = \frac{1}{2} \int_v \left(\{\bar{\varepsilon}\}^T \{\sigma\}\right) dV \tag{8.15}
$$

$$
S_\varepsilon = \frac{1}{2} \int_A \{\varepsilon\}^T \left[D_{FGBC}\right]\{\varepsilon\} dA \tag{8.16}
$$

where

$$
\left[D_{FGBC}\right] = \int_{-h/2}^{h/2} \left[Z\right]^T \left[Q_{FGBC}\right]\left[Z\right] dz.
$$

In this study, FGBC structure is subjected to uniform pressure $q = q_0$ and sinusoidal pressure $q = q_0 \sin\left(\frac{\pi x}{a}\right) \sin\left(\frac{\pi y}{b}\right)$, and the work done can be evaluated as

$$
W = \int_A \{\Phi\}^T \{q\} dA \tag{8.17}
$$

where $\{\Phi\}$ is the displacement field vector.

Further, the energy equations of FGBC sinusoid panel structure are solved using the principle of minimum total potential energy as

$$\delta \Pi = 0 \tag{8.18}$$

where Π is the total potential energy and δ is the variational symbol.

8.2.4 FINITE ELEMENT APPROXIMATIONS

The present model of FGBC sinusoid structure is solved numerically using 2D finite element approximation via nine-node quadrilateral Lagrangian element with 9 degrees-of-freedom per node. For any element, the mid-plane displacement $\{\Phi\}^e$ is represented in nodal form via approximation functions $N(\xi,\eta)$ [32] as

$$\{\Phi\}^e = \sum_{i=1}^{NNE} [N_i]\{\Phi_i\} \tag{8.19}$$

where $\{\Phi_i\}$ denotes the nodal displacements at i^{th} node.

For any element, mid-plane strain vector is expressed as

$$\{\varepsilon\}^e = [\partial]\{\Phi\}^e \tag{8.20}$$

where $[\partial]$ is the differential operator.

The strain energy in elemental form is expressed by utilizing Equation (8.16) as

$$S_\varepsilon^e = \frac{1}{2}\{\Phi\}^{eT}[k]^e\{\Phi\}^e \tag{8.21}$$

where $[k]^e = \int_{-1}^{1}\int_{-1}^{1}[B]^T[D_{FGBC}][B]|J|d\xi d\eta$ is the elemental stiffens matrix. Here $[B] = [\partial]\sum_{i=1}^{NNE}[N_i]$ is the strain-displacement matrix.

Similarly, the external work done for any element is obtained by imposing Equation (8.19) into Equation (8.17), as

$$W^e = \{\Phi\}^{eT}\{f\}^e \tag{8.22}$$

where $\{f\}^e = \int_{-1}^{1}\int_{-1}^{1}[N]^T\{q\}|J|d\xi d\eta$ is the elemental force vector.

The elemental equations of FGBC sinusoid panel structure are obtained by utilizing the principle of minimum total potential energy (Equation 8.18) and presented as

$$[k]^e\{\Phi\}^e = \{f\}^e \tag{8.23}$$

8.3 NUMERICAL RESULTS AND DISCUSSION

The bending responses of the FGBC sinusoid panels are analyzed using a customized computer algorithm, based on the 2D-FEM in conjunction with higher-order kinematics. The present FGBC sinusoid structure is constituted of *Ti* and *HAp* as ceramic materials, respectively, and their properties are mentioned in Table 8.1. First, the present model is tested for different mesh densities to demonstrate the convergence rate, followed by the validation test. Later, a variety of parametric examples are computed and illustrated to highlight the impact of different parameters on the deflection responses of FGBC sinusoid panel structures. For the computational purpose, the following edge constraints are employed throughout in the analysis, if not stated otherwise:

a. Fully simply supported (*SSSS*)

$$v_o = w_o = v_1 = v_2 = v_3 = 0; u_o \neq u_1 \neq u_2 \neq u_3 \neq 0 \text{ at } x = 0, a$$

$$u_o = w_o = u_1 = u_2 = u_3 = 0; v_o \neq v_1 \neq v_2 \neq v_3 \neq 0 \text{ at } y = 0, b$$

b. Fully clamped (*CCCC*)

$$u_o = v_o = w_o = u_1 = u_2 = u_3 = v_1 = v_2 = v_3 = 0 \text{ at } x = 0, a \text{ and } y = 0, b$$

c. Simply and clamped (*SCSC*)

$$v_o = w_o = v_1 = v_2 = v_3 = 0; u_o \neq u_1 \neq u_2 \neq u_3 \neq 0 \text{ at } x = 0, a$$

$$u_o = v_o = w_o = u_1 = u_2 = u_3 = v_1 = v_2 = v_3 = 0 \text{ at } y = 0, b$$

8.3.1 MESH REFINEMENT AND VERIFICATION STUDY

The mesh refinement of FGBC model is accompanied to achieve appropriate mesh density. Here deflection responses of FGBC sinusoid panel structures ($\lambda = 50$, $\chi = 100$, $\phi = 1$, *SSSS*, $\alpha = 0.5$) are computed at different mesh densities (2×2 to 7×7)

TABLE 8.1
Properties of FGBC Constituents [5, 22]

	Properties	
Materials	Young's Modulus *E* (GPa)	Poisson's Ratio (ν)
Hydroxyapatite (*HAp*)	80	0.3
Titanium (*Ti*)	110	0.3

TABLE 8.2

Mesh Refinement of FGBC Sinusoid Panel Structure for Deflection Responses Under Uniform and Sinusoidal Loads

Mesh Size	Homogenization Scheme	Uniform Load	Sinusoidal Load
2 × 2	VRM	0.8403	0.4202
	MT	0.8479	0.4239
3 × 3	VRM	0.6596	0.2995
	MT	0.6656	0.3022
4 × 4	VRM	0.5402	0.2631
	MT	0.5450	0.2654
5 × 5	VRM	0.4853	0.2582
	MT	0.4897	0.2605
6 × 6	VRM	0.4851	0.2576
	MT	0.4895	0.2600
7 × 7	VRM	0.4796	0.2540
	MT	0.4839	0.2563

and presented in Table 8.2. The presented deflection responses using Voigt and Mori–Tanaka schemes are converging at (6 × 6) mesh, and therefore this mesh size is utilized in the forthcoming computations, if not stated otherwise.

For the verification purpose, a cylindrical FGM (Al/ZrO_2) panel with geometric parameters: $h = 0.01$ m, $R = 1$ m, $a = 0.2$ m; and material properties: $E_{Al} = 70$ GPa, $E_{ZrO2} = 151$ GPa, $\nu_{Al} = \nu_{ZrO2} = 0.3$, is considered and compared with the results of Zhao et al. [14]. Here deflection responses of FGM panel are computed under uniform pressure ($q_o = -10^6$ Pa) at various power-law indices and presented in Table 8.3. The computed results of the developed model via TSDT mid-plane kinematics with Voigt and Mori–Tanaka schemes are well aligned with the published results of Zhao et al. [14], which adopted the element free kp-Ritz model with FSDT kinematics.

TABLE 8.3

Central Deflection ($w = w_o/h$) of Clamped Al/ZrO_2 Shells under Uniform Load

α	Present (VRM[1])	Present (MT[2])	Zhao 2009 [14]
0	0.0132	0.0132	0.01347
0.2	0.0148	0.0155	0.01516
0.5	0.0167	0.0176	0.01711
1	0.0187	0.0195	0.01915
2	0.0206	0.0212	0.02102

[1] Voigt's rule-of-mixture.
[2] Mori–Tanaka scheme

8.3.2 Numerical Experimentations

In this section, dimensionless centerline deflections $\left(\bar{w} = w_0/h\right)$ of FGBC sinusoid panel structures are illustrated for different parameters, such as aspect ratio (ϕ), side-to-thickness ratio (χ), power-law index (α), amplitude ratio (λ), and support and loading conditions. For computation purpose, following parametric conditions are utilized, if not stated, otherwise: $\alpha = 0.5$, $\phi = 1$, $\lambda = 50$, $\chi = 100$, $Q = 10^5$ and SSSS.

Tables 8.4 and 8.5 exhibit the dimensionless deflection parameters of simply supported FGBC sinusoid panel at different power-law indices ($\alpha = 0, 0.5, 1, 2, 5, 10, \infty$) under uniform and sinusoidal loads, respectively. In both cases, deflection values are falling with the increment in power-law indices because Ti-rich ($\alpha \to \infty$) FGBC structure is stiffer than the HAp-rich ($\alpha \to 0$) FGBC structure. In case of FGBC structure subjected to uniform loading, deflection responses are found maximum at $(a/3) \leq x \leq (2a/3)$ as shown in Table 8.4, whereas in case of sinusoidal loading,

---

TABLE 8.4
Dimensionless Centerline Deflection of FGBC Sinusoid Panel Structure at Different Power-Law Indices Subjected to Uniform Load for Voigt's and Mori–Tanaka Schemes

$X = x/a$	Micromechanical Scheme	Power-law Indices, α						
		0	0.5	1	2	5	10	∞
X_1	VRM, MT	0	0	0	0	0	0	0
X_2	VRM	0.2332	0.2067	0.197	0.1886	0.1804	0.1761	0.1696
	MT	0.2336	0.2085	0.1986	0.1899	0.1813	0.1768	0.1699
X_3	VRM	0.4017	0.3563	0.3392	0.3242	0.3098	0.3026	0.2922
	MT	0.4023	0.3594	0.3419	0.3564	0.3113	0.3037	0.2926
X_4	VRM	0.5038	0.4472	0.4251	0.4056	0.3872	0.3784	0.3664
	MT	0.5045	0.451	0.4285	0.4083	0.3890	0.3797	0.3669
X_5	VRM	0.5401	0.4798	0.4553	0.4335	0.4133	0.4042	0.3928
	MT	0.5408	0.4839	0.4591	0.4364	0.4151	0.4055	0.3933
X_6	VRM	0.5433	0.4829	0.4576	0.4348	0.4141	0.4053	0.3951
	MT	0.5439	0.4871	0.4614	0.4377	0.4159	0.4065	0.3956
X_7	VRM	0.5457	0.4851	0.4595	0.4364	0.4155	0.4067	0.3969
	MT	0.5463	0.4894	0.4634	0.4393	0.4172	0.4078	0.3973
X_8	VRM	0.5476	0.4867	0.4612	0.4383	0.4175	0.4086	0.3982
	MT	0.5482	0.4909	0.4651	0.4412	0.4192	0.4098	0.3987
X_9	VRM	0.5474	0.4862	0.4614	0.4394	0.4190	0.4098	0.3981
	MT	0.5481	0.4904	0.4653	0.4424	0.4209	0.4111	0.3986
X_{10}	VRM	0.5123	0.4546	0.4322	0.4125	0.3939	0.3849	0.3726
	MT	0.513	0.4585	0.4357	0.4153	0.3957	0.3862	0.3731
X_{11}	VRM	0.4089	0.3627	0.3453	0.3301	0.3155	0.3081	0.2974
	MT	0.4095	0.3658	0.3481	0.3323	0.317	0.3092	0.2979
X_{12}	VRM	0.2374	0.2104	0.2006	0.192	0.1837	0.1793	0.1726
	MT	0.2378	0.2122	0.2021	0.1933	0.1846	0.18	0.1729
X_{13}	VRM, MT	0	0	0	0	0	0	0

TABLE 8.5

Dimensionless Centerline Deflection of FGBC Sinusoid Panel Structure at Different Power-Law Indices Subjected to Sinusoidal Load for Voigt's and Mori–Tanaka Schemes

x/a	Micromechanical Scheme	Power-Law Indices, α						
		0	0.5	1	2	5	10	∞
0	VRM, MT	0	0	0	0	0	0	0
1/12	VRM	0.1008	0.0894	0.0851	0.0814	0.0777	0.0759	0.0733
	MT	0.1010	0.0902	0.0858	0.0819	0.0781	0.0762	0.0734
2/12	VRM	0.1839	0.1632	0.1552	0.1482	0.1416	0.1383	0.1338
	MT	0.1842	0.1646	0.1565	0.1492	0.1422	0.1388	0.1340
3/12	VRM	0.2442	0.2168	0.2060	0.1966	0.1876	0.1834	0.1776
	MT	0.2445	0.2186	0.2077	0.1979	0.1885	0.1840	0.1779
4/12	VRM	0.2773	0.2462	0.2338	0.2228	0.2125	0.2078	0.2017
	MT	0.2776	0.2483	0.2357	0.2243	0.2134	0.2084	0.2019
5/12	VRM	0.2902	0.2578	0.2445	0.2327	0.2219	0.2170	0.2110
	MT	0.2905	0.2600	0.2466	0.2343	0.2228	0.2177	0.2113
6/12	VRM	0.2900	0.2576	0.2443	0.2324	0.2214	0.2167	0.2109
	MT	0.2903	0.2599	0.2464	0.2339	0.2224	0.2173	0.2111
7/12	VRM	0.2783	0.2473	0.2345	0.2230	0.2125	0.2079	0.2024
	MT	0.2786	0.2494	0.2365	0.2245	0.2134	0.2086	0.2026
8/12	VRM	0.2565	0.2278	0.2161	0.2057	0.1961	0.1918	0.1865
	MT	0.2568	0.2298	0.2179	0.2071	0.1969	0.1924	0.1867
9/12	VRM	0.2195	0.1949	0.185	0.1763	0.1681	0.1644	0.1596
	MT	0.2197	0.1966	0.1865	0.1774	0.1689	0.1649	0.1598
10/12	VRM	0.1622	0.144	0.1368	0.1305	0.1245	0.1217	0.1180
	MT	0.1624	0.1453	0.1379	0.1313	0.125	0.1221	0.1181
11/12	VRM	0.0879	0.078	0.0742	0.0708	0.0676	0.0660	0.0639
	MT	0.0880	0.0787	0.0748	0.0712	0.0679	0.0662	0.0640
1	VRM, MT	0	0	0	0	0	0	0

maximum deflections are observed at $x = a/2$ (see Table 8.5). However, the difference between the results based on Voigt's rule-of-mixture and Mori–Tanaka micromechanical schemes are very nominal under both the loading conditions. Therefore, the forthcoming numerical examinations are performed using Voigt model, which is simple, and consequently reduces computational cost.

Figures 8.3 and 8.4 illustrate the effect of aspect ratios (ϕ = 1, 1.5, 2, 2.5) on the dimensionless centerline deflections of FGBC sinusoid panel structures subjected to uniform and sinusoidal load, respectively. Under both the loading conditions, deflection responses are diminishing as the aspect ratio increases. This indicates that FGBC sinusoid panel structure with larger aspect ratios (i.e., ϕ = 2.5) is stiffer than the square panel. In addition, deflection behavior is almost symmetrical about $x = a/2$. It is also observed that deflection responses of FGBC sinusoid panel structures under

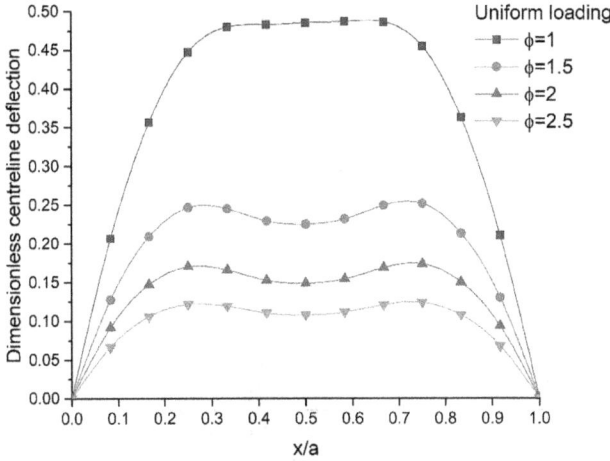

FIGURE 8.3 Dimensionless centerline deflection of FGBC sinusoid panel structure at different aspect ratios subjected to uniform load.

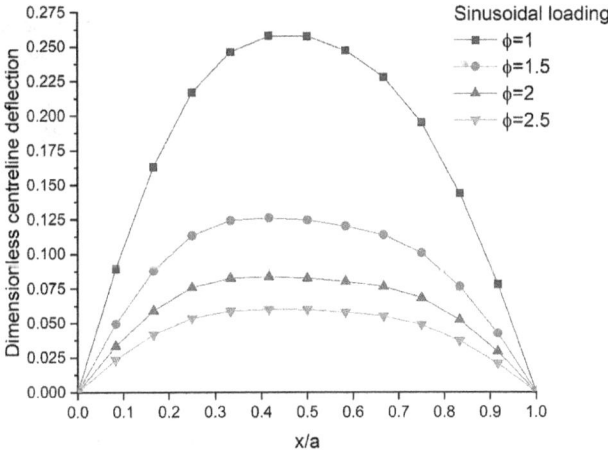

FIGURE 8.4 Dimensionless centerline deflection of FGBC sinusoid panel structure at different aspect ratios subjected to sinusoidal load.

uniform loading (see Figure 8.3) are almost twice of the structure subjected to sinusoidal load (see Figure 8.4).

The effect of side-to-thickness ratios ($\chi = 40, 60, 80, 100$) on the dimensionless deflection parameters of FGBC sinusoid panel structure is demonstrated under uniform and sinusoidal loading conditions and presented in Figures 8.5 and 8.6, respectively. Here, the deflection parameters are increasing with the increase in side-to-thickness ratios. It is because with the increase in side-to-thickness ratios the structure becomes thinner, which is less stiff than the preceding structure. In addition,

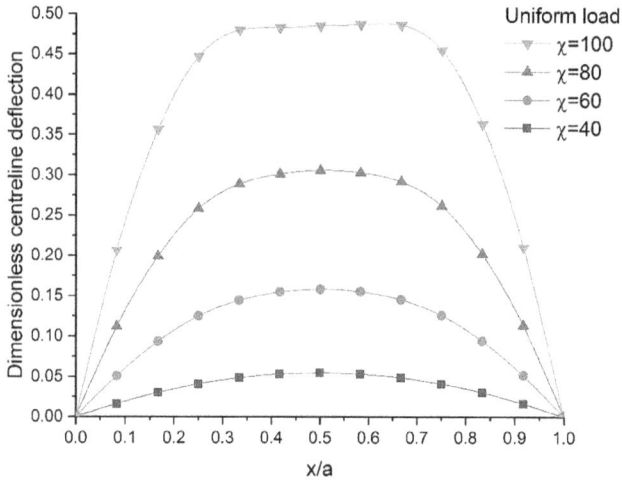

FIGURE 8.5 Dimensionless centerline deflection of FGBC sinusoid panel structure at different side-to-thickness ratio subjected to uniform load.

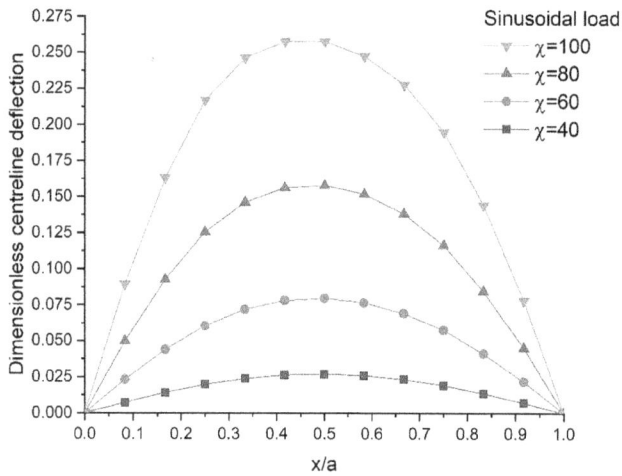

FIGURE 8.6 Dimensionless centerline deflection of FGBC sinusoid panel structure at different side-to-thickness ratio subjected to sinusoidal load.

deflection behavior of FGBC sinusoid panel structures is almost symmetrical about $x = a/2$ under both the loading conditions.

Figures 8.7 and 8.8 demonstrate the significance of amplitude ratios ($\lambda = 40$, 60, 80, 100) on the dimensionless deflection parameters of FGBC sinusoid panel structure under uniform and sinusoidal loads, respectively. Under both loading conditions, deflection behavior of FGBC sinusoid panel structures is almost symmetrical about

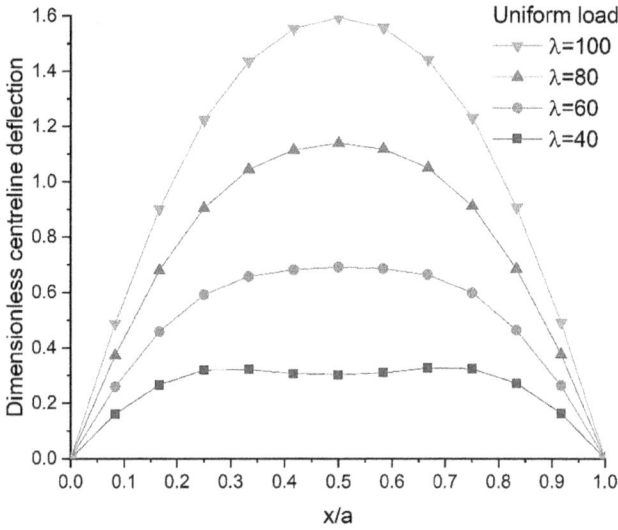

FIGURE 8.7 Dimensionless centerline deflection of FGBC sinusoid panel structure at different amplitude ratio height subjected to uniform load.

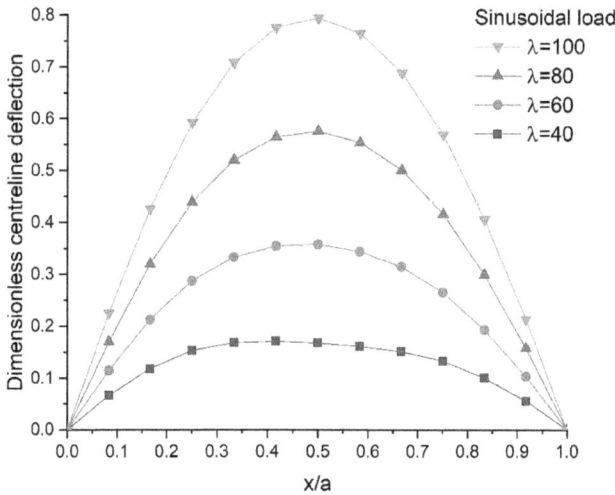

FIGURE 8.8 Dimensionless centerline deflection of FGBC sinusoid panel structure at different amplitude ratio subjected to sinusoidal load.

$x = a/2$. Here, the deflections are increasing along with the amplitude ratios, which indicates that the structural stiffness of sinusoidally curved structure with larger rise is more than the structure with smaller rise.

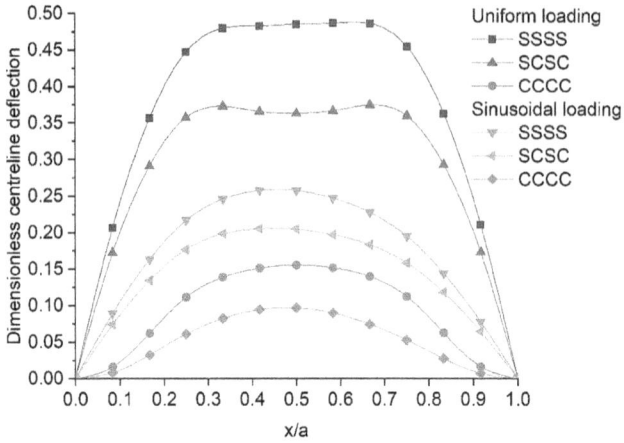

FIGURE 8.9   Dimensionless centerline deflection of FGBC sinusoid panel structure at different support conditions.

Figure 8.9 illustrates the deflection responses of FGBC sinusoid panel structure under different support conditions (SSSS, CCCC, and SCSC). The deflection responses are found maximum in fully simply supported (SSSS) conditions, whereas minimum in case of fully clamped (CCCC) conditions. Figures 8.10 and 8.11 demonstrate the deformation behavior of FGBC sinusoid structures using Voigt's rule-of-mixture and Mori–Tanaka schemes, respectively, under different support (SSSS and CCCC) and loading (uniform and sinusoidal) cases.

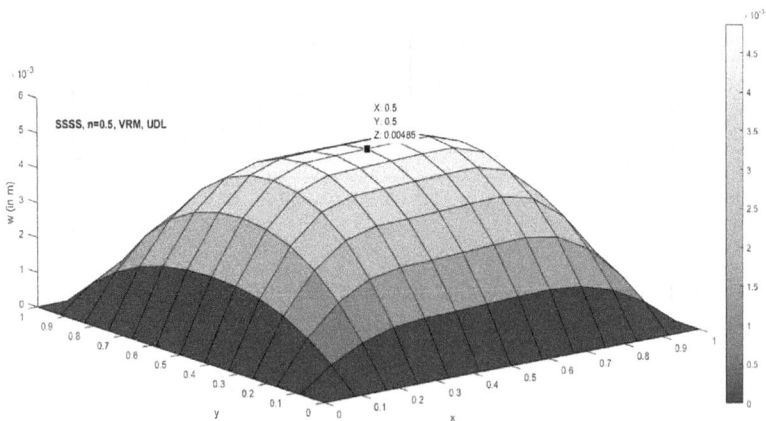

FIGURE 8.10   Deformation behavior of VRM-FGBC sinusoid structures under uniform and sinusoidal pressure.

(Continued)

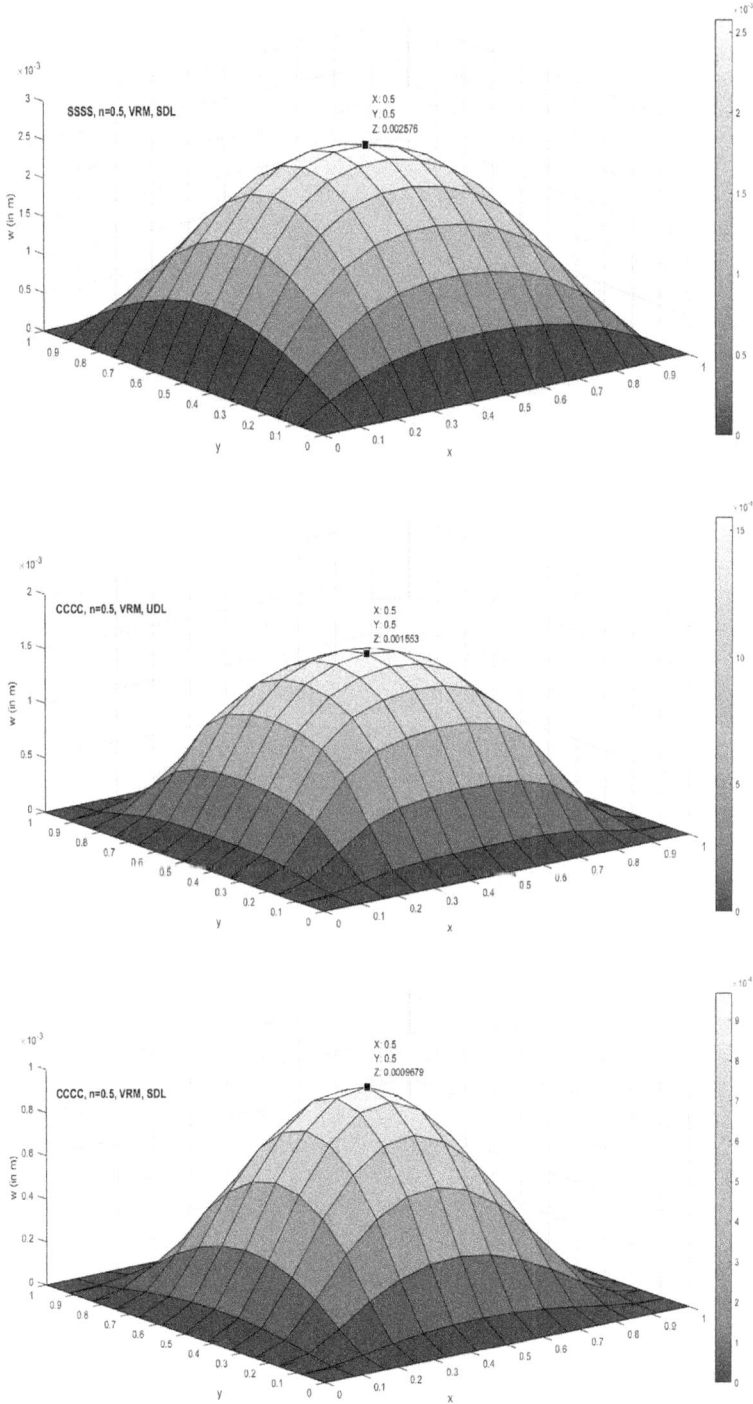

FIGURE 8.10 (Continued)   Deformation behavior of VRM-FGBC sinusoid structures under uniform and sinusoidal pressure.

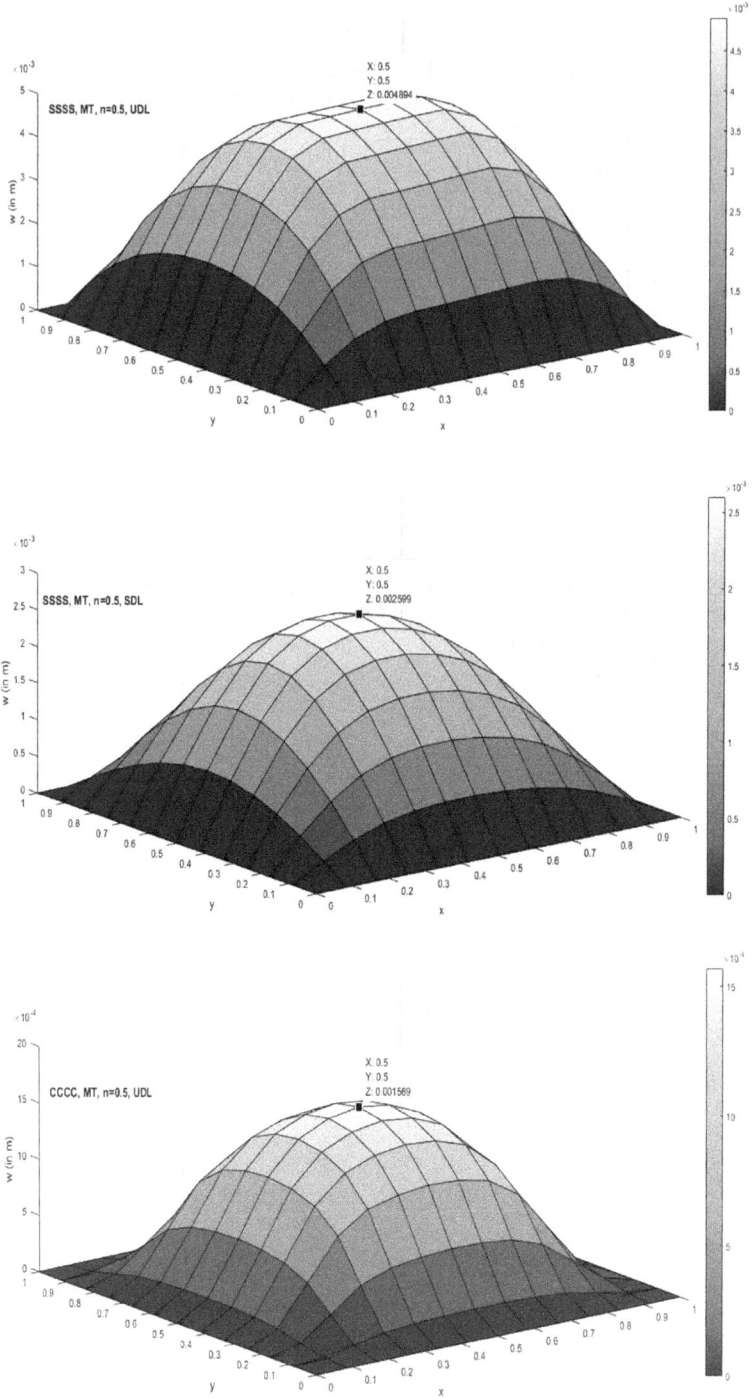

FIGURE 8.11 Deformation behavior of MT-FGBC sinusoid structures under uniform and sinusoidal pressure. *(Continued)*

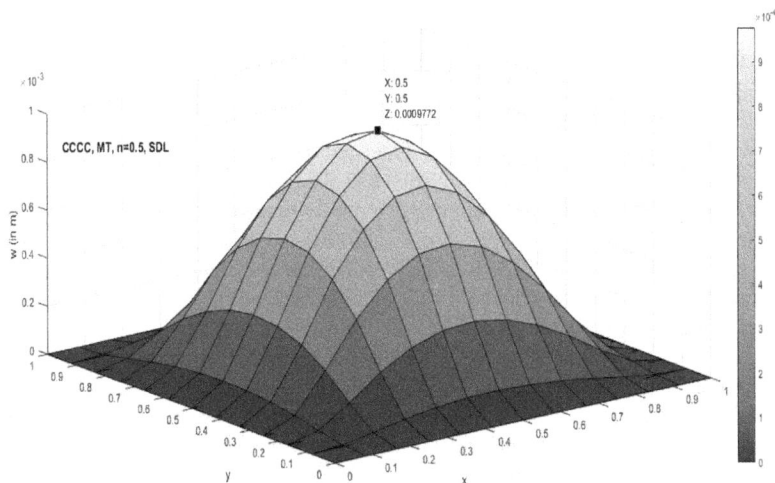

FIGURE 8.11 (Continued) Deformation behavior of MT-FGBC sinusoid structures under uniform and sinusoidal pressure.

8.4 CONCLUSIONS

This is the first time the deformation behaviour of FGBC sinusoid structure is examined under different parametric conditions. Throughout the analysis, *Ti* and *HAp* are utilized as FGBC ingredients, but the heterogeneous material properties are evaluated through Voigt's rule-of-mixture and Mori–Tanaka micromechanical schemes via power-law function. Here the kinematic field of FGBC sinusoid panel structure is developed using TSDT with 9 degrees-of-freedom and general differential curvature equation. The minimum total potential energy principle is adopted to solve the energy equations, whereas the computational results are obtained using 2D isoparametric finite element approximations. The stability and correctness of the discretised FGBC model are confirmed through appropriate mesh refinement and verification tests, respectively. The difference between the deflection responses of FGBC sinusoid panel using Voigt's rule-of-mixture and Mori–Tanaka homogenisation schemes is nominal, thus Voigt scheme can be easily adopted to evaluate the overall material properties. In continuation, a verity of illustrated examples revealed the significance of various parameters, such as aspect ratios, side-to-thickness ratios, amplitude ratios, power-law indices, and the edge supports, on the FGBC sinusoid panel structure under uniform and sinusoidal loads. Moreover, the design implementation of this work may be explored further to various human prosthesis and implants.

ACKNOWLEDGMENT

The authors would like to thank Science and Engineering Research Board, Department of Science and Technology, Government of India (File No. ECR/2016/001829), for the financial support.

REFERENCES

1. F. Watari, A. Yokoyama, M. Omori, T. Hirai, H. Kondo, Biocompatibility of materials and development to functionally graded implant for bio-medical application, *Compos. Sci. Technol.* 64 (2004) 893–908.
2. D.K. Jha, T. Kant, R.K. Singh, A critical review of recent research on functionally graded plates, *Compos. Struct.* 96 (2013) 833–849.
3. H. Mehboob, S. Chang, Application of composites to orthopedic prostheses for effective bone healing: A review, *Compos. Struct.* 118 (2014) 328–341.
4. J. Parthasarathy, B. Starly, S. Raman, A design for the additive manufacture of functionally graded porous structures with tailored mechanical properties for biomedical applications, *J. Manuf. Process.* 13 (2011) 160–170.
5. D. Lin, Q. Li, W. Li, S. Zhou, M.V. Swain, Design optimization of functionally graded dental implant for bone remodeling, *Compos. Part B.* 40 (2009) 668–675.
6. M.A.F. Afzal, P. Kesarwani, K.M. Reddy, S. Kalmodia, B. Basu, K. Balania, Functionally graded hydroxyapatite-alumina-zirconia biocomposite: Synergy of toughness and bio-compatibility, *Mater. Sci. Eng. C* 32(5) (2012) 1164–1173.
7. V.R. Kar, S.K. Panda, Thermoelastic analysis of functionally graded doubly curved shell panels using nonlinear finite element method, *Compos. Struct.* 129 (2015) 202–212.
8. J. Yang, J. Xiong, L. Ma, G. Zhang, X. Wang, L. Wu, Study on vibration damping of composite sandwich cylindrical shell with pyramidal truss-like cores, *Compos. Struct.* 117 (2014) 362–372.
9. S. Xiang, Y.-G. Wang, G.W. Kang, Static analysis of laminated composite plates using *n*-order shear deformation theory and a meshless global collocation method, *Mech. Adv. Mater. Struct.* 22 (2015) 470–478.
10. M. Nejati, R. Dimitri, F. Tornabene, M. Hossein Yas, Thermal buckling of nanocomposite stiffened cylindrical shells reinforced by functionally graded wavy carbon nanotubes with temperature-dependent properties, *Appl. Sci.* 7 (2017) 1223.
11. C. Thurnherr, Y. Mirabito, G. Kress, P. Ermanni, Highly anisotropic corrugated laminates deflection under uniform pressure, *Compos. Struct.* 154 (2016) 31–38.
12. F. Tornabene, N. Fantuzzi, E. Viola, E. Carrera, Static analysis of doubly-curved anisotropic shells and panels using CUF approach, differential geometry and differential quadrature method, *Compos. Struct.* 107 (2014) 675–697.
13. Y. Kiani, A.H. Akbarzadeh, Z.T. Chen, M.R. Eslami, Static and dynamic analysis of an FGM doubly curved panel resting on the Pasternak-type elastic foundation, *Compos. Struct.* 94 (2012) 2474–2484.
14. X. Zhao, Y.Y. Lee, K.M. Liew, Thermoelastic and vibration analysis of functionally graded cylindrical shells, *Int. J. Mech. Sci.* 51 (2009) 694–707.
15. C.L. F. Wang, H.P. Lee, Thermal–mechanical study of functionally graded dental implants with the finite element method, *J. Pak. Med. Assoc.* 67 (2017) 1180–1185.
16. E. Viola, L. Rossetti, N. Fantuzzi, F. Tornabene, Static analysis of functionally graded conical shells and panels using the generalized unconstrained third order theory coupled with the stress recovery, *Compos. Struct.* 112 (2014) 44–65.
17. H.R. Ovesy, S.A.M. Ghannadpour, M. Nassirnia, Post-buckling analysis of rectangular plates comprising Functionally Graded Strips in thermal environments, *Comput. Struct.* 147 (2015) 209–215.
18. M.H. Sherafat, H.R. Ovesy, S.A.M. Ghannadpour, Buckling analysis of functionally graded plates under mechanical loading using higher order functionally graded strip, *Int. J. Struct. Stab. Dyn.* 13 (2013) 1350033.

19. M. Mahmoudi, A.R. Saidi, M.A. Hashemipour, The use of functionally graded dental crowns to improve biocompatibility: A finite element analysis, *Comput. Methods Biomech. Biomed. Engin.* 5842 (2018) 1–8.
20. C. Chang, C. Chen, C. Huang, M. Hsu, Finite element analysis of the dental implant using a topology optimization method, *Med. Eng. Phys.* 34 (2012) 999–1008.
21. A.A. Oshkour, N.A.A. Osman, M.M. Davoodi, Y.H. Yau, F. Tarlochan, W.A.B.W. Abas, M. Bayat, Finite element analysis on longitudinal and radial functionally graded femoral prosthesis, *Int. J. Numer. Meth. Biomed. Engng.* 29 (2013) 1412–1427.
22. F.J. Guild, W. Bonfield, Predictive modelling of the mechanical properties and failure processes in hydroxyapatite- polyethylene (Hapex TM) composite, *J. Mater. Sci. Mater. Med.* 9 (1998) 497–502.
23. S.K. Jalali, R. Yarmohammadi, F. Maghsoudi, Finite element stress analysis of functionally graded dental implant of a premolar tooth, *J Mech Sci Technol.* 30 (2016) 4919–4923.
24. P. Marci, J. Kaiser, L. Horackova, J. Kaiser, T. Zikmund, L. Borak, Micro finite element analysis of dental implants under different loading conditions, *Comput Biol Med.* 96 (2018) 157–165.
25. B.V. Rego, M.S. Sacks, A functionally graded material model for the transmural stress distribution of the aortic valve leaflet, *J. Biomech.* 54 (2017) 88–95.
26. V.A.R. Barão, J.A. Delben, J. Lima, T. Cabral, W.G. Assunção, Comparison of different designs of implant-retained overdentures and fixed full-arch implant-supported prosthesis on stress distribution in edentulous mandible - A computed tomography-based three-dimensional finite element analysis, *J. Biomech.* 46 (2013) 1312–1320.
27. A. Karakoti, V.R. Kar, Deformation characteristics of sinusoidally-corrugated laminated composite panel—A higher-order finite element approach, *Compos. Struct.* 216 (2019) 151–158.
28. H.S. Shen, *Functionally Graded Materials: Nonlinear Analysis of Plates and Shells*, Boca Raton, FL: CRC Press, Taylor & Francis Group (2009).
29. H.S. Shen, Nonlinear bending of functionally graded carbon nanotube-reinforced composite plates in thermal environments, *Compos. Struct.* 91 (2009) 9–19.
30. V.R. Kar, S.K. Panda, Nonlinear free vibration of functionally graded doubly curved shear deformable panels using finite element method, *J. Vib. Control.* 22 (2016) 1935–1949.
31. J.N. Reddy, A general non-linear third-order theory of plates with moderate thickness, *Int. J. Non-Linear Mechanics.* 25 (1990) 677–686.
32. R.D. Cook, D.S. Malkus, M. Plesha, *Concepts and Applications of Finite Element Analysis*, Hoboken, NJ: John Wiley Sons (1989).

9 Stability Behavior of Biocomposite Structures Using 2D-Finite Element Approximation

Kamal Kishore Joshi

National Institute of Technology Jamshedpur,
Jamshedpur, India

Kalinga Institute of Industrial Technology Bhubaneswar,
Bhubaneswar, India

Vishesh Ranjan Kar

NIT Jamshedpur,
Jamshedpur, India

K. Jayakrishna

VIT Vellore,
Vellore, Tamil Nadu, India

Mohamed Thariq Hameed Sultan

University Putra Malaysia,
Serdang, Malaysia

CONTENTS

DOI: 10.1201/9781003158813-9

161

9.1   INTRODUCTION

The last few decades have seen a drastic change in the treatment methodologies and application of materials in the biomedical field. Current research is focused on the development of new biomedical materials to improve their mechanical and physical properties so that they can be specifically used in the area of dental implant and bone replacement. A lot of attention is given to the fabrication and application of advance composite biomaterials such as alloys of titanium, hydroxyapatite (HAP), and partially stabilized zirconia (PSZ).

A number of researchers carried out studies and experiments to design, fabricate, and optimize dental implant for bone remodeling. Lin et al. [1] adopted response surface methodology to optimize functionally graded material (FGM) implant designs. The implants were constructed using CT scan images. Zhang et al. [2] developed a 3D model of skull and cervical spine to investigate the effect of static and dynamic loading conditions on biomechanical responses of head and cervical spine. Watari et al. [3] uses powder metallurgy technique to fabricate Ti/HAP FGM implant by changing concentration gradually along axial direction. Later, tissue response and bone formation in animal implantation were evaluated by using optical microscopy. Monti et al. [4] carried out numerical and experimental analysis to determine the vibration characteristics of biomaterial-based sandwich beam. Parthasarathy et al. [5] examined design and fabrication of cellular structures made of Ti6Al4V using electron beam melting technique to evaluate different mechanical properties. Fang et al. [6] developed CAD-based model to evaluate mechanical properties and structural nonhomogeneity of porous tissue scaffolds. Truninger et al. [7,8] compared the bending moment of restored and internally fixed titanium abutments to zirconia abutments and also analyzed fracture load and fracture pattern for both the abutments. Mehboob et al. [9] reviewed various methods of prosthetic design, types of material used for prostheses and bone healing processes using finite element method to estimate healing of bone fracture. Chung et al. [10] evaluated the flexural strength of resin-based dental composites using two different test methods, namely biaxial flexural test and three-point bending test. Chang et al. [11] developed a finite element model to optimize the shape of dental implants by removing redundant material and redesigned new implant and evaluated its biomechanical properties. Jalali et al. [12] applied a finite element method to determine stress distribution on functionally graded dental implant using COMSOL multi physics software. Mahmoudi et al. [13] determined stresses at cervical region of premolar tooth by implementing null hypothesis in finite element platform based on the assumption that stress at the cervical region is not affected by a FGM crown. Marcian et al. [14] implemented a finite element method to evaluate micro strain and displacement on dental implant subjected to occlusal forces. Bahraminasab et al. [15] adopted a finite element technique and response surface methodology for optimization of stress, micro motion, and wear on functionally graded knee implant. Sadollah et al. [16] developed a multi-optimization model based on genetic and simulated annealing algorithm to optimize mechanical responses of a FGM dental implant. Oshkour et al. [17] employed finite element analysis to determine strain energy and stresses on radial and longitudinal functionally graded femoral prostheses. Tanimoto et al. [18] investigated the effect of silica filler material on bending properties of dental composites by using a finite element method. The proposed model was later on compared with experimental results. Anwar et al. [19] developed a three-dimensional finite element model to determine the effect of implant diameter and length on stress distribution in surrounding bones. Kharazi et al. [20] adopted a 3D finite element method to design metallic bone plate made of polylactic acid matrix and textile bioglass fibre as reinforcement and to optimize volume fraction reinforcement to withstand physiological loads. Wakabayashi et al. [21] discussed the application of nonlinear finite element analysis on different biomechanical properties of dental implant. Moursi et al. [22] investigated the enhancement of biological properties of osteoblast response by incorporating hydroxyapatite (HAP) onto polymethylmethacrylate (PMMA). Teo et al. [23] reported the application and usage of fibrous composite material in orthopedics and dentistry. Hamanaka et al. [24] applied a finite element approach to develop and simulate an orthodontic tooth movement model to determine forces acting on each tooth in sliding. Basha et al. [25] discussed application of natural polymer composite and its recent advancements in fabrication of bone scaffolds. Papini et al. [26] developed a finite element model to compare bio mechanical behavior of synthetic third-generation composite femur (3GCF) to a given sample of cadaveric femur from donor. Chang and his coauthors [27–29] carried out biomechanical analysis of a fractured tibia with different composite structure using a finite element method. Hedia et al. [30] adopted a finite element approach to design a cementless hip stem using concept of bidirectional functionally graded material. Rule of mixture was employed to grade the material in axial and transverse directions. Mehrali et al. [31] summarized the fabrication process and optimization of FGM implants made of different materials, namely titanium (Ti), hydroxyapatite (HAP), zirconia (ZrO_2), and so on.

From the above review, it is clear that very limited effort had been given in the past to analyze the structural behavior of nonhomogeneous biocomposites. Therefore, in the present work, numerical simulation of the biocomposite FG plate is conducted, followed by the analysis to obtain the critical load. In addition, the effects of various parameters such as aspect ratio, thickness ratio, volume fraction index and support condition on the buckling response of biocomposites are investigated.

9.2 MICROMECHANICAL MATERIAL MODELING

9.2.1 EVALUATION OF VOLUME FRACTIONS

A functionally graded plate Ti rich at the bottom and HAP rich at the top is considered, in which the material distribution is assumed to be varying smoothly in accordance with the power-law distribution, whose geometry is shown in Figure 9.1. The volume

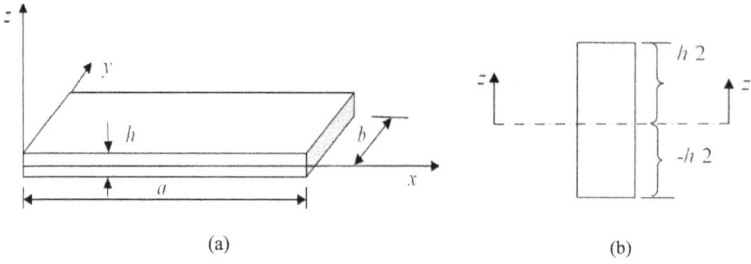

(a)   (b)

FIGURE 9.1 Geometrical description of biocomposite plate structure. (a) Geometrical description. (b) Crosss-sectional view.

fraction of the titanium (V_{Ti}) and HAP (V_{HAP}) by using power-law distribution can be written as

$$V_{Ti} + V_{HAP} = 1 \tag{9.1}$$

$$V_{HAP} = \left(0.5 + \frac{z}{h} \right)^n \tag{9.2}$$

where n represents volume fraction index that lies in the range of $0 \leq n \leq \infty$.

9.2.2 EFFECTIVE MATERIAL PROPERTIES

The effective material properties of a biocomposite FG plate can be evaluated by using the Voigt's model.

$$P = \left(P_c - P_m \right) V_c + P_m \tag{9.3}$$

where P_m and P_c are the properties of metal and ceramic, respectively, and V_c is the volume fraction of the ceramic. Using the above relation, effective material properties, namely Young's modulus (E), density (ρ), and Poisson's ratio (ν) for Ti-HAP based FGM, can be expressed as

$$E = \left(E_{HAP} - E_{Ti} \right) \left(0.5 + \frac{z}{h} \right)^n + E_{Ti} \tag{9.4}$$

$$\rho = \left(\rho_{HAP} - \rho_{Ti} \right) \left(0.5 + \frac{z}{h} \right)^n + \rho_{Ti} \tag{9.5}$$

$$\nu = \left(\nu_{HAP} - \nu_{Ti} \right) \left(0.5 + \frac{z}{h} \right)^n + \nu_{Ti} \tag{9.6}$$

where E_{Ti}, ρ_{Ti}, and ν_{Ti} represent material properties of titanium and E_{HA}, ρ_{HA}, and ν_{HA} represent material properties of hydroxyapatite.

9.3   FINITE ELEMENT FORMULATIONS

9.3.1   KINEMATIC MODEL

The mathematical model for biocomposite FGM is derived by using FSDT mid-plane kinetics. The displacement field at any point along x, y, and z axes can be expressed as

$$\left.\begin{array}{l} u(x,y,z,t) = u_0(x,y,t) + zu_1(x,y,t) \\ v(x,y,z,t) = v_0(x,y,t) + zv_1(x,y,t) \\ w(x,y,z,t) = w_0(x,y,t) + zw_1(x,y,t) \end{array}\right\} \tag{9.7}$$

where z is the thickness coordinate that varies from $-h/2$ to $+h/2$ and t is the time. u_0, v_0, and w_0 are the displacements, and u_1, v_1, and w_1 are the shear rotation in x, y, and z directions.

The strain-displacement relation can be expressed as

$$\{\varepsilon\} = \begin{Bmatrix} \varepsilon_{xx} \\ \varepsilon_{yy} \\ \varepsilon_{zz} \\ \gamma_{xy} \\ \gamma_{xz} \\ \gamma_{yz} \end{Bmatrix} = \begin{Bmatrix} \left(\dfrac{\partial u}{\partial x}\right) \\[2mm] \left(\dfrac{\partial v}{\partial y}\right) \\[2mm] \left(\dfrac{\partial w}{\partial z}\right) \\[2mm] \left(\dfrac{\partial u}{\partial y} + \dfrac{\partial v}{\partial x}\right) \\[2mm] \left(\dfrac{\partial u}{\partial z} + \dfrac{\partial w}{\partial x}\right) \\[2mm] \left(\dfrac{\partial v}{\partial z} + \dfrac{\partial w}{\partial y}\right) \end{Bmatrix} \tag{9.8}$$

It can also rewritten as

$$\{\varepsilon\} = \begin{Bmatrix} \varepsilon_{xx}^o \\ \varepsilon_{yy}^o \\ \varepsilon_{zz}^o \\ \gamma_{xy}^o \\ \gamma_{xz}^o \\ \gamma_{yz}^o \end{Bmatrix} + z \begin{Bmatrix} k_x \\ k_y \\ k_z \\ k_{xy} \\ k_{xz} \\ k_{yz} \end{Bmatrix} \tag{9.9}$$

$$\{\varepsilon\} = [T]\{\bar{\varepsilon}\} \tag{9.10}$$

where $\{\bar{\varepsilon}\} = \{\varepsilon_{xx}^o \ \varepsilon_{yy}^o \ \varepsilon_{zz}^o \ \gamma_{xy}^o \ \gamma_{xz}^o \ \gamma_{yz}^o \ k_x \ k_y \ k_z \ k_{xy} \ k_{xz} \ k_{yz}\}^T$ is the mid-plane strain vector.

9.3.2 CONSTITUTIVE EQUATIONS

The constitutive equations for FG plate can be expressed as

$$\begin{Bmatrix} \sigma_{xx} \\ \sigma_{yy} \\ \sigma_{zz} \\ \tau_{xy} \\ \tau_{xz} \\ \tau_{yz} \end{Bmatrix} = \begin{bmatrix} Q_{11} & Q_{12} & Q_{13} & 0 & 0 & 0 \\ Q_{12} & Q_{22} & Q_{23} & 0 & 0 & 0 \\ Q_{13} & Q_{23} & Q_{33} & 0 & 0 & 0 \\ 0 & 0 & 0 & Q_{66} & 0 & 0 \\ 0 & 0 & 0 & 0 & Q_{55} & 0 \\ 0 & 0 & 0 & 0 & 0 & Q_{44} \end{bmatrix} \begin{Bmatrix} \varepsilon_{xx} \\ \varepsilon_{yy} \\ \varepsilon_{zz} \\ \gamma_{xy} \\ \gamma_{xz} \\ \gamma_{yz} \end{Bmatrix} \tag{9.11}$$

The generalized form of stress-strain relation can be expressed as

$$\{\sigma\} = [Q]\{\varepsilon\} \tag{9.12}$$

where $[Q]$ is the rigidity matrix.

The strain energy for FG plate is given by

$$U = \frac{1}{2} \int_v \{\varepsilon\}^T \{\sigma\} dV \tag{9.13}$$

Now, the above equations can be rewritten as

$$U = \frac{1}{2} \int \left(\{\bar{\varepsilon}\}^T [D]\{\bar{\varepsilon}\} \right) dA \tag{9.14}$$

where $[D] = \int_{-h/2}^{h/2} [T]^T [Q][T] dz$

9.3.3 STRAIN ENERGY DUE TO IN-PLANE LOADING

The strain energy due to in-plane force resultants can be expressed as

$$W_G = \int_A \{\varepsilon_G\}^T [S_G]\{\varepsilon_G\} dA \tag{9.15}$$

where $\{\varepsilon_G\}$ and $[S_G]$ are the in-plane strain vector and force resultant matrix, respectively.

9.3.4   GOVERNING EQUATIONS

The displacement vector for each element having six degree of freedom per node can be expressed as

$$\{\delta\} = \sum_{i=1}^{6} N_i \{\delta_i\} \tag{9.16}$$

where $\{\delta_i\} = [u_{0i} \ v_{0i} \ w_{0i} u_{1i} \ v_{1i} \ w_{1i}]^T$ is the nodal displacement vector at node i. N_i is the shape function for the ith node.

The mid-plane strain vector having nodal displacement vector can be expressed by Equation (9.17).

$$\{\bar{\varepsilon}\} = [B]\{\delta_i\} \tag{9.17}$$

where [B] is the product of differential operators and the shape functions in the strain terms.

By substituting Equation (9.17) in (9.14), elemental strain energy can be expressed by Equation (9.17):

$$U_e = \frac{1}{2} \iint \left(\{\delta_i\}^T [B]^T [D][B]\{\delta_i\} \right) dxdy \tag{9.18}$$

$$U_e = \frac{1}{2} \{\delta_i\}^T [K]_e \{\delta_i\} \tag{9.19}$$

where $[K]_e$ is the elemental stiffness matrix.

Similarly, elemental membrane strain energy can be expressed by Equation (9.20).

$$W_G = \frac{1}{2} \{\delta_i\}^T [K_G]_e \{\delta_i\} \tag{9.20}$$

where K_G is elemental stress stiffness matrix.

The final governing equation of FG plate by using minimum total potential energy principle is given by Equation (9.21),

$$\Delta\Pi = \Delta U - \Delta W = 0 \tag{9.21}$$

where Π is the total potential energy and Δ is the variational symbol.

For buckling analysis, equilibrium equation can be written as

$$\left(\left[K_e\right] + P_{cr}\left[K_g\right]\right)\{\Delta\} = 0 \tag{9.22}$$

where $[K_g]$ and P_{cr} represent geometrical stiffness matrix and critical buckling load factor, respectively.

9.4 STABILITY BEHAVIOR OF BIOCOMPOSITE STRUCTURES

The desired buckling response is calculated using customized code in APDL. The convergence behavior of the present FE model has been established for FG plate subjected to in-plane mechanical loading for different geometrical configuration. Lastly, the present model is validated with previous literature.

9.4.1 CONVERGENCE AND COMPARISON TESTS

Here the convergence and validity of proposed FE model of a FG plate subjected to mechanical loading have been shown by evaluating the critical buckling load for different aspect ratio and mesh size. From the Table 9.1 it is clear that the response is converging well, and hence 12 × 12 mesh size is appropriate for further analysis. For verification purposes, the present model is validated by comparing the desired computed response to the available literature [32] as shown in Figure 9.2, which shows good agreement.

9.4.2 NUMERICAL EXAMPLES

In this example, a FG plate with its top surface ceramic rich (HAP) and bottom surface metal rich (Ti) subjected to uniform uniaxial mechanical loading along x-axis with all four edges simply supported is considered for computing the nondimensional critical buckling load parameter. The material properties are assumed to be same as in previous examples and varying along thickness direction using power-law. The nondimensional critical buckling load parameter is obtained using the relation given below:

$$\overline{P_{cr}} = \frac{P_{cr}a^2}{E_m h^3}$$

TABLE 9.1
Convergence Behavior of Critical Buckling Load with Different Mesh Size

Critical Buckling Load (F_{cr}) (GN/m)	1.896	1.905	1.908	1.91	1.91
Mesh size	4×4	6×6	8×8	10×10	12×12

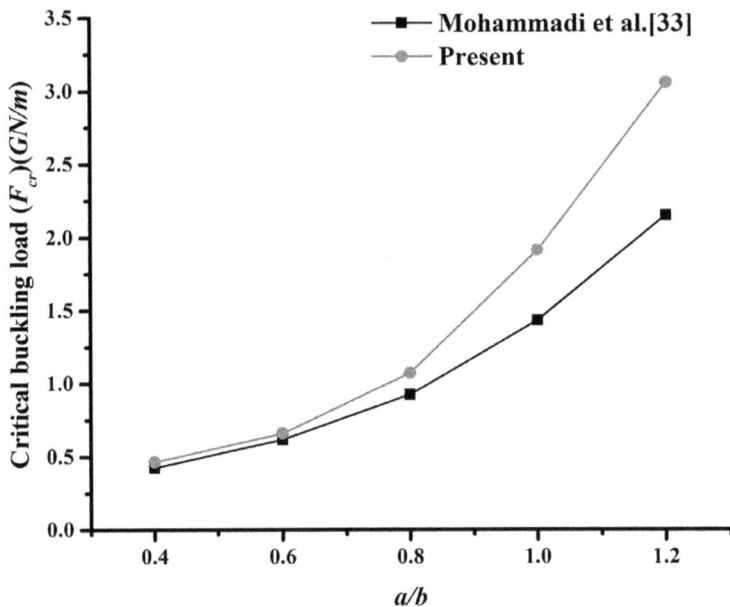

FIGURE 9.2 Comparison of critical buckling load of simply supported FG plate for different aspect ratio ($a = 1$, $a/h = 10$).

9.4.2.1 Effect of Volume Fraction Index on Buckling Strength of FG Plate

The volume fraction index is one of the important factors in the evaluation of material property of FG plate, which affects the stiffness of the FG plate, which in turn affects the buckling strength. Table 9.2 shows the effect of different volume fraction index on critical buckling response parameter of rectangular FG plate ($a/b = 1.5$, $a/h = 50$). The desired response parameter increases with increase in volume fraction index.

9.4.2.2 Effect of Thickness Ratio on Buckling Strength of FG Plate

The geometrical parameters are very important to decide the strength of any structure. In this example variation of critical buckling with four different thickness ratio

TABLE 9.2

Nondimensional Critical Buckling Load Parameter Variation with Different Volume Fraction Index of FG Plate

Mode	N			
	0	0.5	0.8	1
1	5.03	7.57	8.20	8.50
2	5.94	8.94	9.69	10.05
3	10.49	15.77	17.10	17.74
4	12.76	19.14	20.72	21.48
5	14.75	22.16	24.03	24.93

TABLE 9.3
Nondimensional Critical Buckling Load Parameter
Variation with Different Thickness Ratio of FG Plate

Mode	a/h			
	10	20	50	100
1	6.29	7.25	7.57	7.62
2	6.96	8.43	8.94	9.02
3	10.49	14.23	15.77	16.02
4	11.96	17.56	19.14	19.33
5	14.00	19.92	22.16	22.61

TABLE 9.4
Nondimensional Critical Buckling Load Parameter
Variation with Different Aspect Ratio of FG Plate

Mode	a/b			
	1.2	1.4	1.6	1.8
1	4.80	6.51	8.76	11.51
2	5.10	7.82	10.04	12.39
3	8.10	13.59	17.64	22.59
4	11.75	14.20	22.86	27.43
5	16.77	19.91	29.98	37.01

is shown in Table 9.3 for a rectangular FG ($a/b = 1.5$) plate keeping volume fraction index ($n = 0.5$) constant. Results indicate that as the plate becomes thinner, the value of critical buckling load decreases and the critical buckling load parameter increases.

9.4.2.3 Effect of Aspect Ratio on Buckling Strength of FG Plate

In this section, effect of aspect ratio of FG plate ($a/h = 50$, $n = 0.5$) on critical buckling load parameter were investigated. It is clear from the Table 9.4 that the response parameter increases with increase in aspect ratio. This is due to the fact that as aspect ratio increases the stiffness of the plate, the buckling load consequently increases.

9.5   CONCLUSIONS

This study focuses on the stability behavior of biocomposite structures where Ti and HAP are considered as constituents in gradient form. The displacement field is based on the FSDT mid-plane kinematics, whereas equation of motion is derived using principle of minimum potential energy. To obtain the bucking responses, finite element simulations are performed via Block Laczos eigenvalue extraction algorithm in APDL environment. After confirming the consistency and accuracy of the present model, numerous examples have been presented at various sets of parametric conditions. It is found that, with increase in thickness-ratio, aspect ratios, and volume fraction indices, critical buckling load parameter increase.

REFERENCES

1. Lin D, Li Q, Li W, Zhou S, Swain M V. (2009) Design optimization of functionally graded dental implant for bone remodeling. *Compos Part B Eng* 40: 668–675. doi: 10.1016/j.compositesb.2009.04.015
2. Zhang QH, Teo EC, Ng HW, Lee VS (2006) Finite element analysis of moment-rotation relationships for human cervical spine. *J Biomech* 39: 189–193. doi: 10.1016/j.jbiomech.2004.10.029
3. Watari F, Yokoyama A, Omori M, Hirai T, Kondo H, Uo M, Kawasaki T (2004) Biocompatibility of materials and development to functionally graded implant for bio-medical application. *Compos Sci Technol* 64: 893–908. doi: 10.1016/j.compscitech.2003.09.005
4. Monti A, El Mahi A, Jendli Z, Guillaumat L (2017) Experimental and finite elements analysis of the vibration behaviour of a bio-based composite sandwich beam. *Compos Part B Eng* 110: 466–475. doi: 10.1016/j.compositesb.2016.11.045
5. Parthasarathy J, Starly B, Raman S (2011) A design for the additive manufacture of functionally graded porous structures with tailored mechanical properties for biomedical applications. *J Manuf Process* 13: 160–170. doi: 10.1016/j.jmapro.2011.01.004
6. Fang Z, Starly B, Sun W (2005) Computer-aided characterization for effective mechanical properties of porous tissue scaffolds. *CAD Comput Aided Des* 37: 65–72. doi: 10.1016/j.cad.2004.04.002
7. Mühlemann S, Truninger TC, Stawarczyk B, Hämmerle CHF, Sailer I (2014) Bending moments of zirconia and titanium implant abutments supporting all-ceramic crowns after aging. *Clin Oral Implants Res* 25: 74–81. doi: 10.1111/clr.12192
8. Truninger TC, Leutert CR, Sailer TR, Ha CHF (2011) Bending moments of zirconia and titanium abutments with internal and external implant—abutment connections after aging and chewing simulation. 12–18. doi:10.1111/j.1600-0501.2010.02141.x
9. Mehboob H, Chang SH (2014) Application of composites to orthopedic prostheses for effective bone healing: A review. *Compos Struct* 118: 328–341. doi: 10.1016/j.compstruct.2014.07.052
10. Chung SM, Yap AUJ, Chandra SP, Lim CT (2004) Flexural strength of dental composite restoratives: Comparison of biaxial and three-point bending test. *J Biomed Mater Res - Part B Appl Biomater* 71: 278–283. doi: 10.1002/jbm.b.30103
11. Chang CL, Chen CS, Huang CH, Hsu ML (2012) Finite element analysis of the dental implant using a topology optimization method. *Med Eng Phys* 34: 999–1008. doi: 10.1016/j.medengphy.2012.06.004
12. Jalali SK, Yarmohammadi R, Maghsoudi F (2016) Finite element stress analysis of functionally graded dental implant of a premolar tooth. *J Mech Sci Technol* 30: 4919–4923. doi: 10.1007/s12206-016-1011-y
13. Mahmoudi M, Saidi AR, Hashemipour MA, Amini P (2018) The use of functionally graded dental crowns to improve biocompatibility: A finite element analysis. *Comput Methods Biomech Biomed Engin* 21: 161–168. doi: 10.1080/10255842.2018.1431219
14. Marcián P, Wolff J, Horáčková L, Kaiser J, Zikmund T, Borák L (2018) Micro finite element analysis of dental implants under different loading conditions. *Comput Biol Med* 96: 157–165. doi: 10.1016/j.compbiomed.2018.03.012
15. Bahraminasab M, Sahari BB, Edwards KL, Farahmand F, Hong TS, Arumugam M, Jahan A (2014) Multi-objective design optimization of functionally graded material for the femoral component of a total knee replacement. *Mater Des* 53: 159–173. doi: 10.1016/j.matdes.2013.06.050
16. Sadollah A, Bahreininejad A (2011) Optimum gradient material for a functionally graded dental implant using metaheuristic algorithms. *J Mech Behav Biomed Mater* 4: 1384–1395. doi: 10.1016/j.jmbbm.2011.05.009

17. Oshkour AA, Osman NAA, Davoodi MM, Yau YH, Tarlochan F, Wan Abas WAB, Bayat M (2013) Finite element analysis on longitudinal and radial functionally graded femoral prosthesis. *Int J Numer Method Biomed Eng.* 12. doi: 10.1002/cnm.2583
18. Tanimoto Y, Nishiwaki T, Nemoto K, Ben G (2004) Effect of filler content on bending properties of dental composites: Numerical simulation with the use of the finite-element method. *J Biomed Mater Res - Part B Appl Biomater* 71: 188–195. doi: 10.1002/jbm.b.30079
19. El-Anwar MI, El-Zawahry MM (2011) A three dimensional finite element study on dentalimplant design. *J Genet Eng Biotechnol* 9: 77–82. doi: 10.1016/j.jgeb.2011.05.007
20. Kharazi AZ, Fathi MH, Bahmany F (2010) Design of a textile composite bone plate using 3D-finite element method. *Mater Des* 31: 1468–1474. doi: 10.1016/j.matdes.2009.08.043
21. Wakabayashi N, Ona M, Suzuki T, Igarashi Y (2008) Nonlinear finite element analyses: Advances and challenges in dental applications. *J Dent* 36: 463–471.doi: 10.1016/j.jdent.2008.03.010
22. Moursi AM, Winnard A V, Winnard PL, Lannutti JJ, Seghi RR (2002) Enhanced osteoblast response to a polymethylmethacrylate—hydroxyapatite composite. *Biomaterials* 23: 133–144. doi:10.1016/j.compscitech.2003.09.012
23. Fujihara K, Teo K, Gopal R, Loh PL, Ganesh VK (2004) Fibrous composite materials in dentistry and orthopaedics: Review and applications. *Compos Sci Technol* 64: 775–788. doi: 10.1016/j.compscitech.2003.09.012
24. Hamanaka R, Yamaoka S, Anh TN, Tominaga J, Koga Y, Yoshida N Numeric simulation model for long-term orthodontic tooth movement with contact boundary conditions using the finite element method. *Am J Orthod Dentofac Orthop* 152: 601–612. doi: 10.1016/j.ajodo.2017.03.021
25. Yunus Basha R, Sampath Kumar TS, Doble M (2015) Design of biocomposite materials for bone tissue regeneration. *Mater Sci Eng C* 57: 452–463. doi: 10.1016/j.msec.2015.07.016
26. Papini M, Zdero R, Schemitsch EH, Zalzal P (2007) The biomechanics of human femurs in axial and torsional loading: Comparison of finite element analysis, human cadaveric femurs, and synthetic. *ASME* 129. doi: 10.1115/1.2401178
27. Mehboob H, Chang S (2015) Effect of structural stiffness of composite bone plate—scaffold assembly on tibial fracture with large fracture gap. *Compos Struct* 124: 327–336. doi: 10.1016/j.compstruct.2015.01.011
28. Mehboob H, Son D, Chang S (2013) Composite s science and technology finite element analysis of tissue differentiation process of a tibia with various fracture configurations when a composite intramedullary rod was applied. *Compos Sci Technol* 80: 55–65. doi: 10.1016/j.compscitech.2013.02.020
29. Kim H, Kim S, Chang S (2011) Composites: Part B Bio-mechanical analysis of a fractured tibia with composite bone plates according to the diaphyseal oblique fracture angle. *Compos Part B* 42: 666–674. doi: 10.1016/j.compositesb.2011.02.009
30. Hedia HS, Shabara MAN, Ei-Midany TT, Fouda N (2006) Improved design of cementless hip stems using two-dimensional functionally graded materials. *J Biomed Mater Res Partn B.* doi: 10.1002/jbmb
31. Mehrali M, Shirazi FS, Mehrali M, et al. (2013) Dental implants from functionally graded materials. *J Biomed Mater Res - Part A* 101: 3046–3057. doi: 10.1002/jbm.a.34588
32. Mohammadi M, Saidi AR, Jomehzadeh E (2010) A novel analytical approach for the buckling analysis of moderately thick functionally graded rectangular plates with two simply-supported opposite edges. *Proc Inst Mech Eng Part C J Mech Eng Sci* 224: 1831–1841. doi: 10.1243/09544062JMES1804

10 Dynamic Analysis of Sandwich Composite Plate Structures with Honeycomb Auxetic Core

Mrityunjay Kumar and Madan Lal Chandravanshi
Indian Institute of Technology (ISM) Dhanbad,
Dhanbad, India

Mayank K. Ghosh and Vishesh Ranjan Kar
National Institute of Technology Jamshedpur,
Jamshedpur, India

Kamal Kishore Joshi
National Institute of Technology Jamshedpur,
Jamshedpur, India

Kalinga Institute of Industrial Technology Bhubaneswar,
Bhubaneswar, India

CONTENTS

DOI: 10.1201/9781003158813-10

173

10.1   INTRODUCTION

Using the CMT theory, Scarpa and Tomlinson [1] investigated the anisotropic mechanical characteristics of a honeycomb sandwich plate. They discovered that the out-of-plane effect shear moduli of a honeycomb sandwich construction significantly change the geometric parameters of a honeycomb core with a negative cell angle of the wall (Auxetic cell). Additionally, they discovered that when compared with the normal honeycomb core, the negatively angled honeycomb core had a higher density. They looked at the structure's natural frequency with varied cell wall angles and concluded that a well-designed honeycomb cell might considerably improve the dynamic behavior of a honeycomb sandwich structure. The governing equations and fundamental frequencies of sandwich plates were derived using first-order shear deformation theory (FSDT) in their research. In other cases, however, this theory may be incorrect.

Dobyns [2] used an analytical approach to investigate the behavior of laminated composite plates subjected to static and dynamic loads. He calculated the fundamental natural frequency of a generic composite plate with a simply supported boundary condition and investigated the plate's behavior under transient loadings. He also looked at how the plate behaved when these loads were given to different sections of the plate. A survey of recent research on damping in composite materials was conducted by Chandra, Singh, and Gupta [3]. When studying viscoelastic properties, they discovered that the strain energy technique was extremely popular in predicting damping of a composite material. They also pointed out areas that needed more research. A few models, including transverse and shear damping effects, were developed. Limited amount of research has been done on interlaminar tension. It was also necessary to conduct a damping optimization study in a composite structure. The impact of various design factors on the dynamic behavior of a laminated composite plate was studied by Latheswary, Valsarajan, and Sadasiva Rao [4]. The finite element approach was used to conduct the research, which included transient analysis and time harmonic analysis. Damping ratio, width-to-thickness ratio, material anisotropy, fiber orientation, number of layers, and aspect ratio were among the design characteristics investigated. The plate's harmonic response was investigated using various damping ratios. The damping ratio was shown to have a considerable effect on the resonance response, while changes in fiber orientation and width-to-thickness ratio had no effect on the response. Reddy's higher-order shear deformation theory (HSDT) was investigated by Meunier and Shenoi [5]. They used Hamilton's principle to derive the system's equations of motion. The viscoelastic material property was then added to the model, and the results were calculated. Natural frequencies and modal loss factors are two types of loss factors. They discovered during the research that the plate's behavior was significantly influenced by the dynamic properties of material.

In the aircraft industry, carbon fiber composites and aluminum honeycombs are commonly employed. Their sandwich panels, which are made up of two carbon-fiber face sheets and an aluminum honeycomb core, make use of the complementing strength and weight features. Aside from their high specific strength and stiffness, such laminar composite structures can also have much desired energy absorption capacity and good damping qualities [6]. Carbon-fiber and aluminum-honeycomb sandwich structures are extremely attractive due to their superior structural properties, which are significant in many structural applications. They can also be easily produced through lamination, providing designers with new and beneficial structure-property possibilities [7–9]. Many researchers have looked at the structural performance and properties of honeycomb sandwich systems. Petras and Zhu, for example, created failure-mode maps of honeycomb-core sandwich constructions under three-point bending, quasi-static indentation, and low-velocity impact [10,11]. Material qualities, structural configuration, loading distribution, and bonding condition of the face–core interface were found to have a strong relationship with failure mechanisms and failure loads. The impact of cell sizes, core densities, core materials, and face sheet thickness on the damage characteristics of honeycomb sandwich panels were investigated by Kaman and Zhou [12,13]. Arunkumar et al. [14,15] investigated honeycomb sandwich panel vibro-acoustic responses. Sui et al. [16] looked at how honeycomb structures affected total weight and sound transmission loss. The objective of this chapter is to model and simulate the modal analysis of flat honeycomb sandwich panels. First, an eight-nodded element (SHELL281) from ANSYS is used to model the sandwich plate with a honeycomb core. The validity and convergence of the current model are explored in order, and various numerical instances are presented.

10.2 THE PROPERTIES OF HONEYCOMB STRUCTURES AND MATERIALS

In order to examine the behavior of a honeycomb sandwich plate, researchers must have a thorough understanding of the mechanical properties of a honeycomb structure, as this is the foundation of the honeycomb sandwich plate theory. As a result, the honeycomb structure's mechanical qualities are described in this section.

10.2.1 The Effective Properties of Honeycomb Cells

A regular hexagonal honeycomb-shaped layer is shown in Figure 10.1. In-plane stiffness and strength are very low, causing cell walls to bend under in-plane loads. Because axial extension or compression of the cell walls is significantly more difficult to deform, out-of-plane stiffness and strength are higher. Here R is the center distance and t is the thickness of support.

A typical honeycomb cell is shown in Figure 10.2. If the conditions hold, the honeycomb cell will have isotropic in-plane material properties:

1. The honeycomb cell is hexagonal in shape (equal on all sides). The angle between two neighboring sides is 120°. The angle formed by the intersection of two adjacent sides can be specified by $90° + \theta$, resulting in a regular honeycomb of $\theta = 30°$.
2. The cell's wall thickness is the same on every edges. The effective Young's modulus E and the effective shear modulus G will be two independent

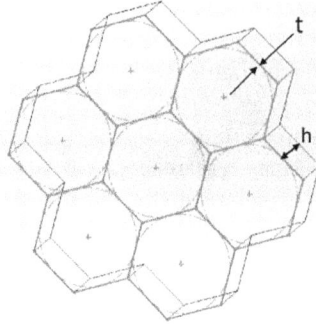

FIGURE 10.1   Regular hexagonal honeycomb structure.

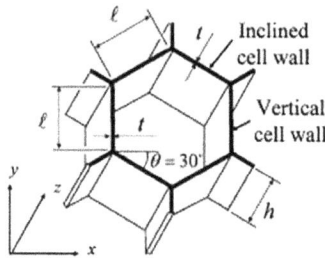

FIGURE 10.2   A regular hexagonal-shaped honeycomb cell as per D. H. Chen (2011).

moduli in this situation. When one or both of the two conditions are not met, the material exhibits anisotropic behavior, resulting in more independent moduli. The parameters needed are Young's modulus E_1 in the X direction, Young's modulus E_2 in the Y direction, the in-plane shear modulus G_{12}, and the Poisson's ratio v_{12}.

The density of the honeycomb is calculated geometrically as illustrated below. The ratio of h to l is defined as α, i.e., $\alpha = h/l$. The ratio of t to l is defined as β, i.e., $\beta = t/l$.

$$\frac{\rho_c}{\rho_m} = \frac{\beta(\alpha+2)}{2\cos\theta(\alpha+\sin\theta)} \tag{10.1}$$

ρ_c and ρ_m here are the densities of the honeycomb and the core material, respectively. There can be an introduction of v_{21} if the honeycomb is subjected to loads in X-Y direction and deformed in a linear elastic way. The reciprocal relation mentioned below can reduce the moduli number by 1.

$$E_1 v_{21} = E_2 v_{12} \tag{10.2}$$

In order to find the Young's modulus, the following methodology has to be followed.

$$M = \frac{Pl\sin\theta}{2}$$

(10.3)

where

$$P = \sigma_1\left(h + l\sin\theta\right)b$$

(10.4)

b is the honeycomb core depth.
The deflection by the wall is given by

$$\delta = \frac{Pl^3\sin\theta}{12E_mI}$$

(10.5)

I is the second moment of inertia of the cell wall.
The strain in the X1 direction is given by

$$\varepsilon_1 = \frac{\delta\sin\theta}{l\cos\theta} = \frac{\sigma_1\left(h + l\sin\theta\right)bl^2\sin^2\theta}{12E_mI\cos\theta}$$

(10.6)

The derivation of the Young's modulus in X1 direction is

$$\frac{E_1}{E_m} = \left(\beta^3\right)\frac{\cos\theta}{\left(\alpha + \sin\theta\right)\sin^2\theta}$$

(10.7)

Similarly for the Y direction, the derivation of moment, the deflection of the wall and the strain is given by

$$M = \frac{Wl\cos\theta}{2}, \ \delta = \frac{Wl^3\cos\theta}{12E_mI}, \ \varepsilon_2 = \frac{\delta\cos\theta}{h + l\sin\theta} = \frac{\sigma_2bl^4\cos^3\theta}{12E_mI\left(h + l\sin\theta\right)}$$

(10.8)

The derivation of the Young's modulus in Y direction is

$$\frac{E_2}{E_m} = \beta^3\frac{\left(\alpha + \sin\theta\right)}{\cos^3\theta}$$

(10.9)

For the hexagon with walls,

$$\frac{E_1}{E_m} = \frac{E_2}{E_m} = 2.3\beta^3$$

(10.10)

Honeycomb layers having a regular hexagon shape will have isotropic in-plane characteristics, as can be seen. Poisson's ratio can be calculated by taking the negative ratio of the strains in the X and Y directions.

$$V_{12} = -\frac{\varepsilon_2}{\varepsilon_1} = \frac{\cos^2 \theta}{(\alpha + \sin \theta) \sin \theta} \tag{10.11}$$

For a regular hexagon, the in-plane Poisson's ratio will be 1. This may not be true, however, in the case of a more sophisticated hypothesis, which will be described later. Knowing v_{12}, v_{21} (which is the reciprocal of v_{12}) may be calculated in the same way. When analyzing the Poisson's ratio, an interesting phenomenon can be observed. That is, when the honeycomb cell's wall angle is negative, the Poisson's ratio is negative, which is extremely rare. When an extensional force is applied in the X or Y directions, the normal in-plane side expands instead of compressing.

For a general honeycomb, as

$$E_1 = E_2, \quad E_1 v_{21} = E_2 v_{12}, \text{ then } \quad v_{12} = v_{21} \tag{10.12}$$

The shear deflection can be given by

$$\gamma = \frac{2u_s}{(h + l \sin \theta)} = \frac{Fh^2 (l + 2h)}{24 E_m I (h + l \sin \theta)} \tag{10.13}$$

and the shear stress is given by

$$\tau = \frac{F}{2lb \cos \theta} \tag{10.14}$$

and the derivation of the shear Modulus is given by

$$\frac{G_{12}}{E_m} = (\beta^3) \frac{(\alpha + \sin \theta)}{(\alpha)^2 (1 + 2\alpha) \cos \theta} \tag{10.15}$$

The out-of-plane properties of a honeycomb structure are more relevant, since they are substantially larger than the in-plane values for hexagonal honeycomb structures. This is because cell walls will be stretched or contracted when loaded in the Z direction. The load is assumed to be applied in the Z direction, or on the surface normal to the Z direction. A honeycomb layer with a low density is assumed. The cell walls are made of the same thickness. To describe the out-of-plane properties, five more moduli are required. The E_3 is the Young's modulus in the Z direction given by

$$\frac{E_3}{E_m} = \frac{\alpha + 2}{2(\alpha + \sin \theta) \cos \theta} \beta = \frac{\rho_H}{\rho_m} \approx \beta \tag{10.16}$$

The Poisson's ratios v_{31} and v_{32} are the same as the core material's Poisson's ratio.

$$v_{31} = v_{32} = v_m \qquad (10.17)$$

Using the reciprocal relation,

$$E_1 v_{31} = E_3 v_{13}, \ E_2 v_{32} = E_3 v_{23} \qquad (10.18)$$

In the case of honeycombs,

$$E_1/E_3 = v_{13}/v_{31} \approx \left(t/l\right)^2 \qquad (10.19)$$

The ratio of cell wall thickness to cell length, $\beta = (t/l)$, is very small, $E_1 << E_3$ as well as

$$v_{13} = v_{23} \approx 0 \qquad (10.20)$$

Using the minimum potential energy and minimum complementary energy theory, the out-of-plane shear moduli was derived. The first theorem gives an upper bound

$$\frac{G_{13}}{G_m} \leq \frac{\cos\theta}{\alpha + \sin\theta}\beta \qquad (10.21)$$

$$\frac{G_{23}}{G_m} \leq \frac{1}{2}\frac{\alpha + 2\sin^2\theta}{\left(\alpha + \sin\theta\right)\cos\theta}\beta \qquad (10.22)$$

For the regular hexagon,

$$\frac{G_{13}}{G_m} = 0.577\beta \qquad (10.23)$$

The second theorem gives a lower bound that is

$$\frac{G_{13}}{G_m} \geq \frac{\cos\theta}{\alpha + \sin\theta}\beta \quad \frac{G_{23}}{G_m} \geq \frac{\alpha + \sin\theta}{\left(1 + 2\alpha\right)\cos\theta}\beta \qquad (10.24)$$

For regular hexagon it becomes

$$\frac{G_{23}}{G_m} = 0.577\beta \qquad (10.25)$$

This is yet another example of isotropic in-plane material properties in a honeycomb structure with regular hexagonal shaped cells.

There is also a refined theory for calculating E_1, E_2, ν_{12}, ν_{21}, G_{12} that comes in handy when (t/l) is greater than 0.2. During the derivation procedure, axial and shear deformation effects are taken into account in this theory. E_1 is given by

$$E_1 = E_m \beta^3 \frac{\cos\theta}{(\alpha + \sin\theta)\sin^2\theta} \frac{1}{1 + (2.4 + 1.5\nu_m + \cot^2\theta)\beta^2} \tag{10.26}$$

and E_2 by

$$E_2 = E_m \beta^3 \frac{\alpha + \sin\theta}{\cos^3\theta} \frac{1}{1 + \left(2.4 + 1.5\nu_m + \tan^2\theta + \dfrac{2\alpha}{\cos^2\theta}\right)\beta^2} \tag{10.27}$$

and ν_{12} by

$$\nu_{12} = E_m \beta^3 \frac{\cos^2\theta}{(\alpha + \sin\theta)\sin\theta} \frac{1 + (1.4 + 1.5\nu_m)\beta^2}{1 + (2.4 + 1.5\nu_m + \cot^2\theta)\beta^2} \tag{10.28}$$

The reciprocal relation can be used to calculate ν_{21}. G_{12} is given by

$$G_{12} = \frac{E_m \beta^3 (\alpha + \sin\theta)}{\alpha^2 \cos\theta} \frac{1}{F} \tag{10.29}$$

where

$$F = 1 + 2\alpha + \beta^2 \left\{ \begin{array}{l} \dfrac{1}{\alpha}(2.4 + 1.5\nu_m)(2 + \alpha + \sin\theta) \\ + \dfrac{\alpha + \sin\theta}{\alpha^2}\left[(\alpha + \sin\theta)\tan^2\theta + \sin\theta\right] \end{array} \right\} \tag{10.30}$$

As most commercial tools, such as ANSYS and ABAQUS, will not enable users to input Poisson's ratios greater than 1 or 0.5, this refined theory is useful. As a result, the moduli offered by the revised theory can be accepted because of the tiny difference if researchers desire to create a model employing the effective qualities to save computing time.

10.3 DERIVATION OF THE GOVERNING EQUATIONS OF A SANDWICH PLATE

10.3.1 ORTHOTROPIC MATERIAL PROPERTIES

Honeycomb's effective characteristics are defined by orthotropic behavior, as we know by now. Orthotropic materials exhibit variable material behavior along orthogonal material directions, a condition known as orthogonal anisotropy.

An orthogonal material's elasticity tensor is defined as

$$[C_{ij}] = \begin{bmatrix} C_{11} & C_{12} & C_{13} & 0 & 0 & 0 \\ & C_{22} & C_{23} & 0 & 0 & 0 \\ & & C_{33} & 0 & 0 & 0 \\ & & & C_{44} & 0 & 0 \\ & sym & & & C_{55} & 0 \\ & & & & & C_{66} \end{bmatrix} \tag{10.31}$$

The stress strain relationship is shown as

$$\{\varepsilon\} = [C]\{\sigma\}. \tag{10.32}$$

Anisotropic elastic material constants can be associated to the compliance matrix [C], which is the inverse of the stiffness matrix, after standardized tensile tests.

$$[C_{ij}] = \begin{bmatrix} \dfrac{1}{E_1} & \dfrac{-v_{12}}{E_1} & \dfrac{-v_{13}}{E_1} & 0 & 0 & 0 \\[2mm] \dfrac{-v_{21}}{E_2} & \dfrac{1}{E_2} & \dfrac{-v_{23}}{E_2} & 0 & 0 & 0 \\[2mm] \dfrac{-v_{31}}{E_3} & \dfrac{-v_{32}}{E_3} & \dfrac{1}{E_3} & 0 & 0 & 0 \\[2mm] 0 & 0 & 0 & \dfrac{1}{G_{13}} & 0 & 0 \\[2mm] 0 & 0 & 0 & 0 & \dfrac{1}{G_{23}} & 0 \\[2mm] 0 & 0 & 0 & 0 & 0 & \dfrac{1}{G_{12}} \end{bmatrix} \tag{10.33}$$

By this we get $\{\varepsilon\} = [Q]\{\sigma\}$. The stiffness matrix can now be defined as

$$[Q] = [C]^{-1}. \tag{10.34}$$

$$[Q_{ij}] = \begin{bmatrix} Q_{11} & Q_{12} & Q_{13} & 0 & 0 & 0 \\ Q_{21} & Q_{22} & Q_{23} & 0 & 0 & 0 \\ Q_{31} & Q_{32} & Q_{33} & 0 & 0 & 0 \\ 0 & 0 & 0 & Q_{44} & 0 & 0 \\ 0 & 0 & 0 & 0 & Q_{55} & 0 \\ 0 & 0 & 0 & 0 & 0 & Q_{66} \end{bmatrix} \tag{10.35}$$

The matrix contains constants that are

$$Q_{11} = E_1\left(1-v_{23}v_{32}\right)/\Delta$$
$$Q_{22} = E_2\left(1-v_{13}v_{31}\right)/\Delta$$
$$Q_{33} = E_3\left(1-v_{12}v_{21}\right)/\Delta$$
$$Q_{12} = \left(v_{21}+v_{31}v_{23}\right)E_1/\Delta = \left(v_{12}+v_{13}v_{32}\right)E_2/\Delta$$
$$Q_{13} = \left(v_{31}+v_{21}v_{32}\right)E_1/\Delta = \left(v_{13}+v_{12}v_{23}\right)E_3/\Delta \qquad (10.36)$$
$$Q_{23} = \left(v_{32}+v_{12}v_{31}\right)E_2/\Delta = \left(v_{23}+v_{21}v_{13}\right)E_3/\Delta$$
$$Q_{44} = G_{12}$$
$$Q_{55} = G_{13}$$
$$Q_{66} = G_{23}$$

where

$$\Delta = 1-v_{12}v_{21}-v_{23}v_{32}-v_{31}v_{13}-2v_{21}v_{32}v_{13} \qquad (10.37)$$

For some reasons, the face sheet of a sandwich plate may be formed of a composite material consisting of various laminae during the design phase. Due to local coordinate systems, as shown in Figure 10.3., these laminae may not share the same direction, implying that the stress imposed on the major geometric direction may be different for each lamina.

In a general coordinate system, a transformation matrix is also used to add a stack of laminae with various orthotropic directions.

A summation of force components on cut free-body diagrams on a stress element in directions specified by rotation of coordinates through an angle in the plane of rotation can be used to construct the transformation matrix. $\{\sigma\} = [T]\{\sigma'\}$ can be

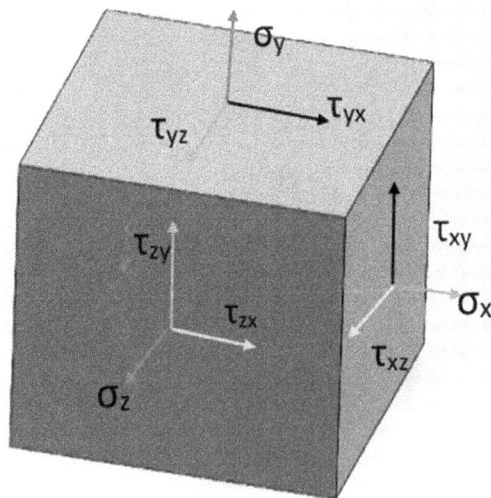

FIGURE 10.3   Stress coordinate system transformation.

used to indicate the transition of stress components from local lamina coordinates X1, X2 to rotating coordinates X, Y. The transformation matrices are as follows:

$$[T] = \begin{bmatrix} m^2 & n^2 & 0 & 0 & 0 & 2mn \\ n^2 & m^2 & 0 & 0 & 0 & -2mn \\ 0 & 0 & 1 & 0 & 0 & 0 \\ 0 & 0 & 0 & m & -n & 0 \\ 0 & 0 & 0 & n & m & 0 \\ -mn & mn & 0 & 0 & 0 & m^2 - n^2 \end{bmatrix}$$ (10.38)

where $m = \cos \theta$ and $n = \sin \theta$. Stress–strain relationship for a laminate in a coordinate system is given as

$$\begin{Bmatrix} \sigma_x \\ \sigma_y \\ \sigma_z \\ \sigma_{xz} \\ \sigma_{yz} \\ \sigma_{xy} \end{Bmatrix}_k = \begin{bmatrix} \bar{Q}_{11} & \bar{Q}_{12} & \bar{Q}_{13} & 0 & 0 & \bar{Q}_{16} \\ & \bar{Q}_{22} & \bar{Q}_{23} & 0 & 0 & \bar{Q}_{26} \\ & & \bar{Q}_{33} & 0 & 0 & \bar{Q}_{36} \\ & & & \bar{Q}_{44} & \bar{Q}_{45} & 0 \\ & & & & \bar{Q}_{55} & 0 \\ & & & & & \bar{Q}_{66} \end{bmatrix}_k \begin{Bmatrix} \varepsilon_x \\ \varepsilon_y \\ \varepsilon_z \\ 2\varepsilon_{xz} \\ 2\varepsilon_{yz} \\ 2\varepsilon_{xy} \end{Bmatrix}_k$$ (10.39)

K here is the kth lamina of the composite.

10.3.2   KINEMATIC DISPLACEMENT AND STRAINS FOR LAMINATE

A linear strain–displacement relationship is adopted in this study. The strain components are defined by derivatives of displacement components for modest strains. The strain tensor components are written in index notation as

$$\varepsilon_{ij} = \frac{1}{2}\left(u_{i,j} + u_{j,i}\right)$$ (10.40)

with *i* and *j* in ranges from 1 to 3. These denotes the direction of components x, y, and z, respectively. The bracket contains the differentiation of the strain component. Elaborating the normal and shear strain component would give

$$\varepsilon_x = \frac{\partial u}{\partial x}$$

$$\varepsilon_y = \frac{\partial v}{\partial y}$$

$$\varepsilon_z = \frac{\partial w}{\partial z}$$ (10.41)

$$\gamma_{xz} = 2\varepsilon_{xz} = \frac{\partial u}{\partial z} + \frac{\partial w}{\partial x}$$

$$\gamma_{yz} = 2\varepsilon_{yz} = \frac{\partial v}{\partial z} + \frac{\partial w}{\partial y}$$

$$\gamma_{xy} = 2\varepsilon_{xy} = \frac{\partial u}{\partial y} + \frac{\partial v}{\partial x}$$

where u, v, and w are the x, y, and z displacements, respectively. Out-of-plane transverse shear strain components γ_{yz} and γ_{xz} exist, while in-plane shear strain component γ_{xy} exists. The displacement through the thickness is considered to be linear in the first-order shear deformation theory, and there is only one element through the thickness. The displacement field only considers one translational and one rotational degree-of-freedom per direction. As a result, the displacement can be expressed as follows:

$$
\begin{aligned}
u(x,y,z) &= u_0(x,y) + z\alpha(x,y) \\
v(x,y,z) &= v_0(x,y) + z\beta(x,y) \\
w(x,y,z) &= w_0(x,y)
\end{aligned}
\tag{10.42}
$$

The rotating terms in classical theory are defined as the derivative of the out-of-plane displacement, $\alpha = -\dfrac{\partial w}{\partial x}$ and $\beta = -\dfrac{\partial w}{\partial y}$. This means that transverse shear deformation isn't taken into account. However, transverse shear deformation is not zero in the first shear deformation hypothesis. It is critical to simulate shear deformation for composites made from a stack of lamina, where certain layers with soft material are sandwiched between tougher layers, in order to accurately reflect the behavior under loading. In addition, as compared to classical theory, first-order shear deformation theory provides more precise stress estimates. The through-thickness normal strain component is considered to be zero using these first-order deformation assumptions. When the displacement field is substituted into the linear strain–displacement relationship, the result is

$$
\begin{aligned}
\varepsilon_x &= \frac{\partial u_0}{\partial x} + z\frac{\partial \alpha}{\partial x} \\
\varepsilon_y &= \frac{\partial v_0}{\partial y} + z\frac{\partial \beta}{\partial y} \\
\varepsilon_z &= 0 \\
\gamma_{xz} &= 2\varepsilon_{xz} = \alpha + \frac{\partial w}{\partial x} \\
\gamma_{yz} &= 2\varepsilon_{yz} = \beta + \frac{\partial w}{\partial y} \\
\gamma_{xy} &= 2\varepsilon_{xy} = \left(\frac{\partial u_0}{\partial y} + \frac{\partial v_0}{\partial x}\right) + z\left(\frac{\partial \alpha}{\partial y} + \frac{\partial \beta}{\partial x}\right)
\end{aligned}
\tag{10.43}
$$

These equations clarify the case of classical theory where transverse shear strain components γ_{yz} and γ_{xz} are zero. The middle surface strain is given by

$$
\varepsilon_{x0} = \frac{\partial u_0}{\partial x}, \quad \varepsilon_{y0} = \frac{\partial v_0}{\partial y}, \quad \varepsilon_{xy0} = \frac{1}{2}\left(\frac{\partial u_0}{\partial y} + \frac{\partial v_0}{\partial x}\right)
\tag{10.44}
$$

The curvatures here are

$$K_x = \frac{\partial \alpha}{\partial x}, \ K_y = \frac{\partial \beta}{\partial y}, \ K_{xy} = \frac{1}{2}\left(\frac{\partial \alpha}{\partial y} + \frac{\partial \beta}{\partial x}\right) \tag{10.45}$$

Using FSDT, the relations can be re written as

$$\begin{Bmatrix} \sigma_x \\ \sigma_y \\ \sigma_{xz} \\ \sigma_{yz} \\ \sigma_{xy} \end{Bmatrix}_k = \left[\bar{Q}\right]_k \begin{Bmatrix} \varepsilon_{x0} + zK_x \\ \varepsilon_{y0} + zK_y \\ \gamma_{xz} \\ \gamma_{yz} \\ 2\left(\varepsilon_{xy0} + zK_{xy}\right) \end{Bmatrix} \tag{10.46}$$

10.3.3 STIFFNESS MATRIX RELATING RESULTANTS FOR A COMPOSITE LAMINATE

Consider stacked laminas turned to composite structure with individual positioning with reference to middle surface as shown in Figure 10.4.

Defining the stress resultant $\{N\}$, stress couple $\{M\}$, and transverse shear resultants $\{Q\}$ per unit width as

$$\begin{Bmatrix} N_x \\ N_y \\ N_{xy} \end{Bmatrix} = \int_{-h/2}^{h/2} \begin{Bmatrix} \sigma_x \\ \sigma_y \\ \sigma_{xy} \end{Bmatrix} dz$$

$$\begin{Bmatrix} Q_x \\ Q_y \end{Bmatrix} = \int_{-h/2}^{h/2} \begin{Bmatrix} \sigma_{xz} \\ \sigma_{yz} \end{Bmatrix} dz \tag{10.47}$$

$$\begin{Bmatrix} M_x \\ M_y \\ M_z \end{Bmatrix} = \int_{-h/2}^{h/2} \begin{Bmatrix} \sigma_x \\ \sigma_y \\ \sigma_{xy} \end{Bmatrix} zdz$$

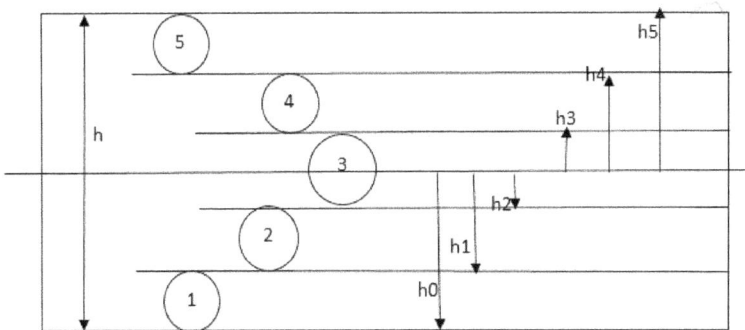

FIGURE 10.4 Composite plate with lamina positions measured from mid surface.

Every coordinate in the transverse direction below the mid surface is negative. Above the midpoint, every coordinate is positive. The stress resultants, stress couples, and transverse shear resultants for a sandwich plate can all be calculated. Each lamina was merged separately and then joined together. The stress–strain relationship can be inserted into the preceding equations. In the case of N laminae, the α, β, and $\bar{Q}$ not being the function of z will be coming out of the integrals, leading to

$$\begin{bmatrix} N_x \\ N_y \\ N_z \end{bmatrix} = \left(\sum_{k=1}^{N} [\bar{Q}] \int_{h_{k-1}}^{h_k} dz \right) \begin{bmatrix} \varepsilon_{x0} \\ \varepsilon_{y0} \\ 2\varepsilon_{xy0} \end{bmatrix} + \left(\sum_{k=1}^{N} [\bar{Q}] \int_{h_{k-1}}^{h_k} z\,dz \right) \begin{bmatrix} K_x \\ K_y \\ 2K_{xy} \end{bmatrix} \qquad (10.48)$$

$$\{N\} = [A]\{\varepsilon_0\} + [B]\{K\} \qquad (10.49)$$

After the integration along the thickness, the matrices A and B would be

$$A_{ij} = \sum_{k=1}^{N} [\bar{Q}_{ij}]_k (h_k - h_{k-1})$$

$$B_{ij} = \frac{1}{2} \sum_{k=1}^{N} [\bar{Q}_{ij}]_k (h_k^2 - h_{k-1}^2) \qquad (10.50)$$

For $i, j = 1, 2, 6$.
Similarly,

$$\{M\} = [B]\{\varepsilon_0\} + [D]\{K\} \qquad (10.51)$$

where

$$D_{ij} = \frac{1}{3} \sum_{k=1}^{N} [\bar{Q}_{ij}]_k (h_k^3 - h_{k-1}^3) \qquad (10.52)$$

The combined stress couple and stress resultant in matrices would be

$$\begin{Bmatrix} N_x \\ N_y \\ N_{xy} \\ M_x \\ M_y \\ M_{xy} \end{Bmatrix} = \begin{bmatrix} A_{11} & A_{12} & A_{16} & B_{11} & B_{12} & B_{16} \\ A_{12} & A_{22} & A_{26} & B_{12} & B_{22} & B_{26} \\ A_{16} & A_{26} & A_{66} & B_{16} & B_{26} & B_{66} \\ B_{11} & B_{12} & B_{16} & D_{11} & D_{12} & D_{16} \\ B_{12} & B_{22} & B_{26} & D_{12} & D_{22} & D_{26} \\ B_{16} & B_{26} & B_{66} & D_{16} & D_{26} & D_{66} \end{bmatrix} \begin{Bmatrix} \varepsilon_{x0} \\ \varepsilon_{y0} \\ \varepsilon_{xy0} \\ K_x \\ K_y \\ 2K_{xy} \end{Bmatrix} \qquad (10.53)$$

where [A] denotes the extensional stiffness matrix that connects in-plane stress resultants to in-plane strain components; [D] represents the flexural stiffness matrix, which connects stress couplings to curvature components; while [B] represents the bending and stretching matrix. The stiffness coefficients are affected by the material qualities as well as the thickness of the laminae.

In the case of transverse shear resultant in relation to shear strains,

$$\begin{bmatrix} Q_x \\ Q_y \end{bmatrix} = \sum_{k=1}^{N} \int_{h_{k-1}}^{h_k} \begin{bmatrix} \sigma_{xz} \\ \sigma_{yz} \end{bmatrix} dz = \sum_{k=1}^{N} \int_{h_{k-1}}^{h_k} \begin{bmatrix} \bar{Q}_{44} \\ \bar{Q}_{45} \end{bmatrix}_k \gamma_{xz} dz + \sum_{k=1}^{N} \int_{h_{k-1}}^{h_k} \begin{bmatrix} \bar{Q}_{45} \\ \bar{Q}_{55} \end{bmatrix}_k \gamma_{yz} dz \quad (10.54)$$

where $\gamma_{xz} = 2\varepsilon_{xz}$ and $\gamma_{yz} = 2\varepsilon_{yz}$. Integrating along the thickness would give

$$\begin{bmatrix} Q_x \\ Q_y \end{bmatrix} = \begin{bmatrix} A_{44} & A_{45} \\ A_{45} & A_{55} \end{bmatrix} \begin{bmatrix} \gamma_{xz} \\ \gamma_{yz} \end{bmatrix} \quad (10.55)$$

where A_{ij} for $i, j = 4, 5$

$$A_{ij} = \sum_{k=1}^{N} (\bar{Q}_{ij})_k [h_k - h_{k-1}] \quad (10.56)$$

The coupling matrix [B] is 0 when the lamina is symmetric about the mid surface. Material differences in two opposed laminae will cause [B] to be nonzero, resulting in a bending-stretching coupling effect, even if the lamina thickness is symmetric about the mid surface. As a result, from the mid surface, the lamina characteristics, fiber orientation, laminae thickness, and location must all be symmetric to allow [B] to become zero.

10.3.4   Stiffness Matrices for a Sandwich Plate

The honeycomb sandwich plate used in this study is a three-layer symmetric component (Figure 10.5), which means the origin of the transverse coordinate axis is at the thickness's midpoint. Above and below this mid-surface laminate are placed in a symmetrical pattern, each with the same thickness.

Detailed derivation of stiffness matrices [A], [B], [D] can be done using the above discussion.

The face sheets of the composite structure are assumed to be isotropic and the core is assumed orthotropic. Here $[\bar{Q}_{ij}] = [Q_{ij}]$, and $[A_{ij}]$ specializes with $N = 3$ and $h_3 = -h_0$ and $h_2 = -h_1$

$$A_{ij} = \sum_{k=1}^{3} (Q_{ij})_k (h_k - h_{k-1})$$

$$= (Q_{ij})_f \left[-\frac{t_c}{2} - \left(-\frac{t_c}{2} - t_f \right) \right] + (Q_{ij})_c \left[\frac{t_c}{2} - \left(-\frac{t_c}{2} \right) \right] + (Q_{ij})_f \left[\left(\frac{t_c}{2} + t_f \right) - \frac{t_c}{2} \right]$$

$$= (Q_{ij})_f (2t_f) + (Q_{ij})_c (t_c) \quad (10.57)$$

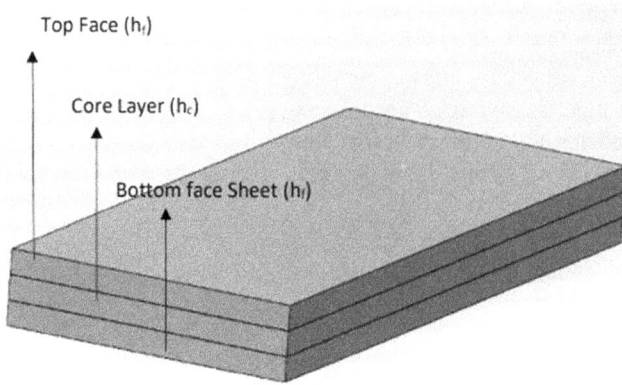

FIGURE 10.5   Sandwich plate along with its dimensions.

where

$$
\begin{aligned}
t_c &= h_2 - h_1 = 2h_2 \\
t_f &= h_3 - h_2 = h_1 - h_0
\end{aligned}
\tag{10.58}
$$

By the Generalized Hooke's Law,

$$
\begin{bmatrix} \varepsilon_{11} \\ \varepsilon_{22} \\ 2\varepsilon_{12} \end{bmatrix} = \frac{1}{E}\begin{bmatrix} 1 & -v & 0 \\ -v & 1 & 0 \\ 0 & 0 & 2(1+v) \end{bmatrix}
\tag{10.59}
$$

and the inverse is

$$
\begin{bmatrix} \sigma_{11} \\ \sigma_{22} \\ \sigma_{12} \end{bmatrix} = \frac{E}{1-v^2}\begin{bmatrix} 1 & v & 0 \\ v & 1 & 0 \\ 0 & 0 & \dfrac{1-v}{2} \end{bmatrix}\begin{bmatrix} \varepsilon_{11} \\ \varepsilon_{22} \\ 2\varepsilon_{12} \end{bmatrix}
\tag{10.60}
$$

Face sheets with $i, j = 1, 2$

$$
\left(Q_{11}\right)_f = \left(Q_{22}\right)_f = \frac{E_f}{1-v_f{}^2}, \left(Q_{12}\right)_f = \frac{vE_f}{1-v_f{}^2}
\tag{10.61}
$$

Shear modulus,

$$
G_f = \frac{E_f}{2\left(1+v_f\right)}
\tag{10.62}
$$

In the case of orthotropic core, $E_c = E_1 = E_2$, $v_c = v_{12} = v_{21}$, $v_{13} = v_{23} \approx 0$, then

$$\left(Q_{11}\right)_c = \left(Q_{22}\right)_c = \frac{E_c}{1-v_c^2}, \left(Q_{12}\right)_c = \frac{v_c E_c}{1-v_c^2} \qquad (10.63)$$

Later,

$$A_{11} = A_{22} = \frac{E_f}{1-v_f^2}\left(2t_f\right) + \frac{E_c}{1-v_c^2}t_c$$

$$A_{12} = \frac{v_f E_f}{1-v_f^2}\left(2t_f\right) + \frac{v_c E_c}{1-v_c^2}t_c \qquad (10.64)$$

$$A_{66} = G_f\left(2t_f\right) + G_{12}\left(t_c\right)$$

The in-plane effective membrane stiffness coefficients $A_{11} = A_{22}$ and A_{12} are determined by the material properties and thickness of the face sheets, as well as the effective characteristics and thickness of the honeycomb core. Honeycomb has an effective in-plane shear modulus of G_{12}, and the composite sandwich plate has an effective in-plane shear stiffness of $(Gt)_{12} \triangleq A_{66}$. The honeycomb sandwich plate's multiscale nature is clear from these equations. Using the simple honeycomb cellular model stated earlier, for regular honeycomb, $v_c = v_{12} = v_{21} = 1$, and cannot be employed, since the sandwich composite's effective stiffness equation will be singular. To overcome this difficulty, we use the refined honeycomb model including effects of transverse shear and axial effects, giving $v_{12} = v_{21} \neq 1$.

The bending stiffness coefficient,

$$D_{ij} = \frac{1}{3}\sum_{k=1}^{N}\rho_k\left(h^3_k - h_{k-1}^3\right)\frac{\partial^2\alpha}{\partial t^2}$$

$$= \frac{1}{3}\left\{2\left(Q_{ij}\right)_f\left[\frac{3}{2}t_c t_f^2 + \frac{3}{4}t_c^2 t_f + t_f^3\right] + \left(Q_{ij}\right)_c\frac{t_c^3}{4}\right\} \qquad (10.65)$$

For $i, j = 1, 2$

$$D_{11} = D_{22} = \frac{1}{3}\left\{\frac{2E_f}{1-v_f^2}\left[\frac{3}{2}t_c t_f^2 + \frac{3}{4}t_c^2 t_f + t_f^3\right] + \frac{E_c}{1-v_c^2}\frac{t_c^3}{4}\right\} \qquad (10.66)$$

$$D_{12} = \frac{1}{3}\left\{\frac{2E_f}{1-v_f^2}\left[\frac{3}{2}t_c t_f^2 + \frac{3}{4}t_c^2 t_f + t_f^3\right] + \frac{v_c E_c}{1-v_c^2}\frac{t_c^3}{4}\right\} \qquad (10.67)$$

$$D_{66} = \frac{1}{3}\left\{2G_f\left[\frac{3}{2}t_c t_f^2 + \frac{3}{4}t_c^2 t_f + t_f^3\right] + G_{12}\frac{t_c^3}{4}\right\} \qquad (10.68)$$

Because the second term involving core characteristics is relatively modest compared to the first term involving face shear materials, it is sometimes overlooked in research. However, terminology involving the basic qualities will be retained in this study.

A_{44} and A_{55} are found similarly:

$$A_{44} = G_f\left(2t_f\right) + G_{13}t_c \tag{10.69}$$

$$A_{55} = G_f\left(2t_f\right) + G_{23}t_c \tag{10.70}$$

For auxetic honeycomb,

$$G_{13} \neq G_{23} \text{ and } A_{44} \neq A_{55} \tag{10.71}$$

The first-order shear deformation theory is used to develop the governing dynamic equations for symmetric sandwich plates. The equations will be derived using stress resultants, stress couples, and stress-strain-displacement relations. The equations are derived using a continuum theory.

The governing equations are:

$$\frac{\partial \sigma_x}{\partial x} + \frac{\partial \sigma_{yx}}{\partial y} + \frac{\partial \sigma_{zx}}{\partial z} + b_x = \rho\frac{\partial^2 u}{\partial t^2}$$

$$\frac{\partial \sigma_{xy}}{\partial x} + \frac{\partial \sigma_y}{\partial y} + \frac{\partial \sigma_{zy}}{\partial z} + b_y = \rho\frac{\partial^2 v}{\partial t^2} \tag{10.72}$$

$$\frac{\partial \sigma_{xz}}{\partial x} + \frac{\partial \sigma_{yx}}{\partial y} + \frac{\partial \sigma_{zx}}{\partial z} + b_z = \rho\frac{\partial^2 w}{\partial t^2}$$

The first part of the equation is taken for ease and integrated along the thickness,

$$\sum_{k=1}^{N}\int_{h_{k-1}}^{h_k}\frac{\partial\left(\sigma_x\right)_k}{\partial x}\,dz + \sum_{k=1}^{N}\int_{h_{k-1}}^{h_k}\frac{\partial\left(\sigma_{yx}\right)_k}{\partial y}\,dz + \sum_{k=1}^{N}\int_{h_{k-1}}^{h_k}\frac{\partial\left(\sigma_{zx}\right)_k}{\partial z}\,dz = \sum_{k=1}^{N}\int_{h_{k-1}}^{h_k}\rho_k\frac{\partial^2 u}{\partial t^2}\,dz \tag{10.73}$$

When the integration and differentiation order is changed in the first two terms, it gives,

$$\frac{\partial}{\partial x}\left[\sum_{k=1}^{N}\int_{h_{k-1}}^{h_k}\left(\sigma_x\right)_k\,dz\right] + \frac{\partial}{\partial y}\left[\sum_{k=1}^{N}\int_{h_{k-1}}^{h_k}\left(\sigma_{yx}\right)_k\,dz\right] + \sum_{k=1}^{N}\left(\sigma_{zx}\right)_k\bigg|_{h_{k-1}}^{h_k} = \left[\sum_{k=1}^{N}\rho_k\left(h_k - h_{k-1}\right)\right]\frac{\partial^2 u_0}{\partial t^2} \tag{10.74}$$

After solving this, the resulting equation provides two shear tractions that are used in the top and the bottom surface. The equation is written as

$$\frac{\partial N_x}{\partial x} + \frac{\partial N_{yx}}{\partial y} + \tau_{top_x} - \tau_{bottom_x} = \rho_t\frac{\partial^2 u_0}{\partial t^2} \tag{10.75}$$

In the case of Y direction,

$$\frac{\partial N_{xy}}{\partial x} + \frac{\partial N_y}{\partial y} + \tau_{top_x} - \tau_{bottom_x} = \rho_t \frac{\partial^2 v_0}{\partial t^2} \tag{10.76}$$

In the case of Z direction,

$$\frac{\partial Q_x}{\partial x} + \frac{\partial Q_y}{\partial y} + p_{top_x} - p_{bottom_x} = \rho_t \frac{\partial^2 w}{\partial t^2} \tag{10.77}$$

Here p_{top} and p_{bottom} are the pressures in Z direction.

The total mass density per unit area for the symmetrical sandwich plate with honeycomb core and face sheets obtained earlier and repeated here is ρ_t in the above equations:

$$\rho_t = \sum_{k=1}^{N} \rho_k \left(h_k - h_{k-1} \right) = \left(2t_f \right) \rho_f + t_c \rho_c \tag{10.78}$$

Moment equations about the mid surface are given by

$$\sum_{k=1}^{N} \int_{h_{k-1}}^{h_k} \frac{\partial (\sigma_x)_k}{\partial x} z dz + \sum_{k=1}^{N} \int_{h_{k-1}}^{h_k} \frac{\partial (\sigma_{yx})_k}{\partial y} z dz + \sum_{k=1}^{N} \int_{h_{k-1}}^{h_k} \frac{\partial (\sigma_{zx})_k}{\partial z} z dz = \sum_{k=1}^{N} \int_{h_{k-1}}^{h_k} \rho_k \frac{\partial^2 \alpha}{\partial t^2} z^2 dz$$

$$\tag{10.79}$$

After integrating the third term by parts,

$$\frac{h}{2} \left(\tau_{top_x} + \tau_{bottom_x} \right) - Q_x \tag{10.80}$$

Here Q_x is the transverse shear resultant and h is the overall plate thickness and the terms at the first are moment about the mid plane. Expressing these moments in terms of stress couples give

$$\frac{\partial M_x}{\partial x} + \frac{\partial M_{xy}}{\partial y} + \frac{h}{2} \left(\tau_{top_x} + \tau_{bottom_x} \right) - Q_x = \frac{1}{3} \sum_{k=1}^{N} \rho_k \left(h^3{}_k - h_{k-1}{}^3 \right) \frac{\partial^2 \alpha}{\partial t^2}$$

$$\frac{\partial M_{xy}}{\partial x} + \frac{\partial M_y}{\partial y} + \frac{h}{2} \left(\tau_{top_y} + \tau_{bottom_y} \right) - Q_y = \frac{1}{3} \sum_{k=1}^{N} \rho_k \left(h^3{}_k - h_{k-1}{}^3 \right) \frac{\partial^2 \beta}{\partial t^2} \tag{10.81}$$

where $I = \frac{1}{3} \sum_{k=1}^{N} \rho_k \left(h^3{}_k - h_{k-1}{}^3 \right)$

The system of equations of the sandwich plate can be stated in terms of five degrees-of-freedom u_0, v_0, α, β, w and using these five dynamic equations and the composite stiffness properties. As previously stated, there is no coupling between in-plane force resultants and curvature, and no coupling between moment resultants and in-plane stresses for the symmetric sandwich plate, i.e., [B] = 0. When the laminates are structured with symmetric layers, this happens. For regular and auxetic honeycomb core with special orthotropic effective properties, $E_c = E_1 = E_2$, $G_{13} = G_{23}$, $\nu_c = \nu_{12} = \nu_{21}$, $\nu_{13} = \nu_{23} \approx 0$ the stress components are

$$N_x = A_{11}\varepsilon_{x0} + A_{12}\varepsilon_{y0}$$
$$N_y = A_{12}\varepsilon_{x0} + A_{22}\varepsilon_{y0}$$
$$N_{xy} = 2A_{66}\varepsilon_{xy0}$$
$$M_x = D_{11}K_{x0} + D_{12}K_{y0}$$
$$M_x = D_{12}K_{x0} + D_{22}K_{y0}$$
$$M_{xy} = 2D_{66}K_{xy0}$$
$$Q_x = A_{55}\left(\alpha + \frac{\partial w}{\partial x}\right)$$
$$Q_y = A_{44}\left(\beta + \frac{\partial w}{\partial y}\right)$$

(10.82)

The governing equations of the transversely loaded sandwich plates after substitution into moment equations:

$$D_{11}\frac{\partial^2 \alpha}{\partial x^2} + D_{66}\frac{\partial^2 \alpha}{\partial y^2} + \left(D_{12}+D_{66}\right)\frac{\partial^2 \beta}{\partial x \partial y} - A_{44}\left(\alpha + \frac{\partial w}{\partial x}\right) = I\frac{\partial^2 \alpha}{\partial t^2}$$
$$\left(D_{12}+D_{66}\right)\frac{\partial^2 \alpha}{\partial x \partial y} + D_{66}\frac{\partial^2 \beta}{\partial x \partial y} - A_{55}\left(\beta + \frac{\partial w}{\partial y}\right) = I\frac{\partial^2 \beta}{\partial t^2}$$
$$A_{44}\left(\frac{\partial \alpha}{\partial x} + \frac{\partial^2 w}{\partial x^2}\right) + A_{55}\left(\frac{\partial \beta}{\partial y} + \frac{\partial^2 w}{\partial y^2}\right) + p(x,y) = \rho_t\frac{\partial^2 w}{\partial t^2}$$

(10.83)

I in the first to equations is the rotational inertia,

$$I = \frac{1}{3}\left\{\rho_f\left[\frac{3}{2}h_c^2 t_f + 3h_c t_f^2 + 2t_f^3\right] + \frac{1}{4}\rho_c h_c^3\right\}$$

(10.84)

In-plane extensions are represented by two equations derived from stress resultant equilibrium equations and in-plane stiffness relations.

$$A_{11}\frac{\partial^2 u_0}{\partial x^2} + A_{66}\frac{\partial^2 v_0}{\partial y^2} + \left(A_{12}+A_{66}\right)\frac{\partial^2 v_0}{\partial x \partial y} = \rho_t\frac{\partial^2 u_0}{\partial t^2}$$
$$\left(A_{12}+A_{66}\right)\frac{\partial^2 u_0}{\partial x \partial y} + A_{66}\frac{\partial^2 v_0}{\partial x^2} + A_{22}\frac{\partial^2 v_0}{\partial y^2} = \rho_t\frac{\partial^2 v_0}{\partial t^2}$$

(10.85)

10.3.5 SIMPLY SUPPORTED SANDWICH PLATE

The cylindrical problem can be expanded to a two-dimensional plate problem. For the sake of simplicity, the two equations for in-plane extension will be ignored. Here are the governing equations,

$$D_{11}\frac{\partial^2\alpha}{\partial x^2} + D_{66}\frac{\partial^2\alpha}{\partial y^2} + (D_{12}+D_{66})\frac{\partial^2\beta}{\partial x\partial y} - A_{44}\left(\alpha+\frac{\partial w}{\partial x}\right) = I\frac{\partial^2\alpha}{\partial t^2}$$

$$(D_{12}+D_{66})\frac{\partial^2\alpha}{\partial x\partial y} + D_{66}\frac{\partial^2\beta}{\partial x^2} + D_{22}\frac{\partial^2\beta}{\partial y^2} - A_{55}\left(\beta+\frac{\partial w}{\partial y}\right) = I\frac{\partial^2\beta}{\partial t^2} \qquad (10.86)$$

$$A_{44}\left(\frac{\partial\alpha}{\partial x}+\frac{\partial^2 w}{\partial x^2}\right) + A_{55}\left(\frac{\partial\beta}{\partial y}+\frac{\partial^2 w}{\partial y^2}\right) + p(x,y) = \rho_t\frac{\partial^2 w}{\partial t^2}$$

The boundary conditions At the edges of the simply supported are
At $x = 0, x = a$

$$w = 0, \text{ and } M_y = D_{12}\frac{\partial\alpha}{\partial x} + D_{22}\frac{\partial\beta}{\partial y} = 0 \qquad (10.87)$$

At $y = 0, y = b$

$$w = 0, \text{and } M_x = D_{11}\frac{\partial\alpha}{\partial x} + D_{12}\frac{\partial\beta}{\partial y} = 0 \qquad (10.88)$$

According to the boundary conditions, the displacement field can be shown as

$$w = \bar{w}e^{iwt} = \left(\sum_{m=1}^{k}\sum_{n=1}^{k}A_{mn}\sin\frac{m\pi x}{a}\sin\frac{n\pi y}{b}\right)e^{iwt} \qquad (10.89)$$

$$\alpha = \bar{\alpha}e^{iwt} = \left(\sum_{m=1}^{k}\sum_{n=1}^{k}B_{mn}\cos\frac{m\pi x}{a}\sin\frac{n\pi y}{b}\right)e^{iwt} \qquad (10.90)$$

$$\beta = \bar{\beta}e^{iwt} = \left(\sum_{m=1}^{k}\sum_{n=1}^{k}C_{mn}\sin\frac{m\pi x}{a}\sin\frac{n\pi y}{b}\right)e^{iwt} \qquad (10.91)$$

That affirms the boundary conditions. For further research, the governing equations can be put in matrix form once more. After substituting the degrees-of-freedom into the governing equations, the matrix form is given below.

$$
\begin{bmatrix}
-\left(D_{11}\dfrac{m^2\pi^2}{a^2}+D_{66}\dfrac{n^2\pi^2}{b^2}+A_{55}\right)+I\omega^2 & -\left(D_{12}+D_{66}\right)\dfrac{mn\pi^2}{ab} & -A_{55}\dfrac{m\pi}{a} \\[2ex]
-\left(D_{12}+D_{66}\right)\dfrac{mn\pi^2}{ab} & -\left(D_{66}\dfrac{m^2\pi^2}{a^2}+D_{22}\dfrac{n^2\pi^2}{b^2}+A_{44}\right)+I\omega^2 & -A_{44}\dfrac{n\pi}{b} \\[2ex]
-A_{55}\dfrac{m\pi}{a} & -A_{44}\dfrac{n\pi}{b} & -\left(A_{55}\dfrac{m^2\pi^2}{a^2}+A_{44}\dfrac{n^2\pi^2}{b^2}-\rho_t\omega^2\right)
\end{bmatrix}
$$

$$
\begin{bmatrix} B_{mn} \\[1ex] C_{mn} \\[1ex] A_{mn} \end{bmatrix}
=
\begin{bmatrix} 0 \\[1ex] 0 \\[1ex] \dfrac{4p}{mn\pi^2}\left(\left(-1\right)^n-1\right)\left(1-\left(-1\right)^m\right) \end{bmatrix}
\tag{10.92}
$$

This section describes the methodology used to conduct the numerical research in the current study:

1. Based on sandwich theory, the honeycomb core was initially equalized as an orthotropic layer. The orthotropic layer's stiffness is believed to be the same as the honeycomb core's. In sandwich theory, the basic assumption is that the core can resist transverse shear deformation and hence has in-plane stiffness. The hexagonal honeycomb core's equivalent orthotropic elastic constants are as follows:

$$
E_x = E_y = \frac{4}{\sqrt{3}}\left[\frac{t}{l}\right]^3 E; \quad G_{xy} = \frac{\sqrt{3}}{2}Y\left[\frac{t}{l}\right]^3 E
$$

$$
G_{xz} = \frac{Y}{\sqrt{3}}\left(\frac{t}{l}\right)G; \quad G_{yz} = \frac{\sqrt{3}}{2}Y\left(\frac{t}{l}\right)E
\tag{10.93}
$$

where E_x, E_y, and G_{xy}, G_{yz}, G_{xz} are the equivalent Young's modulus and shear modulus of core, respectively. E and G are Young's modulus and shear modulus of the core material, respectively. μ is Poisson's ratio, and γ is the corrected coefficient whose value is between 0.4 and 0.6. The honeycomb core modeled as an equivalent orthotropic plate approximation can reduce pre- and postprocessing time to predict vibration response, but the orthotropic approximation's overall global stiffness and distributed mass assumption may not be suitable to accurately predict static stress behavior in certain local places of the sandwich panel.

2. After that, the panel's mid surface is removed, and a geometric model is generated. The shape is then meshed with a four-node quadrilateral layered structural shell element to construct a finite element model (SHELL 181). Three layers make up the finite element model: two FRP rigid layers and

one honeycomb core layer. The elastic characteristics and thickness of each
layer were assigned using settings in the commercial finite element program
ANSYS, which was used to calculate the free vibration response of the
sandwich panel studied in this study.

3. Response for free vibration is obtained by solving the eigenvalue problem
using the following equation:

$$\left(k - \omega_k^2 M\right)\phi_k = 0 \qquad\qquad (10.94)$$

10.4   RESULTS AND DISCUSSION

This part investigates the modal responses of a flat rectangular sandwich plate with
a honeycomb core using ANSYS 18.1. Convergence and validation with existing
literature are investigated to verify the accuracy and dependability of the current
model. For the simply supported at all edges (SSSS) boundary condition, the effect
of face sheet thickness, core height, and cell size on free vibration responses is
determined. The present study makes use of a flat rectangular sandwich plate with
a honeycomb core. Unless otherwise stated, the following material properties, as
reported by Arunkumar et al. [14], are used in the analysis: E = 68 Gpa, $\mu = 0.3$, $\rho =$
2700 kg/m³.

10.4.1   STUDY OF CONVERGENCE AND VALIDATION OF NATURAL FREQUENCIES

The natural frequencies are estimated with increasing mesh sizes in this part to evaluate the convergence behavior of the model utilized in this study. The first step is to
model a flat sandwich rectangular plate with a honeycomb core and all edges simply
supported (SSSS). The plate's parameters are as follows: length = 4 m, width – 2 m,
face sheet thickness = 0.0003 m, core thickness = 0.0144 mm. Figure 10.6 depicts the
variations in natural frequencies for various modes of frequencies. With increasing
mesh sizes, the values of natural frequencies for all types of frequencies appear to be
converging smoothly. Based on this, a 30 × 30 mesh was chosen for further research.
The values of natural frequencies computed are compared with the results published
by Arunkumar et al. [14] and are shown in Table 10.1 to check the trustworthiness of
the methods used in this study. It can be shown that the current process yields result
that are extremely close to the reference. The contour plot for first mode shape and
second mode shape are shown in Figure 10.7 and Figure 10.8, respectively.

10.4.2   INFLUENCE OF FACE SHEET THICKNESS ON NATURAL FREQUENCIES

In the current case, the effect of changing the face sheet thickness on the natural
frequency of a flat rectangular sandwich plate with honeycomb core and all edges
simply supported is explored. In this study Carbon fiber has been used as the material
for face sheet. The value of Young's modulus, shear modulus, density, and Poisson's
ratio are 25 GPa, 4 GPa, 1,900 kg/m³, and 0.20, respectively. The plate's core thickness (hc = 0.0144m) is kept constant. The hexagonal cell is assumed to have a length

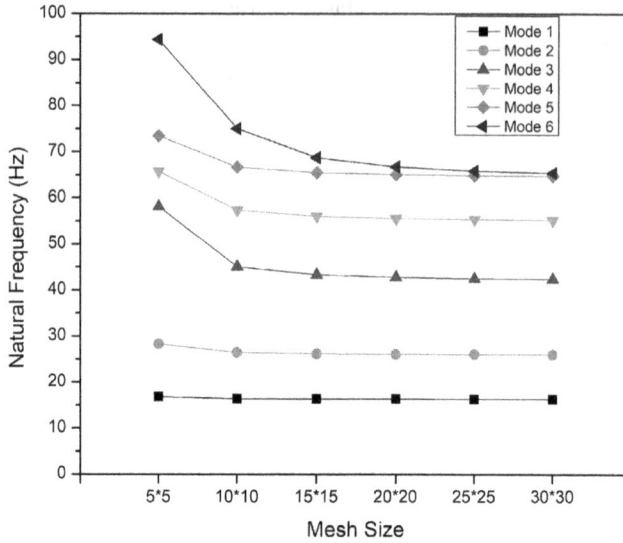

FIGURE 10.6 Convergence of natural frequency.

TABLE 10.1
Validation of Natural Frequency

Natural Frequency (HZ)	Mode 1	Mode 2	Mode 3	Mode 4	Mode 5	Mode 6
Arunkumar et al [11].	16.19	25.81	42.22	55.13	64.55	64.57
Present	16.275	26.021	42.356	55.243	64.789	65.413

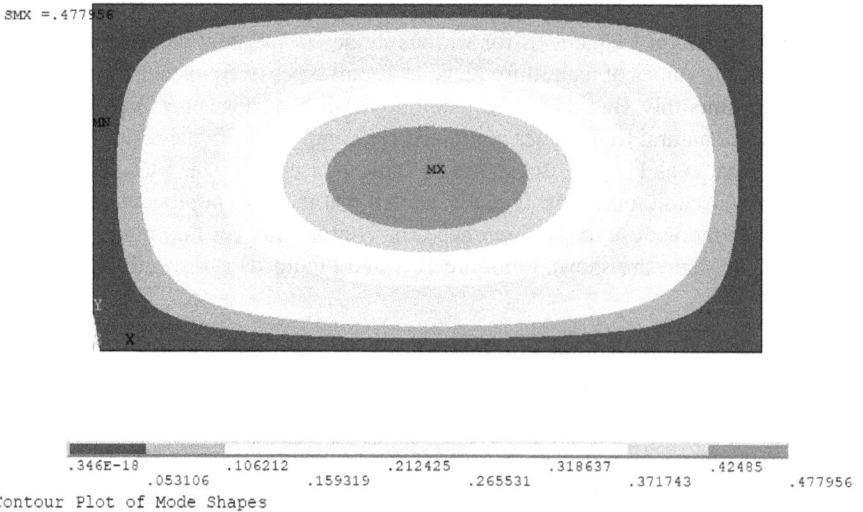

FIGURE 10.7 Contour plot of first mode shape.

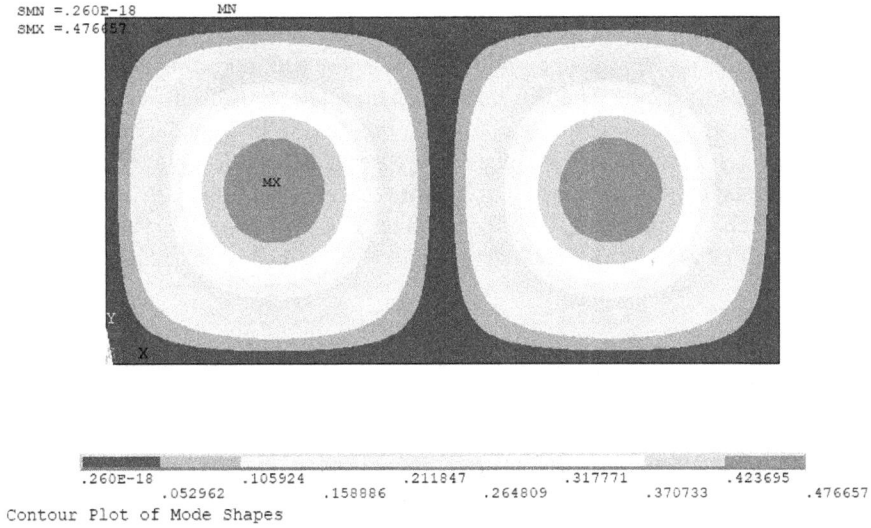

SMN =.260E-18   MN
SMX =.476657

.260E-18   .105924   .211847   .317771   .423695
     .052962   .158886   .264809   .370733   .476657
Contour Plot of Mode Shapes

FIGURE 10.8   Contour plot of second mode shape.

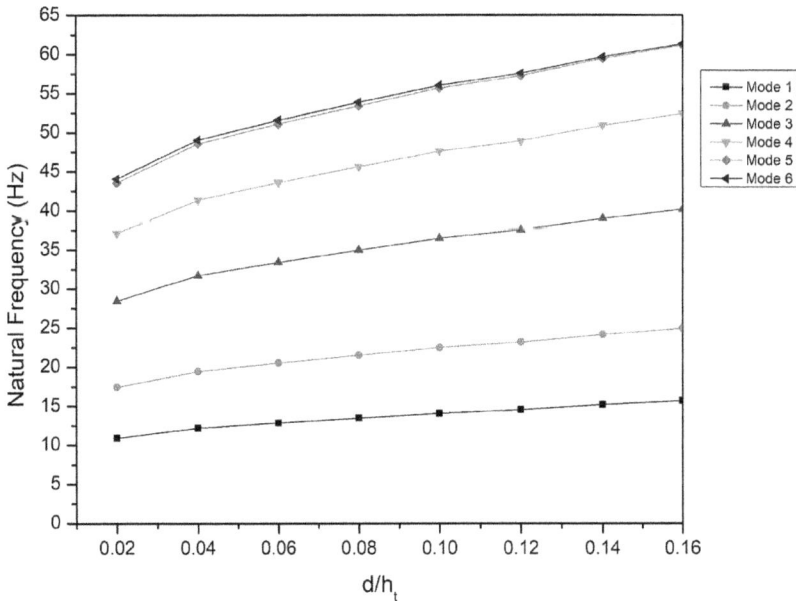

FIGURE 10.9   Influence of face sheet thickness on natural frequency.

of 0.004 m and a thickness of 0.00004 m. From 0 to 0.16, the ratio of face sheet thickness to total height (d/h_t) is changed. Figure 10.9 depicts the fluctuation of natural frequencies for all modes. It can be seen that as the d/h_t ratio increases, the values of natural frequencies increases as well.

10.4.3 Influence of Core Thickness on Natural Frequencies

The fluctuation in the value of natural frequencies with core thickness is investigated in this example. The dimensions of the plate and the material qualities are the same as in the previous example. In this SSSS, the boundary condition is employed. The thickness of the plate's face sheet is kept constant. The proportion of core height to total height (hc/ht) varies between 0.5 and 0.9. Figure 10.10 depicts the change in natural frequency values as a function of the hc/ht ratio. It can be seen that as the hc/ht ratio rises from 0.5 to 0.9, the natural frequency values rise as well.

10.4.4 Influence of Cell Thickness on Natural Frequencies

In this study, the effect of cell size on the values of natural frequencies is investigated. The plate dimensions and material properties are the same as in the preceding example. For this example, total thickness of the plate is 25 mm, height of core is 22 mm, thickness of the face sheet is 1.5 mm, and the cell thickness is varied between 2, 3, 4, 5, and 6 mm. First, the equivalent mechanical characteristics for various cell thicknesses are computed using Eq. (10.93). In this case, the boundary condition is SSSS. The values of natural frequencies vary with cell size, as seen in Figure 10.11. It can be shown that as the cell size increases, the value of natural frequencies decreases.

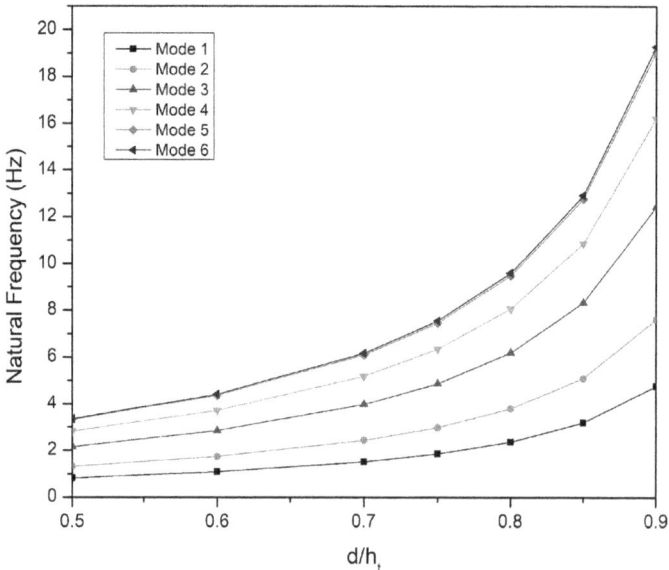

FIGURE 10.10   Influence of core thickness on natural frequency.

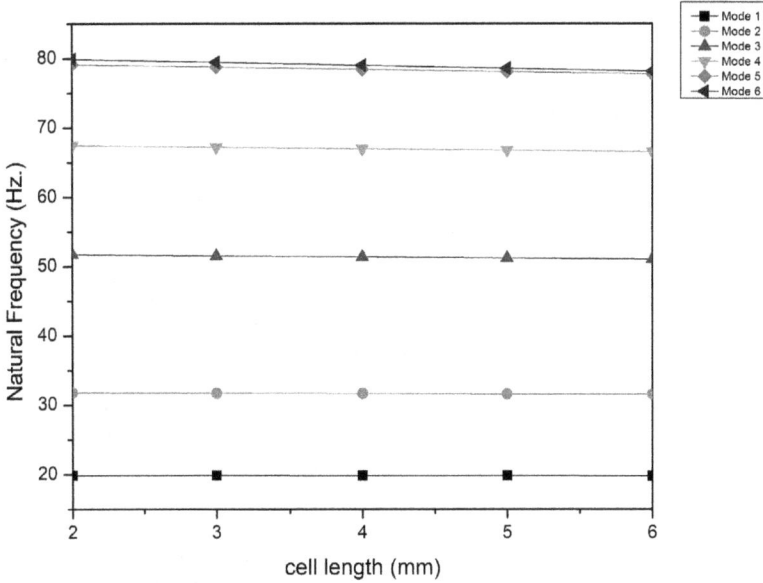

FIGURE 10.11 Influence of cell length on natural frequency.

10.5 CONCLUSIONS

The honeycomb core's unique mechanical characteristic is investigated. The CMT theory, a prominent method for determining effective qualities, is introduced. The density of the honeycomb core is discovered to be much lower than that of the construction material, resulting in significant weight savings in design. While the in-plane moduli are greatly reduced, the out-of-plane moduli remain rather high, demonstrating the honeycomb sandwich plate's high stiffness-to-weight ratio. The honeycomb core has an unusual mechanical property due to its consistent hexagonal shape. The effective Young's modulus in the X and Y directions is the same. Similar results have been obtained in case of Poisson's ratio. As aluminum has a low density and a low price compared to other metals, composite material makers always employ it as the building material for honeycomb sandwich plates. It is also simple to use. As a result, researching aluminum honeycomb sandwich plates is worthwhile.

Using APDL code, the modal responses of a sandwich flat plate with a honeycomb core are investigated in ANSYS 18.1. To ensure the correctness and reliability of the model used for the analysis, the results are first checked against published data. Then, assuming simply supported boundary conditions, the values of natural frequencies are derived for certain dimensions of face sheet thickness, core thickness, and cell thickness. When the ratio of face sheet thickness to total height (d/ht) is altered from 0 to 0.16, the magnitudes of natural frequencies for all modes increases as the d/ht ratio is increased. When the ratio of core thickness to total height (2h/ht) is changed, a similar pattern emerges. Furthermore, the values of natural frequencies are observed to remain constant as the cell size increases.

REFERENCES

[1] Scarpa, F., & Tomlinson, G. (2000). Theoretical characteristics of the vibration of sandwich plates with in-plane negative Poisson's ratio values. *Journal of Sound and Vibration*, 230(1), 45–67.

[2] Dobyns, A. L. (1981). Analysis of simply supported orthotropic plates subject to static and dynamic loads. *AIAA Journal*, 19(5), 642–650.

[3] Chandra, R., Singh, S. P., & Gupta, K. (1999). Damping studies in fiber-reinforced composites–a review. *Composite Structures*, 46(1), 41–51.

[4] Latheswary, S., Valsarajan, K. V., & Sadasiva Rao, Y. V. K. (2004). Dynamic response of moderately thick composite plates. *Journal of Sound and Vibration*, 270(1–2), 417–426.

[5] Meunier, M., & Shenoi, R. A. (2001). Dynamic analysis of composite sandwich plates with damping modelled using high-order shear deformation theory. *Composite Structures*, 54(2–3), 243–254.

[6] Bitzer, T. N. (1997). *Honeycomb technology: materials, design, manufacturing, applications and testing*. Springer Science & Business Media. Heidelberg, Germany.

[7] Göttner, W., & Reimerdes, H. G. (2006). Bending strength of sandwich panels with different cores after impact. In *Fracture of Nano and Engineering Materials and Structures (pp. 1261–1262)*. Springer, Dordrecht.

[8] Burton, W. S., & Noor, A. K. (1997). Assessment of continuum models for sandwich panel honeycomb cores. *Computer Methods in Applied Mechanics and Engineering*, 145(3–4), 341–360.

[9] Foo, C. C., Chai, G. B., & Seah, L. K. (2007). Mechanical properties of Nomex material and Nomex honeycomb structure. *Composite Structures*, 80(4), 588–594.

[10] Guo, J., Guan, Z. D., & Qiu, C. (2016). Experimental and numerical analysis of composite sandwich panel. In *MECHANICS AND MECHANICAL ENGINEERING: Proceedings of the 2015 International Conference (MME2015)* (pp. 718–725).

[11] Zhu, S., & Chai, G. B. (2013). Damage and failure mode maps of composite sandwich panel subjected to quasi-static indentation and low velocity impact. *Composite Structures*, 101, 204–214.

[12] Rao, M.K., & Desai, Y.M. (2004). Analytical solutions for vibrations of laminated and sandwich plates using mixed theory. *Composite Structures*, 63, 361–373.

[13] Zhou, G., Hill, M., Loughlan, J., & Hookham, N. (2006). Damage characteristics of composite honeycomb sandwich panels in bending under quasi-static loading. *Journal of Sandwich Structures and Materials*, 8(1), 55–90.

[14] Arunkumar, M.P., Jagadeesh, M., Pitchaimani, J., Gangadharan, K.V., Lenin Babu, M.C. (2016). Sound radiation and transmission loss characteristics of a honeycomb sandwich panel with composite facings: effect of inherent material damping. *Journal of Sound and Vibration*, 383, 221–232.

[15] Arunkumar, M.P., Pitchaimani, J., Gangadharan, K.V., Lenin Babu, M.C. (2016). Influence of nature of core on vibro acoustic behavior of sandwich aerospace structures. *Aerospace Science and Technology*, 56, 155–167.

[16] Sudhansu, S. Patro, Behera, Ranjan K., Sharma, Nitin, & Joshi, Kamal Kishore. (2021). *Chapter 9 influence of honeycomb core on static and vibration responses of sandwich structures*. Springer Science and Business Media LLC.

[17] Chen, D.H. (2011). Bending deformation of honeycomb consisting of regular hexagonal cells. *Composite Structures*, 93, 736–746.

11 Hygrothermoelastic Responses of Sinusoidally Corrugated Fiber-Reinforced Laminated Composite Structures

Abhilash Karakoti and Vishesh Ranjan Kar

National Institute of Technology Jamshedpur,
Jamshedpur, India

Balakrishnan Devarajan

Applied Materials,
Santa Clara, CA, USA

CONTENTS

DOI: 10.1201/9781003158813-11

11.1 INTRODUCTION

The significance of composite materials development is that the scientists and engineers are able to put strong fibers in exact orientation, in exact place with exact amount of volume fraction to get the desired properties [1]. A major driving force behind the development of composites has been to produce materials with improved specific mechanical properties over existing materials. Generally, fiber-reinforced laminates are piled upon each other to form laminated composite in order to attain assertive properties according to any specific application, and corrugation helps in attaining higher strength when compared to flat panels. Finite element analysis has evolved over time and proved to be an effective tool to analyze the behavior of corrugated laminated composite panels using various numerical techniques.

The effect of moisture and temperature on the deformation of the composite panel is gaining a lot of attention as the composites experiences different environmental conditions that significantly affect the mechanical properties. Upadhyay et al. [2] reported analytical solution where hygrothermomechanical loading is applied on laminated rectangular panel. Higher-order shear deformation theory (HSDT) is used in formulation and governing equation is solved by employing quadratic extrapolation technique and finite double Chebyshev series. Ebrahimi and Barati [3] discussed about the free vibration responses of functionally graded nanobeams under the influence of moisture and temperature by modeling the beam using different beam theories. Alzahrani et al. [4] employed sinusoidal shear deformation theory to obtain governing equation to analyze the bending behavior of the nanoplates when hygrothermomechanical load is applied. The moisture content and temperature field vary throughout the thickness of the plate. Yifeng and Yu [5] employed variational asymptotic method for analysis of composite panel under hygrothermal and mechanical load and compared with the exact solution, classical laminated plate theory (CLPT), and first-order shear deformation theory (FSDT) to show its cost effectiveness over higher-order theories. Kundu and Han [6] developed a mathematical model using total Lagrangian approach to study the bending responses of hygrothermoelastic curved deep and shallow shells, where Newton-Raphson method was employed for solving the equilibrium equation. Lal et al. [7,8] emphasis on the impact of the randomness in the system properties on the buckling and bending of laminated cylindrical and spherical shell, respectively, using HSDT and Von-Karman kinematics. Patel et al. [9] revealed the static and dynamic behavior of thick laminated composite panel under hygrothermal loading. While using an eight-noded C° quadrilateral element having 13 degrees-of-freedom HSDT was employed. Alsubari et al. [10] affirmed a 13-term higher-order shear deformation theory to reveal the bending behavior of cylindrical shell under hygrothermomechanical loading. Shen [11] discussed about the bending of laminated plates under uniform and sinusoidal hygrothermal loading by developing a micromechanical analytical model. Reddy's third-order shear deformation theory is employed to obtain governing equations and perturbation technique is used to obtain load deflection and moment. Tarn and Wang [12] presented exact analysis of torsion and bending behavior of laminated cylindrically anisotropic homogeneous tubes experiencing pressure and bending including the anisotropic elasticity. Shen [13] investigated the post-buckling of laminated plates under hygrothermal loading condition where buckling loads are resolved by perturbation approach and governing equations are derived using HSDT. Upadhyay and Lyons [14] evaluated the large deflection behavior of cylindrical laminates experiencing the hygrothermal loading using Von Karman plate theory. Natrajan et al. [15] investigated the effect of cutout in the center of the laminated plates under hygral and thermal environment on the vibration and buckling behavior using extended finite element method. Nejati et al. [16] emphasized on the functionally graded wavy CNTs reinforced in cylindrical shells under thermal loading. TSDT was employed to obtain governing equations and GDQ is used for parametric study. Lo et al. [17] employed the global-local higher-order theory to develop a four-noded element to analyze the static behavior of laminated plates due to moisture and temperature variation. Panda and Mahapatra [18] analyzed the nonlinear vibration behavior for shallow shell varying different geometric parameters using HSDT under uniform temperature loading. Shariyat [19] proposed a global-local shell theory for bending and buckling analysis of thick shells when thermomechanical load is applied.

It is clear from the above literature that most of the work that reported the bending behavior of laminated composites used FSDT kinematics. Also, only a limited volume of literature is available on bending behavior of corrugated laminated composites under hygrothermal environment using HSDT. The present study looks at a generalized mathematical model for sinusoidally corrugated laminated composite based on HSDT kinematics to investigate the bending behavior under hygrothermal environment. Here the sinusoidally corrugated profile is defined using a general curvature equation. The equilibrium equation is obtained using the total minimum potential energy principle. The bending responses under hygrothermal environment are computed using customized computer code based on the 2D FEM approximation via nine-noded quadrilateral isoparametric Lagrangian element (Q9). Further, the numerical experimentations are transacted for various geometrical and material parameters to exhibit the efficacy of the present model.

11.2   MATHEMATICAL FORMULATIONS

Consider a sinusoidally corrugated laminated composite panel with length a, width b, and wave height h_w ($h_w = C_f a_w$). The corrugated panel consists of N number of layers having uniform thickness h and n_w number of waves are shown in Figure 11.1. Sinusoidal corrugated profile is obtained by introducing the following function:

$$R(x) = \left\{ \frac{-a^2 \left[\dfrac{\psi h_w \cos\left(X^2\right)}{a^2} + 1 \right]^3}{2\psi \sin(X)} \right\} \tag{11.1}$$

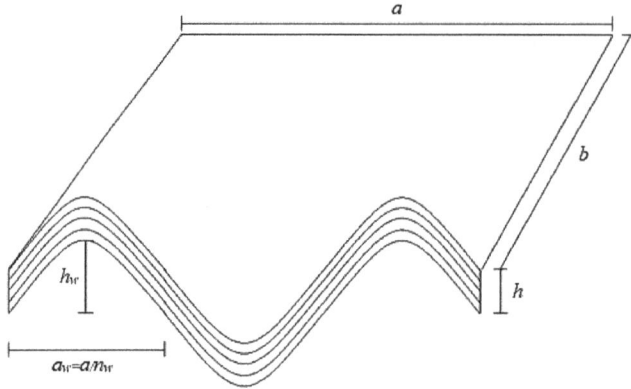

FIGURE 11.1 Sinusoidal corrugated laminated panel geometry.

where

$$X = \frac{\pi n_w x}{a}, \ \psi = \pi^2 h_w n_w^{\ 2} \text{ and } h_w = \frac{C_r a}{n_w}$$

11.2.1 Strain–Displacement Relations

HSDT mid-plane kinematics is employed for ease of calculation, as shear correction factor is not taken into consideration. Mid-plane displacement field is expressed in x, y, and z directions as follows:

$$\left. \begin{aligned} u(x,y,z) &= u_0(x,y) + z u_1(x,y) + z^2 u_2(x,y) + z^3 u_3(x,y) \\ v(x,y,z) &= v_0(x,y) + z v_1(x,y) + z^2 v_2(x,y) + z^3 v_3(x,y) \\ w(x,y,z) &= w_0(x,y) \end{aligned} \right\} \tag{11.2}$$

where (u, v, w) and (u_o, v_o, w_o) indicates the global and mid-plane displacement, respectively. u_1 and v_1 represents the rotations along the transverse normal with respect to y-axis and x-axis, respectively, and other higher-order terms are expanded with the help of Taylor series.

Strain–displacement relation used for modeling the deformation behavior of laminated sinusoidal corrugated panel is presented as

$$\{\varepsilon\} = \begin{Bmatrix} \varepsilon_{xx} \\ \varepsilon_{yy} \\ \gamma_{xy} \\ \gamma_{xz} \\ \gamma_{yz} \end{Bmatrix} = \begin{Bmatrix} \overline{u}_{,x} \\ \overline{v}_{,y} \\ \overline{u}_{,y} + \overline{v}_{,x} \\ \overline{u}_{,z} + \overline{w}_{,x} \\ \overline{v}_{,z} + \overline{w}_{y} \end{Bmatrix} \tag{11.3}$$

From Equations (11.2) and (11.3), the strain–displacement relation for the corrugated panel can be written as

$$\{\varepsilon\} = \begin{Bmatrix} \left[\varepsilon_1^0 + zk_1^1 + z^2k_1^2 + z^3k_1^3\right] \\ \left[\varepsilon_2^0 + zk_2^1 + z^2k_2^2 + z^3k_2^3\right] \\ \left[\varepsilon_6^0 + zk_6^1 + z^2k_6^2 + z^3k_6^3\right] \\ \left[\varepsilon_5^0 + zk_5^1 + z^2k_5^2 + z^3k_5^3\right] \\ \left[\varepsilon_4^0 + zk_4^1 + z^2k_4^2 + z^3k_4^3\right] \end{Bmatrix}$$

(11.4)

The above equation can be rewritten as

$$\{\varepsilon\} = [\mathrm{H}]\{\bar{\varepsilon}\}$$

(11.5)

where $\{\bar{\varepsilon}\} = \begin{Bmatrix} \varepsilon_1^0 & \varepsilon_2^0 & \varepsilon_6^0 & \varepsilon_5^0 & \varepsilon_4^0 & k_1^1 & k_2^1 & k_6^1 & k_5^1 & k_4^1 \\ k_1^2 & k_2^2 & k_6^2 & k_5^2 & k_4^2 & k_1^3 & k_2^3 & k_6^3 & k_5^3 & k_4^3 \end{Bmatrix}^T$ is the mid-plane strain vectors and [H] is the thickness coordinate matrix.

The displacement field here contains higher-order terms. But as the number of laminates increases, the number of unknowns also increases in the mathematical model. This results in increasing the computational cost. Here, Q9 element having nine degrees-of-freedom for every node is adopted for FEM analysis. While employing the FEM, the displacement for each node is asserted in terms of shape function $[N_i]$ as

$$\{\delta^*\} = \begin{bmatrix} u_0 & v_0 & w_0 & u_1 & v_1 & u_2 & v_2 & u_3 & v_3 \end{bmatrix}^T = \sum_{i=1}^{9} [N_i]\{\delta_i\}$$

(11.6)

where $\{\delta_i\} = \begin{bmatrix} u_{0_i} & v_{0_i} & w_{0_i} & u_{1_i} & v_{1_i} & u_{2_i} & v_{2_i} & u_{3_i} & v_{3_i} \end{bmatrix}^T$ represents the nodal displacement vector for i^{th} node, and the shape function is being referred from [20].

Now, the mid-plane strain vector can be written as

$$\{\varepsilon\}_i = [B_i]\{\delta_i\}$$

(11.7)

where $[B_i]$ denotes the strain displacement matrix.

In the present investigation, the bending behavior under uniform thermal and hygral loading is considered for k^{th} orthotropic corrugated lamina with any fiber orientation angle θ can be given by

$$\{\sigma_i\}^k = \left[\bar{Q}_{ij}\right]^k \{\varepsilon_i - \alpha_i \Delta T - \beta_i \Delta C\}^k$$

(11.8)

where $\{\sigma_i\}^k = \{\sigma_1 \ \sigma_2 \ \sigma_6 \ \sigma_5 \ \sigma_4\}^T$ and $\{\varepsilon_i\}^k = \{\varepsilon_1 \ \varepsilon_2 \ \varepsilon_6 \ \varepsilon_5 \ \varepsilon_4\}^T$ are the stress and strain vectors, respectively, for the kth layer, $\left[\bar{Q}_{ij}\right]^k$ is the transferred reduced stiffness matrix for the kth layer, $\{\alpha_i\}^k = \{\alpha_1 \ \alpha_2 \ 0\}^T$ signifies the thermal expansion or contraction coefficient vector, $\{\beta_i\}^k = \{0 \ \beta_2 \ 0\}^T$ is the moisture expansion or contraction coefficient vector, $\Delta T = T - T_0$, where T_0 and T represents the reference and elevated temperatures, respectively, and $\Delta C = C - C_0$, where C_0 and C denotes the reference and elevated moisture concentrations, respectively.

Now, Equation (11.8) can be expanded as

$$
\begin{Bmatrix} \sigma_1 \\ \sigma_2 \\ \sigma_6 \\ \sigma_5 \\ \sigma_4 \end{Bmatrix}^k = \begin{bmatrix} \bar{Q}_{11} & \bar{Q}_{12} & \bar{Q}_{16} & 0 & 0 \\ \bar{Q}_{21} & \bar{Q}_{22} & \bar{Q}_{26} & 0 & 0 \\ \bar{Q}_{16} & \bar{Q}_{26} & \bar{Q}_{66} & 0 & 0 \\ 0 & 0 & 0 & \bar{Q}_{55} & \bar{Q}_{54} \\ 0 & 0 & 0 & \bar{Q}_{45} & \bar{Q}_{44} \end{bmatrix}^k \left(\begin{Bmatrix} \varepsilon_1 \\ \varepsilon_2 \\ \varepsilon_6 \\ \varepsilon_5 \\ \varepsilon_4 \end{Bmatrix}^k - \begin{Bmatrix} \alpha_1 \\ \alpha_2 \\ 2\alpha_{12} \\ 0 \\ 0 \end{Bmatrix}^k \Delta T - \begin{Bmatrix} \beta_1 \\ \beta_2 \\ 2\beta_{12} \\ 0 \\ 0 \end{Bmatrix}^k \Delta C + \begin{Bmatrix} \varepsilon_1 \\ \varepsilon_2 \\ \varepsilon_6 \\ \varepsilon_5 \\ \varepsilon_4 \end{Bmatrix}^k \right)
$$

$$(11.9)$$

In order to obtain the bending behavior of the laminated structure, moisture- and temperature-dependent material properties are taken into consideration.

The strain energy of the panel can be expressed as

$$ U = \frac{1}{2} \int_v \{\varepsilon\}_i^T \{\sigma_i\} dV \tag{11.10} $$

Using the expression of strain vectors and resultant stress from Equations (11.5) and (11.8) and substituting them into Equation (11.10), the strain energy can be expressed as

$$
\begin{aligned}
U &= \frac{1}{2} \int_V \{\varepsilon\}_i^T \left[\bar{Q}\right]\{\varepsilon\}_i \, dV \\
&= \frac{1}{2} \iiint \{\varepsilon\}^T \left[\bar{Q}\right]\{\varepsilon\} \, dx\,dy\,dz \\
&= \frac{1}{2} \int_A \left(\{\varepsilon\}_i^T [\mathrm{D}]\{\varepsilon\}_i \right) dA
\end{aligned} \tag{11.11}
$$

where $[\mathrm{D}] = \sum_{k=1}^{N} \int_{z_{k-1}}^{z_k} [\mathrm{H}]^T [\bar{Q}][\mathrm{H}] \, dz$

By substituting Equation (11.7) into Equation (11.11), the strain energy expression can be rewritten as

$$ U = \frac{1}{2} \int_A \left(\{\delta^*\}_i^T [B]_i^T [D][B]_i \{\delta^*\}_i \right) dA - \{F_{\Delta T} + F_{\Delta C}\}_i \tag{11.12} $$

where $\{\varepsilon\}_i = [B]_i\{\delta^*\}_i$ and $\{F_{\Delta T} + F_{\Delta C}\}_i = [D]\Big[\{\varepsilon\}_i - \{\alpha\}\Delta T - \{\beta\}\Delta C\Big] = \int_A [B]_i^T$

$\{N_{\Delta T} + N_{\Delta C}\}dA.[B]$ is the product form of the differential operator and nodal shape function in the linear strain terms. $\{F_{\Delta T} + F_{\Delta C}\}$ is the hygrothermal load vector due to combined temperature and moisture change.

The work done due to the external applied distributed transverse load p can be expressed as

$$W = \int_A \{\delta\}^T p \, dA \qquad (11.13)$$

The nondimensionalized form of transverse load can be expressed as

$$\bar{p} = \frac{pa^4}{h^4 E_{22}} \qquad (11.14)$$

11.3   GOVERNING EQUATION AND SOLUTION SCHEME

The governing equation for the nonlinear bending analysis of laminated composite curved panel is obtained by minimizing the total energy expression.

$$\delta\Pi = 0 \qquad (11.15)$$

where $\Pi = (U\text{-}W)$

Using Equations (11.18) and (11.19) in Equation (11.21) and applying finite element approximation, the system governing expression can be elaborately expressed as

$$[K]\{\delta\} = \{q\} \qquad (11.16)$$

where $[K]$ is the global stiffness matrices. Equation (11.22) is solved by a direct iterative method, and the following solution steps are executed point-wise [30]:

1. Using FEM step, the elemental stiffness matrices and force vectors are calculated first.
2. The global stiffness and force vector are evaluated by assembling the elemental matrices.
3. By using the static equilibrium equation, bending responses are obtained under various loading condition.

11.4   RESULTS AND DISCUSSIONS

A finite element model is developed for the first time to analyze the deflection of sinusoidally corrugated laminated composite panel under hygrothermomechanical loading. To obtain the bending results, FEM code is developed in MATLAB R2016a environment based on HSDT. Based on the FE model, the bending behaviur of the corrugated panel is obtained for various mesh sizes under hygral and thermal loading. Initially the present model is checked for convergence and validated with the available literature.

The boundary conditions for the analysis are as follows:

a. All edges simply support (SSSS):
$v_o = w_o = v_1 = v_2 = v_3 = 0$ at x = 0, a and $u_o = w_o = u_1 = u_1 = u_3 = 0$ at y = 0, b.
b. All edges clamped (CCCC):
$u_o = v_o = w_o = u_1 = v_1 = u_2 = v_2 = u_3 = v_3 = 0$ for both x = 0, a and y = 0, b.

The central line deflections are nondimensionalised using the relation $W_{Cl} = \dfrac{W_{max}}{h}$.

11.4.1 CONVERGENCE AND VALIDATION

To exhibit the consistency and accuracy of the present corrugated model, the convergence and validation tests are executed in this section. For the convergence test, dimensionless central deflections of corrugated ($n_w = 1$) laminated (0°/90°/0°/90°) composite (a/b = 1, a/h = 50) panel are computed under thermomechanical loading and fully clamped support constraints at various mesh densities (3×3, 4×4, 5×5, 6×6, and 7×7). The material properties of graphite/epoxy are taken from Naidu and Sinha [21] for the computation purpose.

Figures 11.2 and 11.3 represent the dimensionless central deflection of corrugated laminated composite panel under mechanical and thermomechanical load, respectively, for different corrugation ratios (C_r). The obtained results show good convergence rate in both the cases with 6 × 6 mesh size, which is used to obtain the results thereafter.

The present model is also compared with the results of Zhang and Kim [22]. The bending responses of simply supported eight-layered (0°)₈ square composite panel

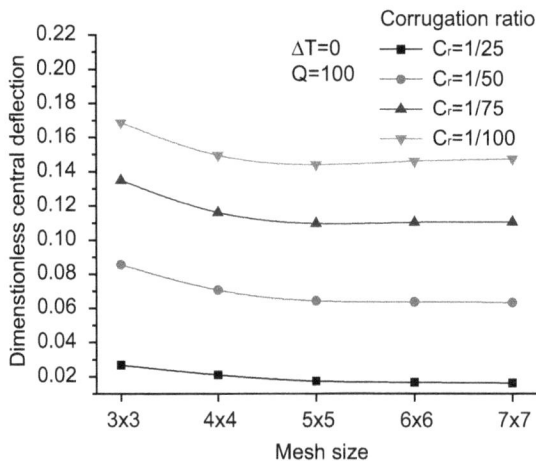

FIGURE 11.2 Central deflection under mechanical load with varying corrugation ratio (C_r).

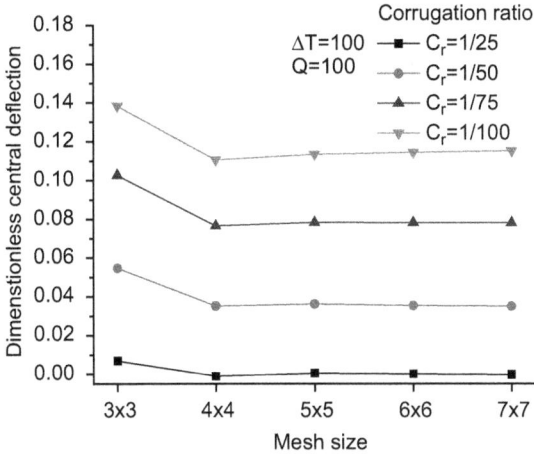

FIGURE 11.3   Central deflection under thermomechanical load while varying the corrugation ratio (C_r).

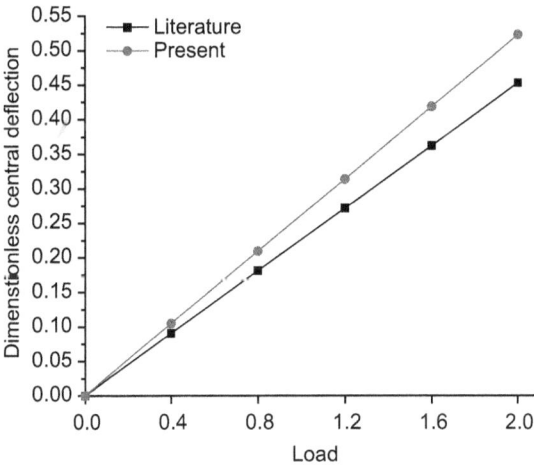

FIGURE 11.4   Comparison of present model with Zhang and Kim [22]).

are obtained under uniform load. Figure 11.4 shows equitable agreement with the literature results, but a nominal difference in the deflection is due to the main reason that HSDT is used to develop the present model with nine degrees-of-freedom; on the other hand, Zhang and Kim [22] used FSDT.

11.4.2   NUMERICAL EXPERIMENTATIONS

In this section, numerical experimentation is presented showing central line deflection of a sinusoidally corrugated laminated composite panel experiencing hygrothermomomechanical load. The effect of various geometric parameters and load conditions

TABLE 11.1

Elastic Properties for Graphite/Epoxy at Various Moisture Concentration (Naidu and Sinha [21])

Elastic Moduli (GPa)	Moisture Concentration H (%)				
	0	0.25	0.5	0.75	1
E_l	130	130	130	130	130
E_t	9.5	9.25	9	8.75	8.5
G_{lt}	6	6	6	6	6

TABLE 11.2

Elastic Properties for Graphite/Epoxy at Various Temperatures (Naidu and Sinha [20])

Elastic Moduli (GPa)	Temperature T (k)					
	300	325	350	375	400	425
E_l	130	130	130	130	130	130
E_t	9.5	8.5	8	7.5	7	6.75
G_{lt}	6	6	5.5	5	4.75	4.5

on the bending behavior of corrugated panel under uniform load is addressed. The material properties remain the same for the parametric studies taken from Naidu and Sinha [21] for all the cases (Tables 11.1, 11.2).

$$E_l = 130\,GPa, \ E_t = 9.5\,GPa, \ G_{lt} = G_{lt} = G_{tz} = 6\,GPa, \ \upsilon_{lt} = 0.3, \ \upsilon_{tl} = \frac{E_t}{E_l}\upsilon_{lt}$$

$\alpha_1 = -0.3 \times 10^{-6}/°\ C, \ \alpha_2 = 28.1 \times 10^{-6}/°\ C, \ \beta_2 = 0.44$

A four-layered $(0°/90°/0°/90°)$ composite panel with three sinusoidally corrugated waves is considered for all the numerical experimentations. The panel is square $(a = b = 1)$ shaped with corrugation ratio $C_r = 1/50$ and side-to-thickness ratio $a/h = 100$. Initially, the temperature (T) and moisture content %H are kept at 400K and 1%, respectively, and all the edges are fully clamped (CCCC).

11.4.2.1 Corrugated Laminated Composite Panel Subjected to Mechanical Load

Figure 11.5 shows the deformation behavior of the corrugated laminated composite panel under uniform pressure $(P_o = 0, 100, 200, 300, 400, 500)$. It is observed that the deflections along the center line $(x, b/2)$ are gradually increasing up to the mid-point $(a/2, b/2)$ and then degrading due to fully clamped supported case at all the panel

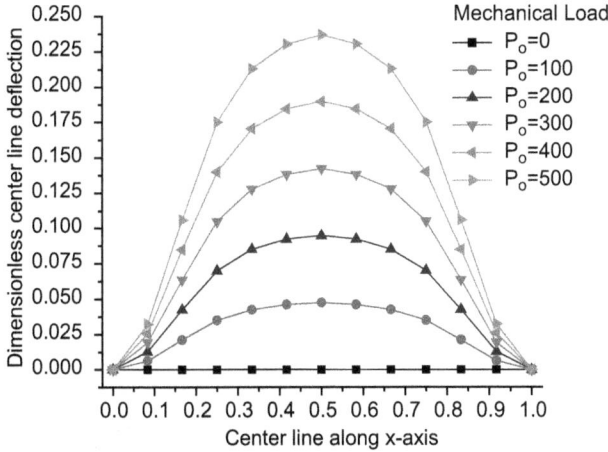

FIGURE 11.5 Deformation under mechanical load.

edges. In addition, the transverse deflections are enhancing with the increment in uniform pressure from $P_o = 0$ to $P_o = 500$.

11.4.2.2 Corrugated Laminated Composite Panel Subjected to Thermal Load

The center line deflection responses of the corrugated composite panel under uniform temperature rise ($\Delta T = 0, 25, 50, 75, 100, 125$) can be seen in Figure 11.6. Deflection along the centerline is increasing and then converging to x-axis in between $x = 0.5833$ and $x = 0.75$. Due to various temperature differences, maximum and

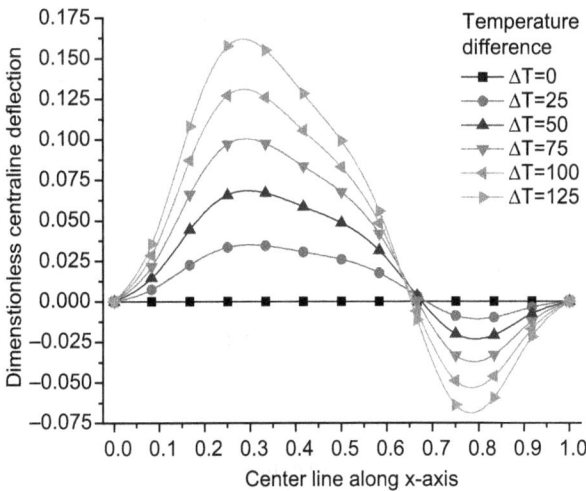

FIGURE 11.6 Deformation under thermal load.

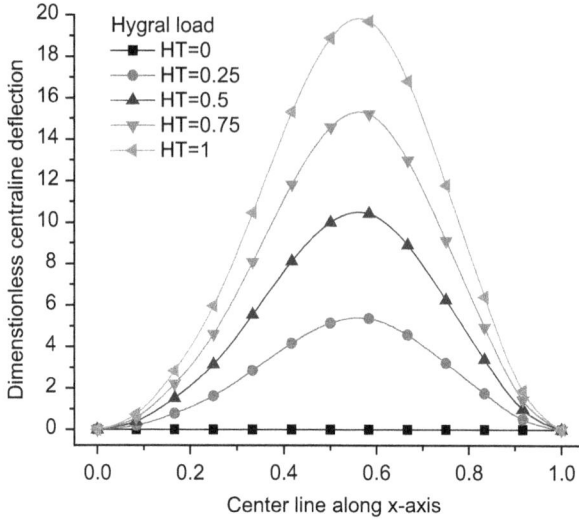

FIGURE 11.7   Deformation under hygral load.

minimum deflections are obtained from $x = 0.25$ to $x = 0.4166$ and $x = 0.75$ to $x = 0.83$, respectively. It may be because of the sinusoidal corrugation ($n_w = 3$).

11.4.2.3   Corrugated Laminated Composite Panel Subjected to Hygral Load

Figure 11.7 shows the deformation behavior of a corrugated panel under uniform hygral load (HT = 0, 0.25, 0.5, 0.75, 1). Deflection is increasing uniformly as the moisture content increases. Deflection increases along the centerline until the mid-point and then degrades. For different moisture content, maximum deflection is obtained in between $x = 0.5$ and $x = 0.58$. Deflection at the end points is zero, as all the edges of the corrugated panel are fully clamped.

11.4.2.4   Corrugated Laminated Composite Panel Subjected to Combined Load

The effects of aspect ratio on the centerline deflection of the sinusoidally corrugated panel subjected to hygrothermomechanical load are demonstrated in Figure 11.8. The maximum central deflection for square ($a/b = 1$) corrugated panel is observed at $x = 0.66$. However, for other panels ($a/b = 1.25, 1.5, 1.75, 2$), deflection parameters are non-monotonous along the centerline. Here maximum and minimum deflections are obtained at $x = 0.25$ and $x = 0.75$, respectively. It is because as aspect ratio increases, the panel will behave like a beam, which may be due to the combined loading.

The centerline deflection responses of corrugated panel on varying the corruga-tion ratio (Cr = 1/25, 1/50, 1/75, 1/100) are shown in Figure 11.9. Significant increase in deflection is noticed with decrease in corrugation ratio, and maximum deflection is obtained at $x = 0.50$ and $x = 0.58$. Lower corrugation ratio signifies lower wave height of the corrugated panel. As the wave height decreases, the panel becomes flat. Thus, corrugated panel with smaller wave height hw have lower stiffness.

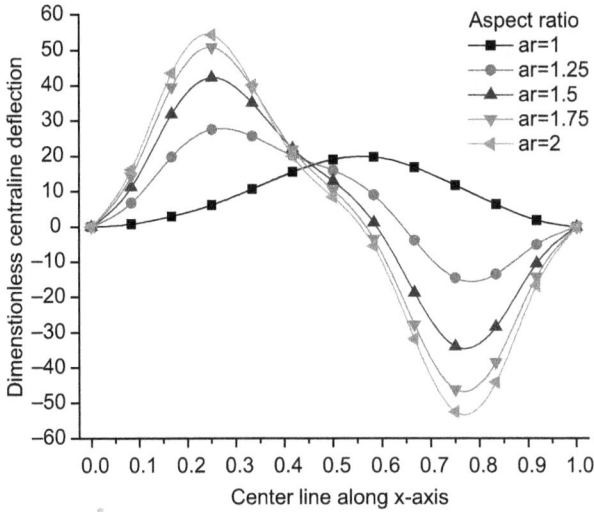

FIGURE 11.8   Centerline deflection with varying aspect ratio.

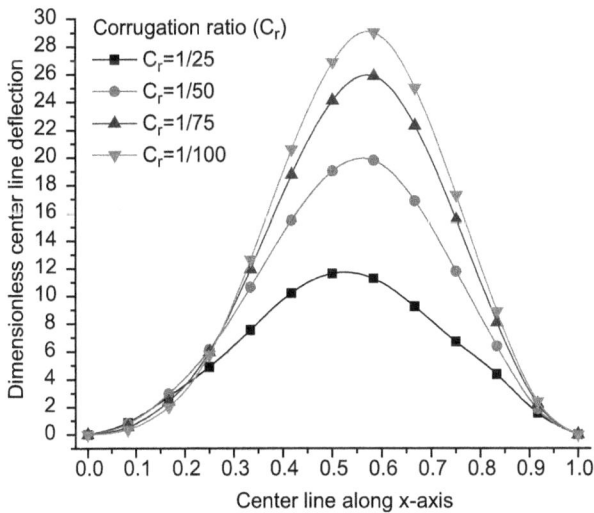

FIGURE 11.9   Centerline deflection with varying corrugation ratio.

The effect of various support conditions on the centerline deflection under combined loading is presented in Figure 11.10. It is noted that lowest and highest deflection is obtained for a fully clamped and simply supported panel. Similar trend in the deflection responses of hinged (HHHH) and clamped (CCCC) panel is obtained with maximum deflection obtained at the mid-point of the central line. But the wavy

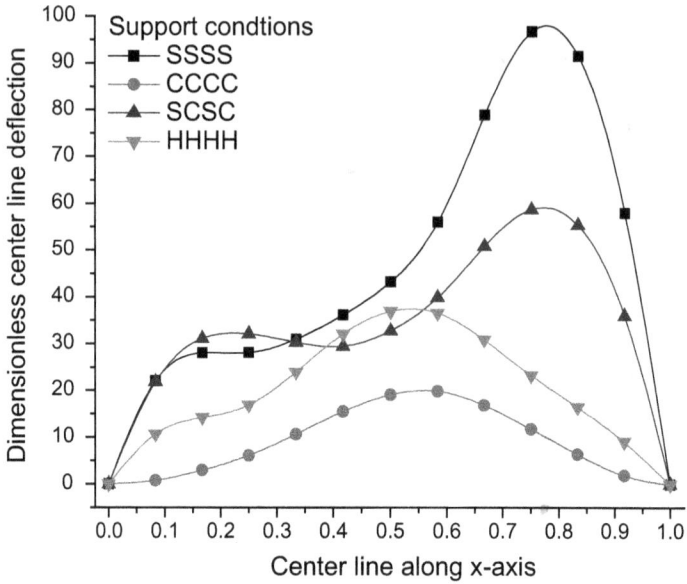

FIGURE 11.10   Centerline deflection under different support conditions.

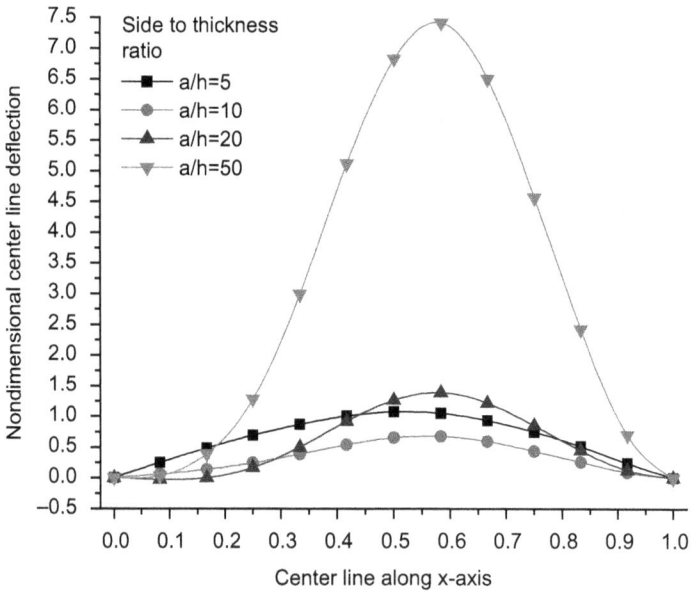

FIGURE 11.11   Centerline deflection with varying side-to-thickness ratio.

pattern can be seen for SCSC and SSSS support condition where maximum deflection is obtained at $x = 0.75$, which may be due to the dominant hygral and thermal load.

Figure 11.11 is showing the effect of thickness ratio for corrugated panel subjected to combined loading. It is observed that the bending responses are higher for thin panel as compared to the thick panels. The maximum deflection is obtained at x = 0.5 and x = 0.58. The peak deflection is shifted away from the mid-point because of the combined loading.

11.5   CONCLUSIONS

In this study, the flexural responses of laminated composite panels are computed under hygrothermal environment. Also, the geometrical configuration of the present composite is considered to be corrugated sinusoidally, which is modeled using the general curvature method. The kinematics of the present model is based on the higher-order shear deformation mid-plane theory with nine degrees-of-freedom. The weak formulation is achieved through the principle of minimum total potential energy, whereas the final form of equilibrium equations is formulated via 2D-isoparametric finite element method in conjunction with Lagrangian elements. The verification of the present results is carried out by comparing with the earlier reported works, which confirms the efficacy of the present model. Further, the numerical parametric examples by taking various sets of conditions including corrugation ratio, edge constraints, length-to-thickness ratio, and side-to-side ratio are presented, which exemplify the significance of these parameters on the deflection responses of the sinusoidally corrugated laminated composites.

ACKNOWLEDGMENT

The authors would like to thank Science and Engineering Research Board, Department of Science and Technology, Government of India (File No. ECR/2016/001829), for the financial support.

REFERENCES

[1] Hull D, Clyne WT. *An Introduction to Composite Material.* 2nd ed. Cambridge: Cambridge University Press, UK, 1996.

[2] Upadhyay AK, Pandey R, Shukla KK. Nonlinear flexural response of laminated composite plates under hygro-thermo-mechanical loading. *Commun Nonlinear Sci Numer Simul* 2010;15:2634–50. doi:10.1016/j.cnsns.2009.08.026.

[3] Ebrahimi F, Barati MR. A unified formulation for dynamic analysis of nonlocal heterogeneous nanobeams in hygro-thermal environment. *Appl Phys A Mater Sci Process* 2016;122:1–14. doi:10.1007/s00339-016-0322-2.

[4] Alzahrani EO, Zenkour AM, Sobhy M. Small scale effect on hygro-thermo-mechanical bending of nanoplates embedded in an elastic medium. *Compos Struct* 2013;105:163–72. doi:10.1016/j.compstruct.2013.04.045.

[5] Yifeng Z, Yu W. A variational asymptotic approach for hygrothermal analysis of composite laminates. *Compos Struct* 2011;93:3229–38. doi:10.1016/j.compstruct.2011.06.003.

[6] Kundu CK, Han JH. Nonlinear buckling analysis of hygrothermoelastic composite shell panels using finite element method. *Compos Part B Eng* 2009;40:313–28. doi:10.1016/j.compositesb.2008.12.001.

[7] Lal A, Singh BN, Kale S. Stochastic post buckling analysis of laminated composite cylindrical shell panel subjected to hygrothermomechanical loading. *Compos Struct* 2011;93:1187–200. doi:10.1016/j.compstruct.2010.11.005.

[8] Lal A, Singh BN, Anand S. Nonlinear bending response of laminated composite spherical shell panel with system randomness subjected to hygro-thermo-mechanical loading. *Int J Mech Sci* 2011;53:855–66. doi:10.1016/j.ijmecsci.2011.07.008.

[9] Patel BP, Ganapathi M, Makhecha DP. Hygrothermal effects on the structural behaviour of thick composite laminates using higher-order theory. *Compos Struct* 2002;56:25–34. doi:10.1016/S0263-8223(01)00182-9.

[10] Alsubari S, Mohamed Ali JS, Aminanda Y. Hygrothermoelastic analysis of anisotropic cylindrical shells. *Compos Struct* 2015;131:151–9. doi:10.1016/j.compstruct.2015.04.035.

[11] Laminated D. *Hygrothermal Effects on the Nonlinear Bending* 2002:493–6.

[12] Tarn JQ, Wang YM. Laminated composite tubes under extension, torsion, bending, shearing and pressuring: A state space approach. *Int J Solids Struct* 2001;38:9053–75. doi:10.1016/S0020-7683(01)00170-6.

[13] Shen HS. Hygrothermal effects on the postbuckling of shear deformable laminated plates. *Int J Mech Sci* 2001;43:1259–81. doi:10.1016/S0020-7403(00)00058-8.

[14] 2016 PC Upadhayay.pdf n.d.

[15] Natarajan S, Deogekar PS, Manickam G, Belouettar S. Hygrothermal effects on the free vibration and buckling of laminated composites with cutouts. *Compos Struct* 2014;108:848–55. doi:10.1016/j.compstruct.2013.10.009.

[16] Nejati M, Dimitri R, Tornabene F, Hossein Yas M. Thermal Buckling of Nanocomposite Stiffened Cylindrical Shells Reinforced by Functionally Graded Wavy Carbon Nanotubes with Temperature-Dependent Properties. *Appl Sci* 2017;7:1223. doi:10.3390/app7121223.

[17] Lo SH, Zhen W, Cheung YK, Wanji C. Hygrothermal effects on multilayered composite plates using a refined higher order theory. *Compos Struct* 2010;92:633–46. doi:10.1016/j.compstruct.2009.09.034.

[18] Panda SK, Mahapatra TR. Nonlinear finite element analysis of laminated composite spherical shell vibration under uniform thermal loading. *Meccanica* 2014;49:191–213. doi:10.1007/s11012-013-9785-9.

[19] Shariyat M. A general nonlinear global-local theory for bending and buckling analyses of imperfect cylindrical laminated and sandwich shells under thermomechanical loads. *Meccanica* 2012;47:301–19. doi:10.1007/s11012-011-9438-9.

[20] Cook RD, Malkus DS, Plesha M. *Concepts and Applications of Finite Element Analysis.* John Wiley Sons 1989:31–59. doi:10.1115/1.3264300.

[21] Naidu NVS, Sinha PK. Nonlinear finite element analysis of laminated composite shells in hygrothermal environments. *Compos Struct* 2005;69:387–95. doi:10.1016/j.compstruct.2004.07.019.

[22] Zhang YX, Kim KS. Geometrically nonlinear analysis of laminated composite plates by two new displacement-based quadrilateral plate elements. *Compos Struct* 2006;72:301–10. doi:10.1016/j.compstruct.2005.01.001.

12 Flexural Behavior of Shear Deformable FGM Composites with Corrugation
Higher-Order Finite Element Approximation

Abhilash Karakoti and Vishesh Ranjan Kar

National Institute of Technology Jamshedpur,
Jamshedpur, India

Hamid M. Sedighi

Shahid Chamran University of Ahvaz,
Ahvaz, Iran

CONTENTS

DOI: 10.1201/9781003158813-12

12.1 INTRODUCTION

Functionally graded materials (FGM) can resist high temperature and have high fracture toughness. Because of these unique properties FGM is preferred over conventional metals and composites in many industrial applications such as spacecraft, pressure vessels, aircraft, nuclear reactors, etc. [1]. FGM is a heterogenous material at microscopic level, whose properties are graded smoothly from one surface to another. Metal/alloy and ceramic material is one of the favorites used for gradation. The smooth gradation of the material also presents delamination, which is one of the major causes of failures in laminated composites.

Most of the studies are focused on the computation of deflection responses of flat or curved FG composite panels [2, 3]. However, the relative paucity of the literature in the field of corrugated FGM panel has drawn the interests of some researchers. Corrugated structures have enhanced mechanical properties perpendicular to the direction of the corrugation, which is possible without adding any external elements [4]. Only change in the geometry of the panel serves the purpose.

Numerous literature is available in which deflection responses of FGM panels having different geometries such as flat, cylindrical, singly and doubly curved, etc. are investigated by grading material properties smoothly. Kar and Panda (2016) investigated the effect of various parameters on the bending behavior of different functionally graded curved panel (elliptical, hyperbolic, and cylindrical) in combination with thermomechanical loading. Kheirikhah and Babaghasabha (2016) investigated bending and buckling behavior of corrugated sandwich plates using ANSYS under uniaxial load. Li et al. (2018) obtained the linear and nonlinear displacements by generalized differential quadrature method for two dimensional FG beam and examined the nonlinear bending behavior using the Euler–Bernoulli beam kinematic theory. Thai et al. (2017) used the isogeometric analysis to show the post-buckling behavior of FG microplates under thermal and mechanical loads using Reddy third-order shear deformation plate theory where material properties are considered as temperature dependent or temperature independent. Li et al. (2018) examined the bending behavior of simply supported FG sandwich plates with FG soft core under distributed load using Navier approach. Lv and Liu (2017) used minimum total potential energy principle to examine the impact of material uncertainties on nonlinear bending behavior of FG nanobeam. Houari et al. (2013) developed a new higher-order shear and normal deformation theories where they distributed the transverse displacement into bending, shear, and thickness stretching part, which doesn't require any shear correction factor and number of unknowns was reduced from six to four. Nejati et al. (2016) evaluated the static and free vibration responses of carbon nanotubes reinforced into FG conical shells under uniform loading, material properties were determined by four parameter power law, and generalized differential quadrature method (GDQ) was used for numerical solution. Viola et al. (2014) calculated tangential and normal stresses in FG conical shells and panels using unconstrained third-order shear deformation theory (UTSDT), and differential equations were solved by GDQ technique. Viola et al. (2016) used UTSDT to signify the effect of material parameters on the static response of FG spherical shells and panels under uniform loadings, and GDQ method was implemented to solve the governing equations.

Over the years researchers have mainly focused on the bending and vibration analysis of FGM panels. However, limited literature is available on geometrical consideration of corrugation. In this chapter, bending behavior of FGM panels having sinusoidal corrugation is investigated using higher-order shear deformation theory (HSDT) incorporating the corrugation effects. The material properties of the corrugated FGM panels are varied continuously according to the Voigt's model via power-law distribution. A general curvature equation is used to introduce the sinusoidal corrugation, and governing equations are obtained using minimum total potential energy principle. Nine-noded quadrilateral isoparametric Lagrangian element is employed for 2D approximation, and a computarized code developed in MATLAB environment is used to obtain the bending solutions of corrugated panels. Further, the bending analysis is carried out for different material and geometric parameters to show the adequacy of the present model.

12.2   MATHEMATICAL FORMULATION

In the present analysis, a metal-ceramic corrugated FG panel is considered. Here the FG panel with gradation is sinusoidally corrugated having one corrugation ($n_w = 2$) with uniform thickness h, length a, breadth b, and wave height h_w, which is shown in Figure 12.1. Sinusoidal form of the corrugated panel is introduced by employing general curvature equation [4]:

$$R_x = \frac{-a^2 \left(\dfrac{\gamma^2 n_w^4}{a^4} \cos\left(\dfrac{n_w \pi x}{a} \right)^2 + 1 \right)^3}{2\gamma n_w^3 \pi^2 \sin\left(\dfrac{n_w \pi x}{a} \right)} \tag{12.1}$$

where $\gamma = \dfrac{h_w a}{n_w}$ indicates the corrugation ratio.

12.2.1   EFFECTIVE MATERIAL PROPERTIES

Here, as we move from one surface to the next, properties are changing smoothly from top to bottom according to the power-law distribution (Shen, 2009), such that the top surface consists of metal and the bottom surface consists of ceramics.

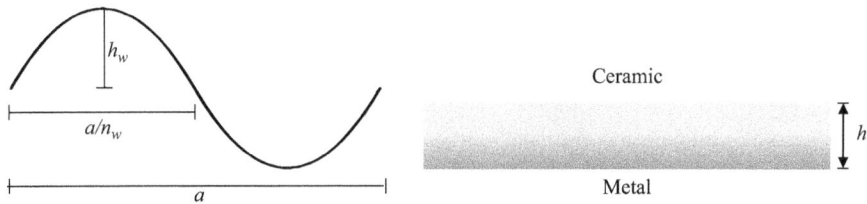

FIGURE 12.1   Parametric description of sinusoidally corrugated FGM panel.

By Voigt's model we have obtained the following expression that gives the effective material property of corrugated FG panel as follows (Gibson et al., 1995):

$$P = (P_m - P_c)V_{fm} + P_c \qquad (12.2)$$

where P_m and P_c are the metal and ceramic material properties, respectively, and V_{fm} is the volume fraction of the metal that is the top surface, which can be evaluated by power-law distribution (Shen, 2009) as follows:

$$V_{fm} = \left(\frac{z}{h} + \frac{1}{2}\right)^{\alpha} \qquad (12.3)$$

where power law index (α) must be within the limits ($0 \leq \alpha < \infty$) and helps in describing the material property along the thickness of the panel. We can vary the power law index with respect to top or bottom layer to get different material profiles.

Now, with the help of Equations (12.2) and (12.3), we can obtain effective material properties like Young's modulus (E) and Poisson's ratio (v):

$$E = (E_c - E_m)\left(\frac{z}{h} + \frac{1}{2}\right)^{\alpha} + E_m \qquad (12.4)$$

$$v = (v_c - v_m)\left(\frac{z}{h} + \frac{1}{2}\right)^{\alpha} + v_m \qquad (12.5)$$

where E_m and v_m represents the properties of metal and E_c and v_c represents the properties of ceramics.

12.2.2 DISPLACEMENT FIELD

HSDT mid-plane kinetics is used for the corrugated FG panel to derive the mathematical model [5]:

$$\left.\begin{aligned} u &= u_0 + zu_1 + z^2u_2 + z^3u_3 \\ v &= v_0 + zv_1 + z^2v_2 + z^3v_3 \\ w &= w_0 \end{aligned}\right\} \quad -\frac{h}{2} \leq z \leq \frac{h}{2} \qquad (12.6)$$

where z is the thickness coordinate varies from $-h/2$ to $+h/2$. u_0, v_0, and w_0 are the displacements in x, y, and z axes, respectively. u_1 and v_1 are the rotations about the y and x axes, respectively. Other higher-order terms u_2, u_3, v_2, and v_3 are obtained by the Taylor series expansion, defined in the mid plane.

12.2.3   STRAIN–DISPLACEMENT RELATIONS

The strain-displacement relation can be expressed as follows:

$$\{\varepsilon\} = \left\{ \begin{array}{c} \varepsilon_{xx} \\ \varepsilon_{yy} \\ \gamma_{xy} \\ \gamma_{xz} \\ \gamma_{yz} \end{array} \right\} = \left\{ \begin{array}{c} \left(\dfrac{\partial u}{\partial x} \right) \\ \left(\dfrac{\partial v}{\partial y} \right) \\ \left(\dfrac{\partial u}{\partial y} + \dfrac{\partial v}{\partial x} \right) \\ \left(\dfrac{\partial u}{\partial z} + \dfrac{\partial w}{\partial x} \right) \\ \left(\dfrac{\partial v}{\partial z} + \dfrac{\partial w}{\partial y} \right) \end{array} \right\} \tag{12.7}$$

Strain displacement relation of Equation (12.7) can also be expressed as

$$\{\varepsilon\} = \left[T \right] \{\varepsilon^{o}\} \tag{12.8}$$

where $\{\varepsilon o\} = \{\varepsilon_{xx}{}^{0}\,\varepsilon_{yy}{}^{0}\,\varepsilon_{xy}{}^{0}\,\varepsilon_{xz}{}^{0}\,\varepsilon_{yz}{}^{0}\,k_{xx}\,k_{yy}\,k_{xy}\,k_{xz}\,k_{yz}\,l_{xx}\,l_{yy}\,l_{xy}\,l_{xz}\,l_{yz}\,m_{xx}\,m_{yy}\,m_{xy}\,m_{xz}\,m_{yz}\}^{T}$ is the mid-plane strain vector.

12.2.4   CONSTITUTIVE RELATION

The constitutive relation for the FG panel is expressed as

$$\left\{ \begin{array}{c} \sigma_{xx} \\ \sigma_{yy} \\ \tau_{xy} \\ \tau_{xz} \\ \tau_{yz} \end{array} \right\} = \begin{bmatrix} \dfrac{1}{(1-v^{2})} & \dfrac{v}{(1-v^{2})} & 0 & 0 & 0 \\ \dfrac{v}{(1-v^{2})} & \dfrac{1}{(1-v^{2})} & 0 & 0 & 0 \\ 0 & 0 & \dfrac{1}{2(1+v)} & 0 & 0 \\ 0 & 0 & 0 & \dfrac{1}{2(1+v)} & 0 \\ 0 & 0 & 0 & 0 & \dfrac{1}{2(1+v)} \end{bmatrix} \left\{ \begin{array}{c} \varepsilon_{xx} \\ \varepsilon_{yy} \\ \gamma_{xy} \\ \gamma_{xz} \\ \gamma_{yz} \end{array} \right\} \tag{12.9}$$

Equation (12.9) can also be expressed as

$$\{\sigma\} = [Q]\{\varepsilon\} \tag{12.10}$$

where $[Q]$ is the stiffness matrix.

12.2.5 STRAIN ENERGY

The strain energy of the corrugated FG panel can be expressed as

$$U = \frac{1}{2}\int_v \{\varepsilon\}^T \{\sigma\} dV \tag{12.11}$$

now the above equations can be evaluated as

$$U = \frac{1}{2}\int \left(\{\varepsilon\}^T [D]\{\varepsilon\}\right) dA \tag{12.12}$$

where $[D] = \int_{-h/2}^{h/2} [T]^T [Q][T] dz$

12.2.6 WORK DONE

When we apply a uniform load of q, we get the relation for work done as

$$W = \int_A \{\delta\}^T \{q\} dA \tag{12.13}$$

12.3 FINITE ELEMENT FORMULATION

With nine degrees-of-freedom per node, the displacement vector for each element is expressed as

$$\{\delta\} = \sum_{i=1}^n N_i \{\delta_i\} \tag{12.14}$$

where $\{\delta_i\} = [u_{0i}\ v_{0i}\ w_{0i}\ u_{1i}\ v_{1i}\ u_{2i}\ v_{2i}\ u_{3i}\ v_{3i}]^T$ is the nodal displacement vector at node i. N_i is the shape function for the ith node [6].

The mid-plane strain vector having nodal displacement vector can be written as

$$\{\varepsilon\} = [B]\{\delta_i\} \tag{12.15}$$

where [B] is the product of differential operators and the shape functions in the linear strain terms.

The final governing equation of corrugated FG panel by using minimum total potential energy principle is given as

$$\delta \prod = \delta U - \delta W = 0 \tag{12.16}$$

where $\prod$ is the total potential energy and δ is the variational symbol.

Equilibrium equation for static analysis is obtained from Equations (12.12–12.16) as

$$\left[K \right]\{\delta\} = \{q\} \tag{12.17}$$

where $[K]$ is the global stiffness matrix and $\{\delta\}$ is the global displacement vector.

12.4   RESULTS AND DISCUSSION

To analyze the bending behavior of the corrugated FG pane,l a homemade code is developed in a MATLAB environment. The present model is validated with the available literature, and parametric studies are carried out to show the effect of corrugation.

The material properties of aluminium and zirconia are given in Table 12.1. The boundary conditions that are utilized throughout the analysis are shown in Table 12.2 where S, C, and F stand for simply supported, clamped, and free end, respectively. Uniform and sinusoidal load is taken into consideration to obtain the centreline deflection.

TABLE 12.1

Properties of the FGM Constituents

Materials	Young's Modulus E (GPa)	Poisson's Ratio (ν)	Density ρ (Kg/m³)
		Properties	
Aluminum (Al)	70	0.3	2707
Zirconia (ZrO₂)	151	0.3	3000

TABLE 12.2

Support Conditions

CCCC	$u_0 = v_0 = w_0 = u_1 = v_1 = u_2 = v_2 = u_3 = v_3 = 0$ at $x = 0$, a and $y = 0$, b
SSSS	$v_0 = w_0 = v_1 = v_2 = v_3 = 0$ at $x = 0$, a; $u_0 = w_0 = u_1 = u_2 = u_3 = 0$ at $y = 0$, b
SCSC	$v_0 = w_0 = v_1 = v_2 = v_3 = 0$ at $x = 0$, a
	$u_0 = v_0 = w_0 = u_1 = v_1 = u_2 = v_2 = u_3 = v_3 = 0$ at $y = 0$, b
SFSF	$v_0 = w_0 = v_1 = v_2 = v_3 = 0$ at $x = 0$, a

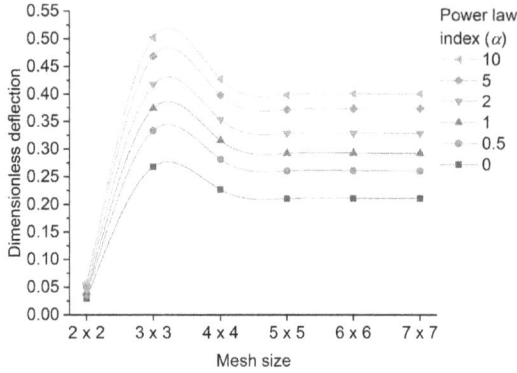

FIGURE 12.2   Deflection responses for corrugated FG panel under uniform load for different mesh size ($a = b = 1$, $n_w = 1$, CCCC, $\gamma = 1/50$, $a/h = 50$).

12.4.1   Convergence Behavior of Corrugated FG Panel

Deflection responses of a FG composite panel can be seen in Figure 12.2 for differnet mesh densities. The square-shaped FG panels are subjected to uniform loading, and all edges of the panels are fully clamped. It is noticed that the deflection responses are identical after a mesh size of 5×5 for all the cases considered. Hence, a 6×6 mesh would be suitable for further numerical computations.

12.4.2   Validation with FG Cylindrical Shell

Bending responses obtained from the present model are compared with the available literature to show its efficacy. Material properties are taken same as considered for convergence test. The results are obtained for the FG cylindrical shell and compared with the results of Zhao et al. [7] as shown in Table 12.3. The FG cylindrical panel is subjected to uniform loading with varying powerlaw index ($\alpha= 0, 0.2, 0.5, 1, 2, 5$) and support conditions. The published results are based on the FSDT having five

TABLE 12.3
Central Deflection of FG Cylindrical Panel Subjected to Uniform Loading with Varying Power-Law Index and Support Conditions

Boundary Condition	CCCC		CFCF	
power-Law Index (*n*)	Present	Ref- Zhao	Present	Ref- Zhao
0	0.0132	0.01347	0.0242	0.02778
0.2	0.0148	0.01516	0.0271	0.0313
0.5	0.0167	0.01711	0.0305	0.03535
1	0.0187	0.01915	0.0342	0.03956
2	0.0206	0.02102	0.0377	0.04333
5	0.0224	0.02289	0.0414	0.04703

degrees-of-freedom as compared to HSDT with nine degrees-of-freedom that it utilized for present results.

12.4.3 NUMERICAL EXPERIMENTATIONS

Validation and convergence studies for the present model revealed that implementation of HSDT produces efficient results to analyze the bending behavior. Hence, the proposed model is explored to solve numerous examples for various design parameters ($\lambda=a/h$, a_r, γ, n_w, support conditions) to illustrate their significance on the deflection responses. In general, simply supported FG corrugated panel is subjected to uniform load, and other design parameters are taken as $a = b = 1$, $\alpha = 0.5$, $\lambda = 50$, $n_w = 3$, and $\gamma = 1/50$ if not defined otherwise. The centreline deflection is nondimensionalised by using the formula:

$$w_{ND} = w_{centreline}\left(\frac{E_m h^3 \times 10}{q_0 a^4}\right) \qquad (12.18)$$

Tables 12.4 and 12.5 indicate the centreline dimensionless deflection of sinusoidal corrugated FG composite panel for different side-to-thickness ratios ($\lambda = 10, 20, 30, 40,$ and 50) under uniform and sinusoidal loads, respectively. It can be stated that the deflection responses are increasing as the ratio increases in both the cases of loading. But lower deflection responses are obtained for FG panel subjected to sinusodal loading.

TABLE 12.4
Centreline Deflection of Corrugated FG Panel Subjected to Uniform Load with Side-to-Thickness Ratio

	Uniform Load			
x/a	$\lambda = 10$	$\lambda = 20$	$\lambda = 30$	$\lambda = 40$
0	0	0	0	0
0.0833	0.7258	1.1813	1.4515	1.5923
0.1667	1.3708	2.2138	2.6808	2.8902
0.25	1.8817	3.0179	3.6041	3.8209
0.3333	2.2523	3.5801	4.2083	4.3785
0.4166	2.4892	3.9233	4.5526	4.6661
0.5	2.5747	4.045	4.6731	4.7667
0.5833	2.4917	3.9276	4.5583	4.6728
0.6667	2.2566	3.5875	4.2184	4.3905
0.75	1.8867	3.0266	3.616	3.8353
0.8333	1.3751	2.2214	2.6911	2.9029
0.9167	0.7282	1.1857	1.4576	1.5998
1	0	0	0	0

TABLE 12.5
Centreline Deflection of Corrugated FG Panel Subjected to Sinusoidal Load with Side-to-Thickness Ratio

x/a	$\lambda = 10$	$\lambda = 20$	$\lambda = 30$	$\lambda = 40$
	Sinusoidal Load			
0	0	0	0	0
0.0833	0.3396	0.5525	0.6755	0.7361
0.1667	0.6495	1.0476	1.2637	1.3558
0.25	0.9029	1.4448	1.7205	1.8176
0.3333	1.0912	1.7301	2.0298	2.1073
0.4166	1.212	1.9053	2.2082	2.2603
0.5	1.2525	1.9622	2.2634	2.305
0.5833	1.2045	1.8925	2.1908	2.2395
0.6667	1.0778	1.7074	1.9991	2.0704
0.75	0.8872	1.4179	1.6839	1.7735
0.8333	0.6352	1.0235	1.2312	1.3166
0.9167	0.3314	0.5385	0.6563	0.7127
1	0	0	0	0

Figures 12.3 and 12.4 demonstrate the centreline dimensionless deflection of sinusoidally corrugated FG panel with different aspect ratio ($a_r = 1, 1.5, 2, 2.5$) when subjected to the uniform and sinusoidal loads, respectively. Among all the considered cases, maximum deflection responses are obtained for a square panel ($a_r = 1$), and

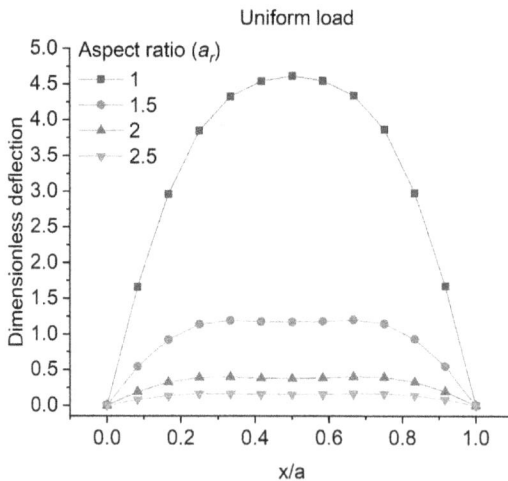

FIGURE 12.3 Centreline deflection of corrugated FG panel subjected to uniform load with varying aspect ratio.

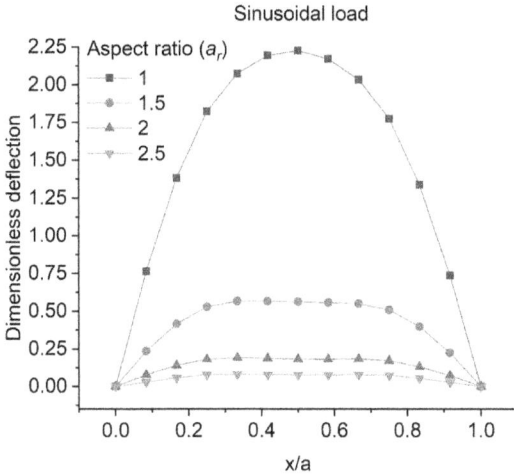

FIGURE 12.4 Centreline deflection of corrugated FG panel subjected to sinusoidal load with varying aspect ratio.

deflection responses are reducing as the aspect ratio is increasing. In general we know that a square panel is stiffer, but here a square panel is showing maximum deflection because of nondimensionalization of deflection responses of all the cases.

Figure 12.5 and 12.6 is demonstrating the nondimensionalised centerlinie deflection for sinusoidally corrugated FG panel with different corrugation ratios ($\gamma = 1/10$, 1/20, 1/40, 1/50) under uniform and sinusoidal load, respectively. In all the cases,

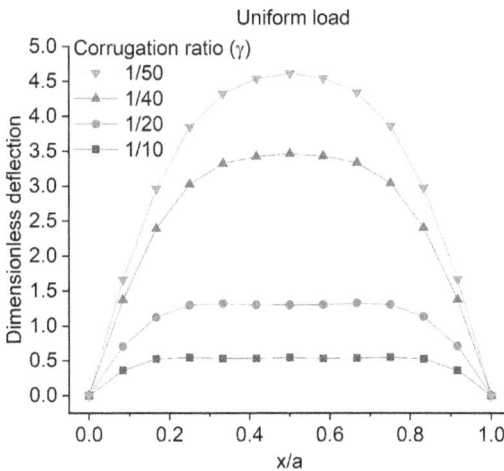

FIGURE 12.5 Centreline deflection of corrugated FG panel subjected to uniform load with varying corrugation ratio.

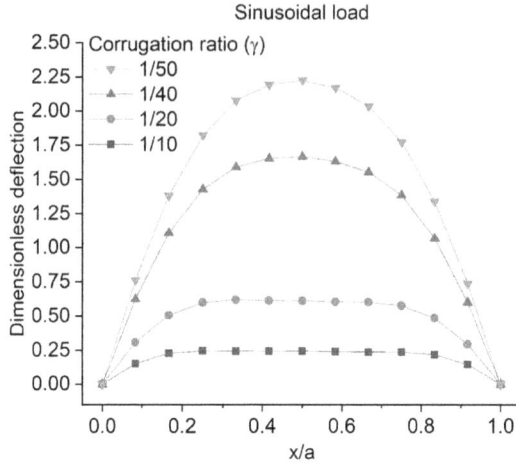

FIGURE 12.6 Centreline deflection of corrugated FG panel subjected to sinusoidal load with varying corrugation.

deflection responses are increamenting as the corrugation is decreasing. It is because wave height ($h_w = (\gamma \times a)/n_w$) is directly proportional to the corrugation ratio, which implies that panels with higher wave height will be stiffer.

Figure 12.7 and 12.8 shows the centreline deflection of sinusoidally corrugated FG panel under uniform and sinusoidal loading, respectively. It is obvious from the figures that the deflection responses increases as the power-law indix ($\alpha = 0$, 0.5, 1, 2, 5, 10, inf) is increasing. The result agrees with the fact that ceramic material is much stiffer compared to the conventional metals, and thus the corrugated FG panels

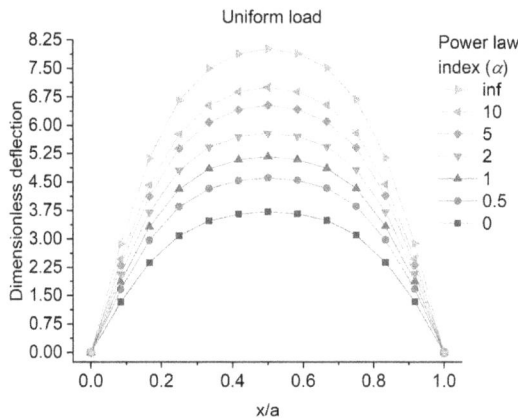

FIGURE 12.7 Centreline deflection of corrugated FG panel subjected to uniform load with varying power-law index.

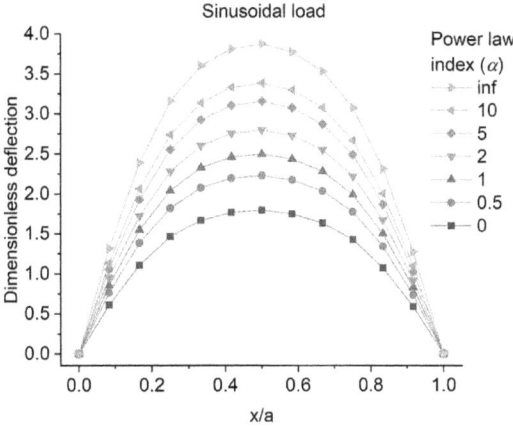

FIGURE 12.8   Centreline deflection of corrugated FG panel subjected to sinusoidal load with varying power law index.

with lower power-law indices are stiffer as they have ceramic-rich properties when compared with the corrugated panel with higher power-law indices having metal-rich properties.

Figures 12.9 and 12.10 demonstrate the centreline deflection of corrugated FG panel under uniform and sinusoidal load with increasing number of corrugations (1, 2, 3, 4), respectively. Here each corrugation consists of two sinusoidal waves (for one corrugation, $n_w = 2$). It is evident from both figures that the deflection responses

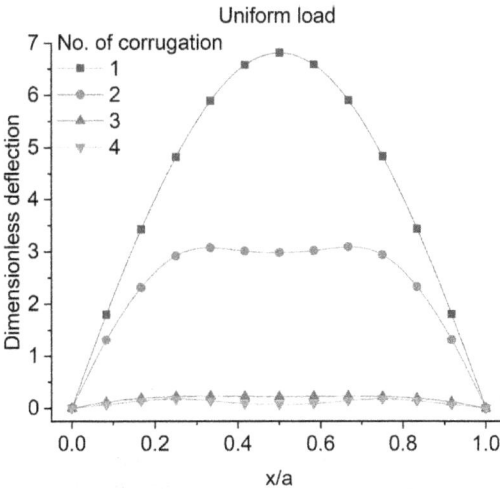

FIGURE 12.9   Centreline deflection of corrugated FG panel subjected to uniform load with increasing number of corrugation.

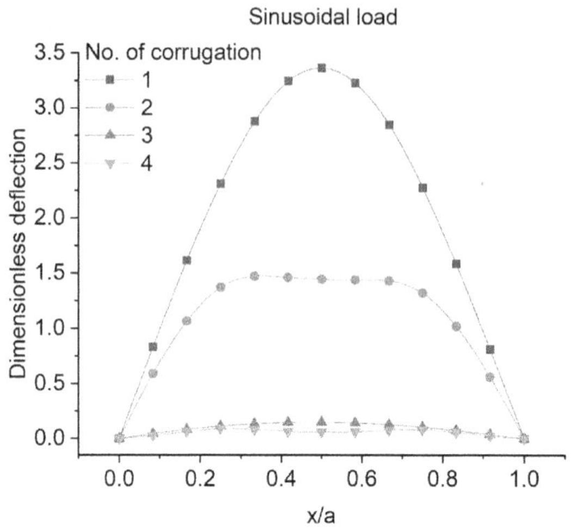

FIGURE 12.10 Centreline deflection of corrugated FG panel subjected to sinusoidal load with increasing number of corrugations.

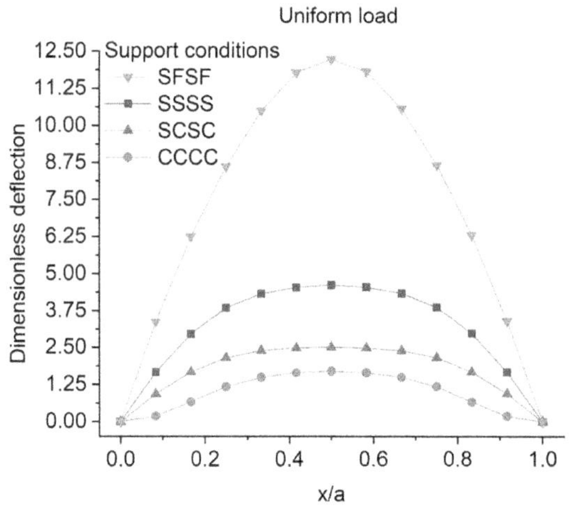

FIGURE 12.11 Centreline deflection of corrugated FG panel subjected to uniform load with different support conditions.

are inversely proportion to the number of corrugation. Thus, the higher number of corrugations will produce higher stiffness for a corrugated FG panel.

Figures 12.11 and 12.12 demonstrate the centreline deflection of corrugated FG panels with different support conditions (*SFSF, SSSS, SCSC, CCCC*). Among all the cases, panels with *SFSF* support condition exhibit maximum deflection, and fully clamped (*CCCC*) corrugation FG panels exhibit minimum deflection.

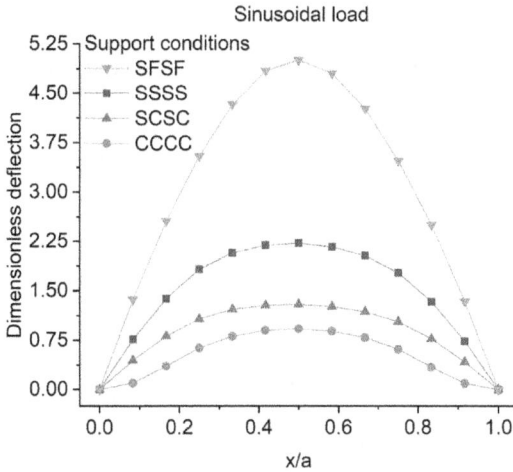

FIGURE 12.12   Centreline deflection of corrugated FG panel subjected to sinusoidal load with different support conditions.

12.5   CONCLUSION

A sinusoidally corrugated FG composite panel is modeled by using HSDT kinematics with nine degrees-of-freedom through a customized code to observe the bending behavior. The sinusoidal corrugation is introduced by using a general curvature equation. Governing equations are obtained by minimum total potential energy principle, and the panel is subjected to uniform and sinusoidal load. The proposed model is compared with the available literature, showing good agreement. For the first time the effect of corrugation on the deflection responses of the FG composite panel is examined. Thus, the following remarks can be confirmed from the numerical illustration:

1. The deflection responses of the sinusoidally corrugated FG panel under sinusoidal load are much lower as compared to the panels under uniform load.
2. Corrugated FG panel exhibits high deflection responses for high side-to-thickness ratio λ, aspect ratio a_r, and power law index α.
3. Corrugated FG panels with higher wave height h_w and number of corrugations will be stiffer.
4. Among different support conditions, minimum deflection responses are obtained when the panel is fully clamped.

ACKNOWLEDGMENT

The authors would like to thank Science and Engineering Research Board, Department of Science and Technology, Government of India (File No. ECR/2016/001829), for the financial support.

REFERENCES

[1] J. Yang, K.M. Liew, Y.F. Wu, S. Kitipornchai, Thermo-mechanical post-buckling of FGM cylindrical panels with temperature-dependent properties, *Int. J. Solids Struct.* 43 (2006) 307–324. doi:10.1016/j.ijsolstr.2005.04.001.

[2] V.R. Kar, S.K. Panda, Thermoelastic analysis of functionally graded doubly curved shell panels using nonlinear finite element method, *Compos. Struct.* 129 (2015) 202–212. doi:10.1016/j.compstruct.2015.04.006.

[3] V.R. Kar, T.R. Mahapatra, S.K. Panda, Effect of different temperature load on thermal postbuckling behavior of functionally graded shallow curved shell panels, *Compos. Struct.* 160 (2017) 1236–1247. doi:10.1016/j.compstruct.2016.10.125.

[4] A. Karakoti, V.R. Kar, Deformation characteristics of sinusoidally-corrugated laminated composite panel – A higher-order finite element approach, *Compos. Struct.* (2019). doi:10.1016/j.compstruct.2019.02.097.

[5] J.N. Reddy, A Simple Higher-Order Theory for Laminated Composite Plates, *J. Appl. Mech.* 51 (1984) 745. doi:10.1115/1.3167719.

[6] R.D. Cook, D.S. Malkus, M. Plesha, *Concepts and Applications of Finite Element Analysis*, John Wiley Sons. (1989) 31–59. doi:10.1115/1.3264300.

[7] X. Zhao, Y.Y. Lee, K.M. Liew, Thermoelastic and vibration analysis of functionally graded cylindrical shells, *Int. J. Mech. Sci.* 51 (2009) 694–707. doi:10.1016/j.ijmecsci.2009.08.001.

13 Multiscale Analysis of Laminates Printed by 3D Printing Fused Deposition Modeling Method

Taha Sheikh and Kamran Behdinan

University of Toronto,
Toronto, Canada

CONTENTS

13.1 INTRODUCTION

Composite materials are generally understood as a combination of more than one material. These materials maintain different physical or chemical phases, resulting in improved properties when compared to their constituent components [1]. The continuous phase is termed *matrix* while the discontinuous phase is called *reinforcement*. Composite laminates are structures composed by the arrangement of different layers hierarchically. These laminas can be of the same material or different. The reason for such layer-by-layer arrangements lies behind the advanced properties of the composite laminate such as high strength, improved fracture properties, and many more.

Like these laminates, scientists have investigated the structures produced by 3D printing and observed similarities [2, 3]. FDM is an additive manufacturing method that 3D prints the objects directly in a layer-by-layer way as shown in Figure 13.1 [4]. In FDM, a polymer filament is forced into an extruding nozzle, where after melting,

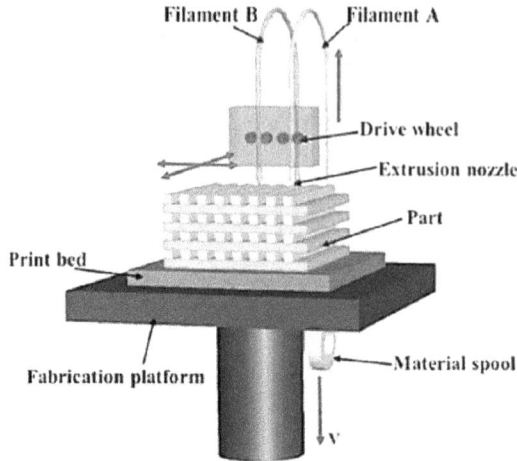

FIGURE 13.1 The FDM printing process in a layer-by-layer deposition process. Reprinted from Composites part B: Engineering, Volume 143, 2018, T. D. Ngo, A. Kashani, G. Imbalzano, K.T.Q. Nguyen, D. Hui, Additive manufacturing (3D printing): A review of materials, methods, applications and challenges, Copyright (2018), with permission from Elsevier.

the viscous polymer is printed over a bed. Once solidified, the next layer is printed over the previous layer. With further analysis, researchers have reported experimental analysis of mechanical properties of the FDM printed structures and verified their behavior to be similar to the laminated composite structures. This experimental investigation is also validated by applying the classical laminate theory (CLT) for the analysis of FDM printed structures. The printed laminates suffer from inferior mechanical properties because of anisotropy and porosity in the final structure [5–8]. From the analysis of these structures, it was reported that the mechanical properties of the FDM laminate differ from the material filament used [9]. The reason for this difference is that during the layer-by-layer printing, the microstructure of the laminate changes. For further structural analysis of printed laminate, this variation must be taken into consideration. Therefore, the stiffness matrix of the printed structure must have the final properties to capture its mechanical behavior. This requires efficient characterization of these printed parts to calculate the final mechanical properties.

The parts fabricated using FDM processes still suffer from some serious limitations. These printed parts—neat polymers (rigid), their composites, or nanocomposites—mainly suffer dimensional inaccuracies and poor mechanical properties [10]. These arise because of the inclusion of microvoids and voids due to the positive air gap between the adjacent layers [11, 12]. Also, the anisotropy arises because of poor interfacial adhesion of different layers. This results in loss of strength in z-direction in comparison to the other planes when printed [5, 13]. In the past, scientists have conducted experimental characterization to predict the mechanical properties of the FDM-printed structures [10–12, 14]. It was reported from these investigations that the FDM process parameters such as the layer height, infill density, or raster orientation affect these properties [10–12, 14]. Several studies were performed to capture how the process parameters affect the mechanical properties.

The paradigm of these investigations shifted toward numerical simulations because the experimental studies are financially restrictive. Scientists have developed numerical and computational models to characterize the mechanical and other properties with the effect of the parameters [11, 12]. Martinez et al. [9] performed a numerical simulation study for FDM-printed composite structures. The ply level method was used and ABAQUS software was employed to simulate the FEM model. These models assume the raster is infused into holes (space between two rasters) as filler, this way assuming the acrylonitrile butadiene styrene (ABS) part as an orthotropic composite material. The method does not account for parameters affecting the mechanical strength of the part. Domingo-Espin et al. [15] studied the effect of build orientation using FEA analysis for polycarbonate material. This was a simple FEA simulation on a macro scale without incorporating any small-scale properties. Anisotropy was not considered in this study, and the error against experimental validation was less than 8%. Qattawi et al. [16] reported the FEA analysis of PLA printed part to analyze the effect of voids. They incorporated apparent density and used holes per area to model accurately printed structures. The analysis was still based on several assumptions such as modeling of material property as filament property, and throughout the layer isotropy was considered, which is not the actual case.

Another numerical modeling analysis was based on classical laminate theory (CLT). This assumed that the FDM part is fabricated using layer by layer and the whole structure is assumed to be a composite structure. Lie et al. [17] in their work incorporate the density of the void to take the effect of air gaps. Casavola et al. use the model to predict Young's modulus of parts fabricated in different stacking sequences. The CLT-based models fail to apply for parts printed with positive air gaps (the space between two adjacent layers) [18].

To answer these challenges, multiscale methods have been employed to capture the effect of anisotropy and heterogeneity introduced in the FDM-printed part [11, 19, 20]. These methods are also capable of capturing the effect of the process parameters on the mechanical properties of the part. The work proposes the use of a multiscale method to calculate the stiffness matrix of the part that can efficiently capture the mechanical properties. The work also analyzes the effect of FDM process parameters on the mechanical properties of the part. The results are validated by comparing these with the results obtained from the ANSYS material designer module, showing the efficiency of the proposed work.

13.2 METHODOLOGY

13.2.1 MATHEMATICAL HOMOGENIZATION OF THE RVE

The FDM part is printed by depositing layers over each layer. Each layer has a thickness h and is printed with raster orientation θ. This orientation could be the same or different for each layer. We can refer to the single layer as lamina and the depositing filament as fiber. The entire printed structure can be referred to as composite

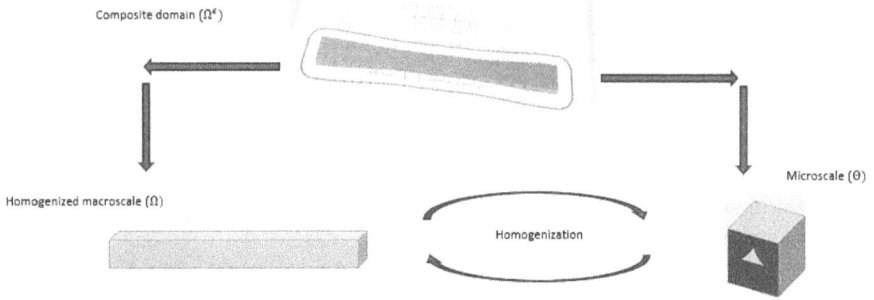

FIGURE 13.2   Macro- and microscale discretization of the printed composite structure.

laminate. The analysis is based on discretization of the printed structure into micro- and macroscale structures, and homogenization is applied to calculate the mechanical properties efficiently. The printed structure is assumed to be a composite domain (Ω^e) composed of the microscale (Θ) and macroscale (Ω) structures as shown in Figure 13.2.

The macroscale structure is composed of RVE, the smallest volume element that is periodically repeated. This periodically repeating unit captures macroscale effective properties. The investigation involves analyzing the RVE of the FDM part with different layer heights, raster orientation (θ) with solid (100%) infill. The constitutive model for FDM laminate is developed using numerical homogenization [20]. Elastic properties of RVE are calculated by the multiscale volume average method (VAM). The printed structure is assumed to be periodic, therefore periodic boundary conditions are applied for six linear strains. The direction of the raster is considered as direction 1, vertical to this is direction 2, and the transverse direction to 1 is direction 3. The stress–strain relationship of the RVE can be described by the generalized Hooke's law as [19]

$$
\begin{bmatrix} \bar{\sigma}_{11} \\ \bar{\sigma}_{22} \\ \bar{\sigma}_{33} \\ \bar{\sigma}_{12} \\ \bar{\sigma}_{13} \\ \bar{\sigma}_{23} \end{bmatrix} = \begin{bmatrix} C_{11} & C_{12} & C_{12} & 0 & 0 & 0 \\ C_{21} & C_{22} & C_{23} & 0 & 0 & 0 \\ C_{31} & C_{32} & C_{33} & 0 & 0 & 0 \\ 0 & 0 & 0 & C_{44} & 0 & 0 \\ 0 & 0 & 0 & 0 & C_{55} & 0 \\ 0 & 0 & 0 & 0 & 0 & C_{66} \end{bmatrix} \begin{bmatrix} \bar{\varepsilon}_{11} \\ \bar{\varepsilon}_{22} \\ \bar{\varepsilon}_{33} \\ \bar{\gamma}_{12} \\ \bar{\gamma}_{13} \\ \bar{\gamma}_{23} \end{bmatrix}
\tag{13.1}
$$

where $\bar{\sigma}$, $\bar{\varepsilon}$, and $\bar{\gamma}$ are the average normal stress, normal linear strain, and shear stress, respectively. These properties are averaged over the volume of RVE. The relationship between the strain and displacement can be expressed as [21]

$$
\bar{\varepsilon}_{ij} = \frac{\partial u_i}{\partial i}
\tag{13.2}
$$

$$
\bar{\gamma}_{ij} = \frac{\partial u_i}{\partial j} + \frac{\partial u_j}{\partial i}
\tag{13.3}
$$

$$\bar{\sigma}_{ij} = \frac{1}{V_{RVE}} \iiint \sigma_{ij} d1 d2 d3$$

where i and $j = 1, 2, 3$, u is the displacement, and V_{RVE} is the volume of RVE. To calculate the elastic constants from stiffness matrix elements (C_{ij}) values, the RVE is superimposed with six separate strain boundary conditions [20]. Therefore, after applying the boundary conditions and keeping the rest of the strain values as 0, the corresponding direction element of stiffness matrix in Equation (13.1) is evaluated using relations:

$$C_{ij} = \frac{\bar{\sigma}_{ij}}{\bar{\varepsilon}_{ij}} \tag{13.4}$$

The compliance matrix is determined from the inverse of stiffness matrix (Equation 13.1) as

$$
\begin{bmatrix} \bar{\varepsilon}_{11} \\ \bar{\varepsilon}_{22} \\ \bar{\varepsilon}_{33} \\ \bar{\gamma}_{12} \\ \bar{\gamma}_{13} \\ \bar{\gamma}_{23} \end{bmatrix} =
\begin{bmatrix}
{1}/{E_1} & {-\upsilon_{21}}/{E_2} & {-\upsilon_{31}}/{E_3} & 0 & 0 & 0 \\
{-\upsilon_{12}}/{E_1} & {1}/{E_2} & {-\upsilon_{32}}/{E_3} & 0 & 0 & 0 \\
{-\upsilon_{13}}/{E_1} & {-\upsilon_{23}}/{E_2} & {1}/{E_3} & 0 & 0 & 0 \\
0 & 0 & 0 & {1}/{G_{12}} & 0 & 0 \\
0 & 0 & 0 & 0 & {1}/{G_{13}} & 0 \\
0 & 0 & 0 & 0 & 0 & {1}/{G_{23}}
\end{bmatrix}
\begin{bmatrix} \bar{\sigma}_{11} \\ \bar{\sigma}_{22} \\ \bar{\sigma}_{33} \\ \bar{\sigma}_{12} \\ \bar{\sigma}_{13} \\ \bar{\sigma}_{23} \end{bmatrix} \tag{13.5}
$$

where E_1, E_2, E_3 are the Young's moduli in the corresponding directions, G_{12}, G_{13}, G_{23} are the shear moduli along the corresponding planes, and υ_{12}, υ_{13}, υ_{23} are the Poisson's ratios along the corresponding planes. Equation (13.5) gives the elastic properties of the RVE of the full-scale printed structure as shown in Figure 13.2. To capture the effect of raster orientation, the matrix in Equation (13.1) is transformed in the horizontal plane. The reason behind this is that in FDM printing the raster orientation only has an effect along the horizontal plane. The transformation matrix can be written as [22]

$$
[T] =
\begin{bmatrix}
l_1^2 & m_1^2 & n_1^2 & l_1 m_1 & l_1 n_1 & n_1 m_1 \\
l_2^2 & m_2^2 & n_2^2 & l_2 m_2 & l_2 n_2 & n_2 m_2 \\
l_3^2 & m_3^2 & n_3^2 & l_3 m_3 & l_3 n_3 & n_3 m_3 \\
2l_1 l_2 & 2m_1 m_2 & 2n_1 n_2 & l_1 m_2 + l_2 m_1 & l_1 n_2 + l_2 n_1 & m_1 n_2 + m_2 n_1 \\
2l_1 l_3 & 2m_1 m_3 & 2n_1 n_3 & l_1 m_3 + l_3 m_1 & l_1 n_3 + l_3 n_1 & m_1 n_3 + m_3 n_1 \\
2l_3 l_2 & 2m_3 m_2 & 2n_3 n_2 & l_2 m_3 + l_3 m_2 & l_2 n_3 + l_3 n_2 & m_2 n_3 + m_3 n_2
\end{bmatrix} \tag{13.6}
$$

where l, m, n are the direction cosines with raster angle θ, $l_1 = \cos\theta$, $l_2 = -\sin\theta$, $l_3 = 0$, $m_1 = \sin\theta$, $m_2 = \cos\theta$, $m_3 = 0$, $n_1 = 0$, $n_2 = 0$, and $n_3 = 1$. The transformed compliance matrix can be written as

$$\left[S'\right] = \left[\bar{T}\right]\left[S\right]\left[\bar{T}\right]^T$$

where $\left[S'\right]$ and $[S]$ are the transformed and original compliance matrix, respectively. The obtained elastic properties using the present formulation of RVE are homogenized over the macroscale structure. This gives us the advantage of analyzing the RVE only, not the full structure like CLT theory, to obtain the elastic properties of FDM-printed laminate.

Polylactic acid (PLA) is chosen as the raw material for the proposed study [23]. The effective properties of the FDM part are confirmed by validating the proposed study with the results derived from Ansys material designer module simulations.

13.2.2 DESIGN OF THE RVE

The microscale model as RVE, which resembles the cross-section of FDM printed laminate as shown in Figure 13.2, is designed in Ultimaker Cura, a slicing software. Following this, the RVE corresponding to the layer height of 0.4 mm, 0.2 mm, and 0.1 mm is designed for 0°. To study the effect of raster angle, 0.4 mm RVE is created for 0° and 0°/90° raster angle in Solidworks and meshed in Ansys workbench as shown in Figure 13.3. The infill density for both the RVEs is taken as 100%.

13.2.3 COMPUTATIONAL ANALYSIS OF THE RVE

The elastic properties of RVEs are determined using the formulation presented in the previous section. The RVEs were assigned the properties of PLA filament [23]. The multiscale analysis is done in the Ansys workbench by applying the strain boundary condition on RVEs. The methodology for this method is averaging the stresses over

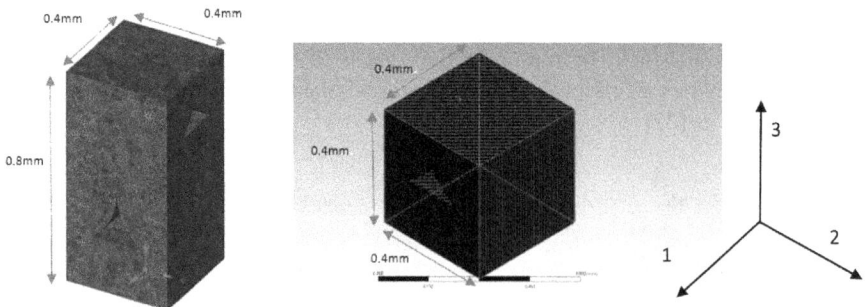

FIGURE 13.3 RVE with layer height 0.4 mm for raster angle 0°/90° and 0° meshed in Ansys workbench.

the RVE volume and then deriving the orthotropic stiffness matrix upon applying the straining boundary conditions [24]. To obtain the convergence, the mesh is refined along the corner edges of RVE.

The constitutive stiffness matrix for layer height of 0.4 mm both for 0°/90° and 0° raster angle is obtained as

$$C = \begin{bmatrix} 3931.72 & 1685.89 & 1546.29 & 0 & 0 & 0 \\ & 3930.12 & 1558.59 & 0 & 0 & 0 \\ & & 3435.53 & 0 & 0 & 0 \\ & \text{Symmetric} & & 980.72 & 0 & 0 \\ & & & & 951.37 & 0 \\ & & & & & 994.50 \end{bmatrix} \text{MPa}$$

$$C = \begin{bmatrix} 4722.97 & 2144.59 & 1891.42 & 0 & 0 & 0 \\ & 4186.85 & 1769.70 & 0 & 0 & 0 \\ & & 3484.24 & 0 & 0 & 0 \\ & \text{Symmetric} & & 1113.21 & 0 & 0 \\ & & & & 1001.32 & 0 \\ & & & & & 1014.58 \end{bmatrix} \text{MPa}$$

13.3 RESULTS AND DISCUSSIONS

The multiscale analysis of FDM-printed laminate over the corresponding RVE gives the elastic properties tabulated in Table 13.1. The obtained results are validated

TABLE 13.1

Comparison of Mechanical Properties of Proposed Study with ANSYS Material Designer Module

| Elastic Properties | 0.4 mm Layer Height (0°) | | | 0.4 mm Layer Height (0°/90°) | | |
	Proposed Study	Material Designer Ansys	Error %	Proposed Study	Material Designer Ansys	Error %
E_1 (MPa)	3269.91	3096.59	5.60	2923.56	2720.26	7.47
E_2 (MPa)	2908.78	2740.81	6.13	2935.10	2720.18	7.90
E_3 (MPa)	2468.57	2329.61	6.00	2577.33	2382.19	8.19
G_{12} (MPa)	1113.21	1060.34	4.98	951.37	937.28	1.50
G_{13} (MPa)	1001.32	959.36	4.37	980.72	930.19	5.43
G_{23} (MPa)	1014.58	962.96	5.36	994.50	930.17	6.91
v_{12}	0.350	0.319	9.80	0.304	0.303	0.33
v_{13}	0.340	0.271	25.5	0.310	0.278	11.51
v_{23}	0.334	0.333	0.30	0.318	0.317	0.31

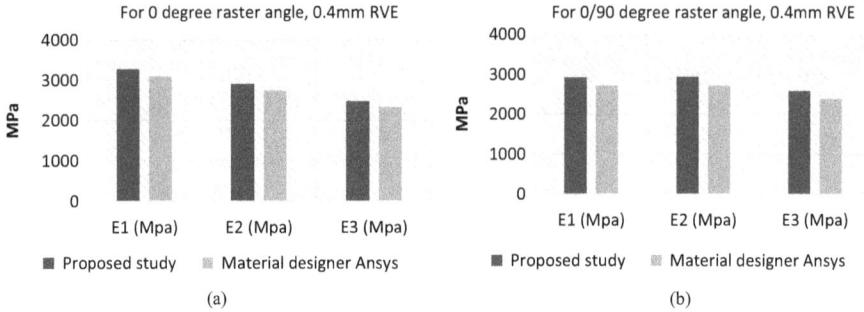

FIGURE 13.4 Comparing Young's modulus calculated from the proposed analysis with Ansys material designer module for (a) 0° and (b) 0°/90° raster angle.

with the homogenized properties determined from Ansys material designer module simulations.

The elastic properties obtained from the multiscale analysis are in good agreement with the Ansys material designer module as shown in Table 13.1.

Table 13.1 gives us the information on the efficiency of this method, which shows an error $\approx 6\%$ for Young's modulus and $\approx 5\%$ for shear modulus for 0° raster angle. For 0°/90° raster angle, the error is reported as $\approx 8\%$ for Young's modulus and $\approx 6\%$ for shear modulus. The reason for these errors is because of the assumption of calculating the stress value as the average over the elements (RVE), which does not consider the effect of the triangular voids as shown in Figure 13.3. The effect of raster orientation can be understood from the values of Table 13.1, and it can be confirmed that the FDM-printed laminate is stronger in the raster direction and behave orthotropic for 0° raster angle and nearly isotropic for 0°/90°. The high percentage error in the values of v_{13} could be because of the poor adhesion between the layers in the corresponding plane. Figure 13.4 shows the graphical comparison of Young's modulus of 0.4 mm layer height RVE for 0° and 0°/90° raster angles, respectively. The deviation in the values could be because of the presence of microvoids in the macro scale because of manufacturing, which is not incorporated in this method.

To study the effect of layer height elastic properties of three different RVEs, 0.1 mm, 0.2 mm, and 0.4 mm layer height are investigated. Figure 13.5 shows that as the height is increased from 0.1 mm to 0.4 mm, the mechanical strength in terms of Young's modulus and shear modulus of the laminate increases correspondingly. This agrees well with the study presented in the literature [25]. The multiscale analysis of the RVE at the micro scale is homogenized over the macro scale, and the elastic constants of RVE are assumed as effective elastic constants of FDM-printed laminate. This eliminates the analysis of FDM laminates as a whole, making analysis easy and simple by homogenizing RVE. The proposed study can capture the effect of positive air gaps and investigate the effect of raster orientation and layer height that CLT fails to capture.

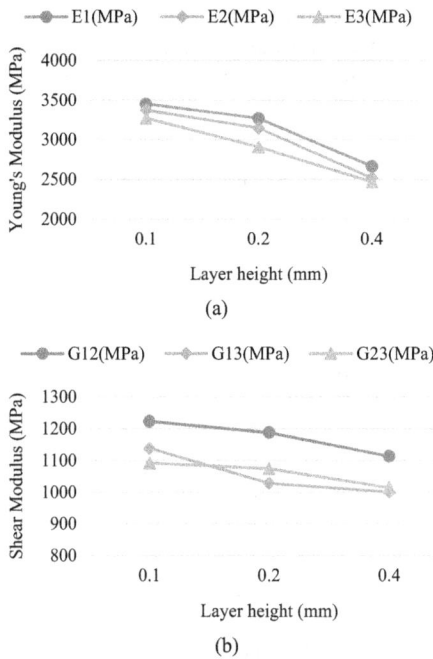

FIGURE 13.5   Variation of (a) Young's modulus and (b) shear modulus with layer height for 0°.

13.4 CONCLUSION

This chapter presented the multiscale analysis of the FDM-printed structures that are considered as laminates. The isotropic PLA filament was used for printing the laminate, but the final constitutive properties differed from the filament material. To compute the elastic properties of the macro scale, the properties of RVE were homogenized. The study was further extended to analyze the effect of process parameters—layer height and raster orientation—on the final mechanical properties: Young's modulus, shear modulus, and Poison's ratio. The predictions derived from the proposed method were validated against the elastic properties calculated by Ansys material designer module simulations. The results were found to be in good agreement with the Ansys simulation results. The following points are drawn from the proposed study:

- Proposed multiscale analysis effectively predicts the constitutive elastic constant matrix of the FDM printed laminates.
- The effect of the raster orientation reported that the FDM printed laminate is stronger in the raster direction and behaves orthotropic for 0° raster angle and nearly isotropic for 0°/90°.

- The layer height is another important process parameter that is reported to have an inverse effect on the mechanical properties of the laminate. The laminate mechanical properties increase when printed with small layer heights.

The proposed study could be extended for further analysis of other process parameters and different filament materials. The study also finds an extension to investigate the viscoelastic properties and fracture behavior of these FDM-printed laminates.

ACKNOWLEDGMENT

The funding received for this research from The National Science and Engineering Research Council of Canada (NSERC under grant RGPIN-217525) as well as the support from the Advanced Research Laboratory Multifunctional Lightweight Structures (ARL-MLS), University of Toronto, are gratefully acknowledged.

REFERENCES

[1] Taylor P, Kessler MR. Polymer matrix composites: A perspective for a special issue of polymer reviews polymer matrix composites: A perspective *Polym Rev* 2012:37–41. https://doi.org/10.1080/15583724.2012.708004

[2] Rodríguez JF, Thomas JP, Renaud JE. Mechanical behavior of acrylonitrile butadiene styrene (ABS) fused deposition materials. Experimental investigation. *Rapid Prototyp J* 2001. https://doi.org/10.1108/13552540110395547

[3] Casavola C, Cazzato A, Moramarco V, Pappalettere C. Orthotropic mechanical properties of fused deposition modelling parts described by classical laminate theory. *JMADE* 2016;90:453–458. https://doi.org/10.1016/j.matdes.2015.11.009

[4] Ngo TD, Kashani A, Imbalzano G, Nguyen KTQ, Hui D. Additive manufacturing (3D printing): A review of materials, methods, applications and challenges. *Compos Part B Eng* 2018;143:172–196. https://doi.org/10.1016/j.compositesb.2018.02.012

[5] Stansbury JW, Idacavage MJ. 3D printing with polymers: Challenges among expanding options and opportunities. *Dent Mater* 2016. https://doi.org/10.1016/j.dental.2015.09.018

[6] Torrado AR, Shemelya CM, English JD, Lin Y, Wicker RB, Roberson DA. Characterizing the effect of additives to ABS on the mechanical property anisotropy of specimens fabricated by material extrusion 3D printing. *Addit Manuf* 2015. https://doi.org/10.1016/j.addma.2015.02.001

[7] Torrado AR, Roberson DA. Failure analysis and anisotropy evaluation of 3D-printed tensile test specimens of different geometries and print raster patterns. *J Fail Anal Prev* 2016. https://doi.org/10.1007/s11668-016-0067-4

[8] Chuang KC, Grady JE, Draper RD, Shin ESE, Patterson C, Santelle TD. Additive manufacturing and characterization of ultem polymers and composites. *CAMX 2015 - Compos. Adv. Mater. Expo*, 2015.

[9] Martínez J, Diéguez JL, Ares E, Pereira A, Hernández P, Pérez JA. Comparative between FEM models for FDM parts and their approach to a real mechanical behaviour. *Procedia Eng*, 2013. https://doi.org/10.1016/j.proeng.2013.08.230

[10] Jaisingh Sheoran A, Kumar H. Fused Deposition modeling process parameters optimization and effect on mechanical properties and part quality: Review and reflection on present research. *Mater Today Proc* 2020;21:1659–1672. https://doi.org/10.1016/j.matpr.2019.11.296

[11] Vyavahare S, Teraiya S, Panghal D, Kumar S. Fused deposition modelling: A review. *Rapid Prototyp J* 2020;26:176–201. https://doi.org/10.1108/RPJ-04-2019-0106

[12] Cuan-Urquizo E, Barocio E, Tejada-Ortigoza V, Pipes RB, Rodriguez CA, Roman-Flores A. Characterization of the mechanical properties of FFF structures and materials: A review on the experimental, computational and theoretical approaches. *Materials (Basel)* 2019;16. https://doi.org/10.3390/ma12060895

[13] Mohamed OA, Masood SH, Bhowmik JL. Optimization of fused deposition modeling process parameters: A review of current research and future prospects. *Adv Manuf* 2015. https://doi.org/10.1007/s40436-014-0097-7

[14] Penumakala PK, Santo J, Thomas A. A critical review on the fused deposition modeling of thermoplastic polymer composites. *Compos Part B Eng* 2020;201. https://doi.org/10.1016/j.compositesb.2020.108336

[15] Domingo-Espin M, Puigoriol-Forcada JM, Garcia-Granada AA, Llumà J, Borros S, Reyes G. Mechanical property characterization and simulation of fused deposition modeling Polycarbonate parts. *Mater Des* 2015. https://doi.org/10.1016/j.matdes.2015.06.074

[16] Alafaghani A, Qattawi A, Alrawi B, Guzman A. Experimental optimization of fused deposition modelling processing parameters: A design-for-manufacturing approach. *Procedia Manuf* 2017. https://doi.org/10.1016/j.promfg.2017.07.079

[17] Li L, Sun Q, Bellehumeur C, Gu P. Composite modeling and analysis for fabrication of FDM prototypes with locally controlled properties. *J Manuf Process* 2002. https://doi.org/10.1016/S1526-6125(02)70139-4

[18] Magalhães LC, Volpato N, Luersen MA. Evaluation of stiffness and strength in fused deposition sandwich specimens. *J Brazilian Soc Mech Sci Eng* 2014. https://doi.org/10.1007/s40430-013-0111-1

[19] Liu X, Shapiro V. Homogenization of material properties in additively manufactured structures. *CAD Comput Aided Des* 2016;78:71–82. https://doi.org/10.1016/j.cad.2016.05.017

[20] Somireddy M, Czekanski A, Singh CV. Development of constitutive material model of 3D printed structure via FDM. *Mater Today Commun* 2018. https://doi.org/10.1016/j.mtcomm.2018.03.004

[21] Fish J. Practical multiscaling. vol. 3. 2013.

[22] Anoop MS, Senthil P. Homogenisation of elastic properties in FDM components using microscale RVE numerical analysis. *J Brazilian Soc Mech Sci Eng* 2019;41:1–16. https://doi.org/10.1007/s40430-019-2037-8

[23] Hikmat M, Rostam S, Ahmed YM. Investigation of tensile property-based Taguchi method of PLA parts fabricated by FDM 3D printing technology. *Results Eng* 2021;11:100264. https://doi.org/10.1016/j.rineng.2021.100264

[24] Somireddy M, Czekanski A, Singh CV. Development of constitutive material model of 3D printed structure via FDM. *Mater Today Commun* 2018. https://doi.org/10.1016/j.mtcomm.2018.03.004

[25] Chokshi H, Shah DB, Patel KM, Joshi SJ. Experimental investigations of process parameters on mechanical properties for PLA during processing in FDM. *Adv Mater Process Technol* 2021:1–14. https://doi.org/10.1080/2374068x.2021.1946756

14 Flexural Behavior of Carbon Nanotube-Reinforced Composites with Multiple Cutouts

Shyam Kumar Chaudhary, Vishesh Ranjan Kar, and Karunesh Kumar Shukla

National Institute of Technology Jamshedpur, Jamshedpur, India

CONTENTS

14.1 INTRODUCTION

Composite materials are defined as those combining two or more materials on a microscopic scale. They are widely used because of their advantageous properties such as stiffness, strength, low weight, corrosion resistance, thermal properties, fatigue life, and wear resistance. Composites have two constituent elements, namely fiber and matrix. The fibers are used in modern composites because of their high specific mechanical properties compared to those of traditional bulk materials. Carbon and graphite are the common fiber materials used by many weight-sensitive industries over the past few decades. Matrix acts as a bonding element that protects fiber from external break or damage. The main function of the matrix is to distribute and transfer the load to the fibers or reinforcements. Metal, ceramic, and polymer are the commonly used material for the matrix phase. Transformation of load depends on the bonding interface between the reinforcement and matrix. Bonding depends on the types of reinforcement and matrix and the fabrication technique. Functionally graded material (FGM) is a new kind of advanced composite material in which the constituents are gradually changed with respect to the spatial coordinate over the volume, resulting in a consistent change in the properties of the material (see Figure 14.1). The overall properties of functionally graded material are exclusive and dissimilar from any of the individual materials that form it. Nowadays a wide range of FGM applications are used in engineering field, and this trend is likely to continue and extend to other industries as the costs of manufatruring and processing go down with further technological advencements. This study presents an overview of fabrication processes area of application.

Material properties depend on the spatial position in the structure. The materials can be designed for specific function and applications. The properties that may be designed/controlled for desired functionality include chemical, mechanical, thermal, and electrical. They provide the ability to control deformation, dynamic response, wear, and corrosion, as well as the ability to design for different complex environments, and to eliminate stress concentrations.

Carbon nanotubes (CNT) are advanced materials known for their high mechanical strength. The functionally graded carbon nanotube–reinforced composite (FGCNT) structures have vast applications under different environment conditions due to their excellent mechanical, thermal, and chemical properties. In addition, stiffness of any structure depends critically on the geometry of the structure.

Carbon nanotubes were invented in 1991 by Sumio Iijima. These are thick sheet of graphite rolled into a tube with a diameter of 1 nm. CNT properties depend on how the nanotubes are rolled and include tensile strength and Young's modulus. Graphene is the name given to a single layer of carbon atoms densely packed into a benzene-ring structure, and is widely used to describe properties of many carbon-based materials, including graphite, large fullerenes, nanotubes, etc.

Carbon nanotubes have very high stiffness and axial strength due to carbon-carbon sp^2 bonding. They are the best stiff material [1] having Young's modulus of 1.4 TPa and tensile strength well above 100 GPa. CNT are the best thermal conductors having thermal conductivity at least twice that of diamond and showing super-conducting properties at low temperature. They possess extremely low electrical resistance an can carry the highest current density measured as high as 10^9 A/cm^2.

These are the few methods of production of carbon nanotubes:

1. Carbon-arc discharge method
2. Laser ablation method
3. Chemical vapor deposition (CVD) method
4. Flame synthesis and Smalley's high-pressure carbon monoxide (HIP-CO) process.

Functionally graded carbon nanotube (FGCNT) composites show continuous improvements in properties such as thermomechanical, lightweight, stiffness-to-weight ratio, high strength, fatigue life, weight-to-strength ratio, vibration absorption capability, and electrical conductivity [2,3]. These properties provide avenues for development of ever more advanced technologies in the future.

Functionally graded materials (FGMs) are composites involving spatially varying volume fraction of constituent materials, thus providing a graded microstructure and macro properties. CNT is an effective reinforcement material in the case of composite material developments, owing to its good physical and chemical properties. Guided by the concept of functionally graded (FG) materials, a class of new emerging composite materials, the FGCNT–reinforced composite, has been proposed, making use of CNTs as reinforcements in a functionally graded pattern. These have many applications:

a. FGCNT is used as a reactor shield in nuclear reactors to reduce chemical corrosion and thermal stress, as well as TBCs in combustion chambers.
b. FGCNT is used in manufacturing the components of propulsion systems, submarines, and in aerospace structures to withstand aero-thermal loads.
c. FGCNT is used in cutting tools, drills, machining of wear-resistant materials, mining, and geothermal drilling.

14.2 TYPES OF COMPOSITE

CNTs have three unique geometrical arrangements of carbon atoms. Thes can be categorized by how a graphene sheet is wrapped into a tube form. Because physical and mechanical properties of CNT depend on its atomic arrangement, these forms are armchair, chiral, and zig-zag.

14.2.1 TYPE OF CNT ACCORDING TO NUMBER OF TUBES

  i. A single-walled CNT is considered as a cylinder with only one wrapped graphene sheet.
 ii. A double-walled CNT is a collection of the two concentric SWCNT.
iii. A multi-walled CNT is a collection of more than two concentric SWCNT.

14.2.2 TYPE OF CNT ACCORDING TO DISTRIBUTION

1. FGCNT UD
2. FGCNT-X
3. FGCNT-O
4. FGCNT-V
    1. **FGCNT UD:** The CNT volume fraction is uniformly distributed throughout the structure known as FGCNT UD. See Figure 14.1.
    2. **FGCNT-X**: The CNT volume fraction is distributed diagonally throughout the structure known as FGCNT -X. See Figure 14.2.
    3. **FGCNT-O**: The CNT volume fraction is maximum distributed at center. See Figure 14.3.

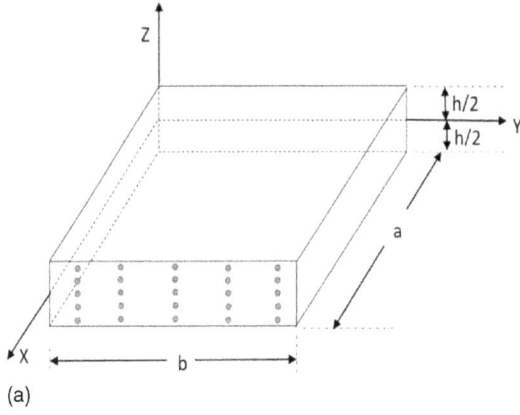

(a)

FIGURE 14.1   FGCNT-UD.

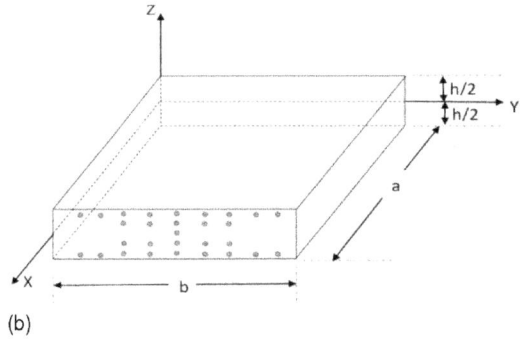

(b)

FIGURE 14.2   FGCNT-X.

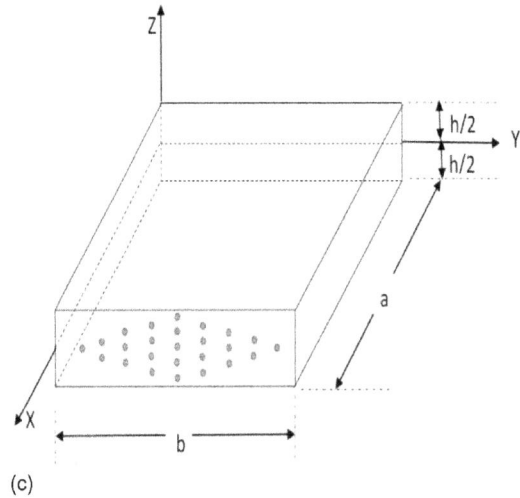

(c)

FIGURE 14.3   FGCNT-O.

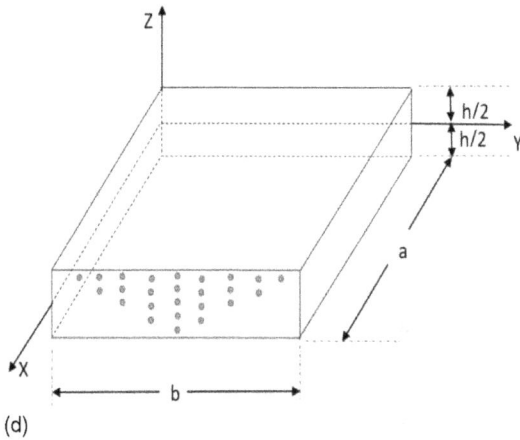

(d)

FIGURE 14.4   FGCNT-V.

4. **FGCNT-V**: The CNT volume fraction distribution gradually increases as the length increases and found to be maximum at the center; as the length increases, the volume fraction decreases. See Figure 14.4.

Future applications demand materials having extraordinary mechanical, thermal, and chemical properties that must sustain a wide range of environmental conditions and at the same time be available at reasonable prices. CNT-reinforced functionally graded composite materials (FGCM) are expected to be the new-generation materials having a wide range of unexplored potential applications in various technological areas such as aerospace, military, energy, automobile, medicine, chemical, construction, and manufacturing. They can be used as gas adsorbents, templates, actuators, catalyst supports, probes, chemical sensors, nano-pipes, nano-reactors, etc.

Functionally graded materials (FGMs) are a new generation of engineered materials that are gaining interest in recent years. FGMs were initially designed as thermal barrier materials for aerospace structural applications and fusion reactors [4]. FGMs also found application in structural components operating under extremely high-temperature [5]. FGMs are the composite materials in which the content of reinforcement is gradually varied in some direction to achieve gradient in properties. Due to graded variation in the content of constituent materials, the properties of FGMs undergo appreciable and continuous change from one surface to another, thus eliminating interface problems and diminishing thermal stress concentrations. The application of the concept of FGMs to Metals Matrix Composites (MMCs) has led to the development of components designed with the purpose of employing selective reinforcement in certain regions where enhanced properties like Young's modulus, strength, and wear resistance are required [6].

14.3 APPLICATION OF FGCNT-REINFORCED COMPOSITES

Applications of FGM [7] that have recently been reported include the following:

1. CNT-reinforced functionally graded prosthesis joint increasing adhesive strength and reducing pain.
2. CNT-reinforced functionally graded polyester-calcium phosphate materials for bone replacement with a controllable in vitro polyester degradation rate
3. CNT-reinforced functionally graded TBCs for combustion chambers.
4. CNT-reinforced functionally graded piezoelectric actuators.
5. CNT-reinforced functionally graded reactor shield in nuclear reactors to reduce chemical corrosion and thermal stress.
6. CNT-reinforced functionally graded tools and dies will enable better thermal management, possess better wear resistance, reduce scrap, and improve process productivity. FGMs also find application as furnace liners and thermal shielding elements in microelectronics.

14.4 RESEARCH IN FGCNT-REINFORCED COMPOSITE

The reported studies on the analyses of FGCNT composite panels depict that the FGCNT composite panels were examined without considering the cutouts. Based on the authors' knowledge, no study has reported yet on the deformation behavior of FGCNT-reinforced composite plate with perforations under thermomechanical loading. In the present study, the maximum deflection responses of perforated FGCNT-reinforced composite panel are examined under thermomechanical loading. The FGCNT-reinforced composite panel is comprised of SWCNT as fiber and PmPV as matrix materials. Here, four different types of FGCNT plates are considered by varying the gradation type of SWCNT fibers through the thickness. To evaluate the overall material properties of FGCNT plate, extended rule of mixture is employed. The finite element solutions are obtained in ANSYS parametric design language (APDL) platform using eight-noded serendipity elements with six degrees-of-freedom per node. Finally, the effects of different geometrical parameters (thickness ratios, aspect ratios), gradation type, volume fractions, temperature, and support conditions on the FGCNT-reinforced composite perforated plates are demonstrated through numerous numerical illustrations.

Vodenitcharova and Zhang [8] examined the bending behavior of a single-walled CNT (SWCNT) reinforced composite plate using Airy's stress function. Shen [9] investigated the nonlinear bending of a simply supported FGCNT composite plate using higher-order shear deformation theory (HSDT) and von-Karman nonlinearity. Murmu and Pradhan [10] investigated the stability behavior of a SWCNT-embedded Timoshenko beam using nonlocal elasticity. Ayatollahia et al. [11] used molecular model via finite element method (FEM) to examine the nonlinear mechanical properties of armchair and zigzag SWCNTs. Zhu et al. [12] analyzed the free vibration and bending responses of FG-CNTRC panel using first-order shear deformation theory (FSDT) kinematics. Yas and Samadi [13] utilized Hamilton's principal and generalized differential quadrature method to study the stability and vibration behavior of FGCNT composite plates. Bian et al. [14] examined the bending and free vibration responses of functionally graded plate embedded with piezoelectric layer using piezoelasticity method. Bousahla et al. [15] proposed a new trigonometric higher-order theory to obtain the bending responses of FG plate. Alibeigloo and Emtehani [16] obtained the deflection responses of FGCNT composite plate using three-dimensional elasticity theory. Zhang et al. [17] analyzed the static and dynamic behavior of FGCNT cylindrical panel using FSDT kinematics. Mehrabadi and Aragh [18] used HSDT kinematics to examine the bending behavior of FG-CNTRC cylindrical panel. Mehar et al. [19] examined the dynamic responses of FGCNT composite panel using HSDT kinematics.

14.5   EFFECTIVE MATERIAL PROPERTIES

In this study, the temperature-dependent material properties are considered. The FGCNT-reinforced composite plate is comprised of SWCNT of armchair (10, 10) with configuration $L = 9.26$ nm, $R = 0.68$ nm, $h = 0.067$ nm; as fiber material and PmPV as matrix material. The temperature-dependent material properties of SWCNT material are mentioned is Table 14.1 and of matrix material are as follows:

$$E^m = (3.51 - 0.0047T) GPa, \, \alpha^m = 45(1 + 0.0005\Delta T) \times 10^{-6}/K,$$
$$\rho_m = 1500 \, \text{kg/m}^3 \text{ and } \upsilon_m = 0.34$$

where $\Delta T = T - T_0$ and $T_0 = 300 \, K$.

Here four different types of gradation patterns along the thickness direction of the plate structure are considered (see Figure 14.5), namely uniformly distributed CNT fiber through the thickness (UD), volume fraction of CNT fiber (V_{CNT}) increasing gradually from bottom to top surfaces (FG-V), V_{CNT} is maximum at the middle and minimum at bottom and top surfaces (FG-O), and V_{CNT} is maximum at the bottom and top surfaces and minimum at the middle (FG-X). The effective volume fraction of CNT fiber can be evaluated for different gradation type as mentioned in Equation (14.1). The extended rule-of-mixture is adopted to evaluate overall mechanical and thermal properties of FGCNT reinforced composite material, as mentioned in Equations (14.2–14.8) [12].

TABLE 14.1
Temperature-Dependent Material Properties of SWCNT [12]

Temperature (K)	E_{11}^{CNT} (TPa)	E_{22}^{CNT} (TPa)	G_{12}^{CNT} (TPa)	α_{11}^{CNT} (10^{-6}/K)	α_{22}^{CNT} (10^{-6}/K)	ρ^{CNT} (kg/m³)	v_{12}^{CNT}
300	5.6466	7.0800	1.9445	3.4584	5.1682	1400	0.175
500	5.5308	6.9348	1.9643	4.5361	5.0189	1400	0.175
700	5.4744	6.8641	1.9644	4.6677	4.8943	1400	0.175

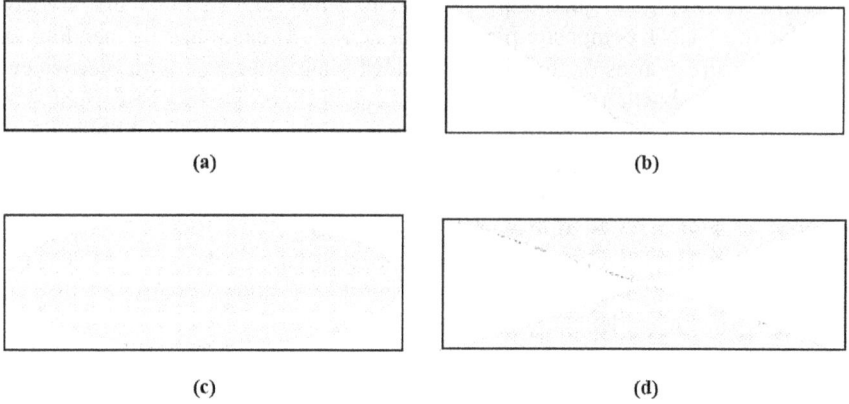

(a)

(b)

(c)

(d)

FIGURE 14.5 Distribution patterns of FGCNT reinforced composite panel (a) UD, (b) FG-V, (c) FG-O, and (d) FG-X.

$$
\left.
\begin{aligned}
V_{CNT} &= V_{CNT}^* && \left(UDCNTRC \right) \\
V_{CNT}(z) &= \left(1 + \frac{2z}{h} \right) V_{CNT}^* && \left(FG - VCNTRC \right) \\
V_{CNT}(z) &= 2 \left(1 - \frac{2z}{h} \right) V_{CNT}^* && \left(FG - OCNTRC \right) \\
V_{CNT}(z) &= 2 \left(\frac{2|z|}{h} \right) V_{CNT}^* && \left(FG - XCNTRC \right)
\end{aligned}
\right\}
\tag{14.1}
$$

$$
E_{11} = \zeta_1 V_{CNT} E_{11}^{CNT} + V_m E^m \tag{14.2}
$$

$$
\frac{\zeta_2}{E_{22}} = \frac{V_{CNT}}{E_{22}^{CNT}} + \frac{V_m}{E^m} \tag{14.3}
$$

$$
\frac{\zeta_3}{G_{12}} = \frac{V_{CNT}}{G_{12}^{CNT}} + \frac{V_m}{G^m} \tag{14.4}
$$

$$
\upsilon_{12} = V_{CNT} \upsilon_{12}^{CNT} + V_m \upsilon^m \tag{14.5}
$$

$$
\rho = V_{CNT} \rho^{CNT} + V_m \rho^m \tag{14.6}
$$

$$
\alpha_{11} = V_{CNT} \alpha_{11}^{CNT} + V_m \alpha^m \tag{14.7}
$$

TABLE 14.2

Effectiveness Parameter of SWCNT (10,10) [12]

V_{CNT}	ζ_1	ζ_2	ζ_3
0.11	0.149	0.934	0.934
0.14	0.150	0.941	0.941
0.17	0.149	1.381	1.381

$$\alpha_{22} = \left(1 + \upsilon_{12}^{CNT}\right) V_{CNT} \alpha_{22}^{CNT} + \left(1 + \upsilon^m\right) V_m \alpha^m - \upsilon_{12}\alpha_{11} \qquad (14.8)$$

where $(E_{11},\ E_{22},\ G_{12},\ \upsilon_{12})$, $(\alpha_{11},\ \alpha_{22})$, and (ρ) represent the effective elastic properties, coefficient of thermal expansion, and mass density of the FGCNT-reinforced composite material, respectively.

$$V_{CNT} = \frac{w_{CNT}}{w_{CNT} + \left(\dfrac{\rho^{CNT}}{\rho^m}\right) - \left(\dfrac{\rho^{CNT}}{\rho^m}\right) w_{CNT}}$$ denotes the volume fraction of CNT fiber,

w_{CNT} is the mass fraction of the CNT fiber, whereas superscripts m and CNT represent matrix and fiber materials, respectively. ζ is the effective parameter of SWCNT as mentioned in Table 14.2.

14.6   NUMERICAL MODELING OF PERFORATED PLATE

In the present study, a rectangular FGCNT reinforced composite plate of sides $(a \times b)$ is considered with $(n \times n)$ number of circular holes of radius $r = a/5(n+1)$ as shown in Figure 14.6. Here, all circular holes are equispaced with distance $c = a/5(n + 1)$ and $d = b/5(n + 1)$ along x and y directions, respectively.

The present model is discretized using eight-noded quadrilateral serendipity shell element (SHELL281) that is based on the FSDT mid-plane kinematics with six degrees-of-freedom, i.e., three translations (u_0, v_0, w_0) and three rotations (ϕ_x, ϕ_y, ϕ_z) in x, y, z directions. The mid-plane displaced field can be written as

$$\left. \begin{array}{l} u\left(x,y,z\right) = u_0\left(x,y\right) + z\phi_x\left(x,y\right) \\ v\left(x,y,z\right) = v_0\left(x,y\right) + z\phi_y\left(x,y\right) \\ w\left(x,y,z\right) = w_0\left(x,y\right) + z\phi_z\left(x,y\right) \end{array} \right\} \qquad (14.9)$$

The displacements (u_0, v_0, w_0) can be expressed in terms of shape functions (N_i) as

$$\delta = \sum_{i=1}^{j} N_i \delta_i$$

where $(u_{0i}\ v_{0i}\ w_{0i}\ \phi_{xi}\ \phi_{yi}\ \phi_{zi})$ are the nodal displacement.

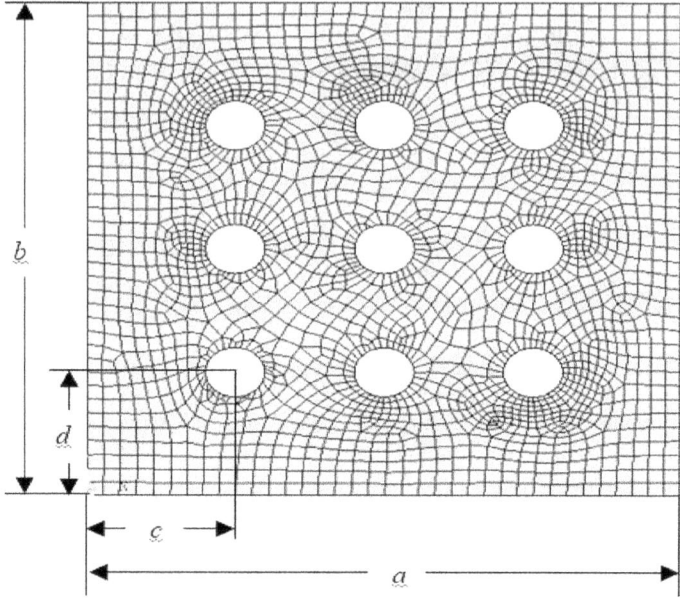

FIGURE 14.6 Geometrical description of perforated FGCNT reinforced composite modes.

The mid-plane displacements can be expressed in terms of shape functions N_i as

$$\left.\begin{aligned}
u_0 &= \sum_{i=1}^{i=8} N_i u_{0_i}; \quad \phi_x = \sum_{i=1}^{i=8} N_i \phi_{x_i} \\
v_0 &= \sum_{i=1}^{i=8} N_i v_{0_i}; \quad \phi_y = \sum_{i=1}^{i=8} N_i \phi_{y_i} \\
w_0 &= \sum_{i=1}^{i=8} N_i w_{0_i}; \quad \phi_z = \sum_{i=1}^{i=8} N_i \phi_{z_i}
\end{aligned}\right\} \tag{14.10}$$

where $(u_{0i}\ v_{0i}\ w_{0i}\ \phi_{xi}\ \phi_{yi}\ \phi_{zi})$ are the nodal displacements. The shape functions for side nodes at $i = 1$ to 4 and mid-side nodes at $I = 5$ to 8, in natural coordinates $(\xi\text{-}\eta)$, and shape functions can be expressed as

$$\left.\begin{aligned}
N_1 &= \frac{1}{4}(1-\xi)(1-\eta)(-\xi-\eta-1); \quad N_5 = \frac{1}{2}(1-\xi^2)(1-\eta) \\
N_2 &= \frac{1}{4}(1+\xi)(1-\eta)(\xi-\eta-1); \quad N_6 = \frac{1}{2}(1+\xi)(1-\eta^2) \\
N_3 &= \frac{1}{4}(1+\xi)(1+\eta)(\xi+\eta-1); \quad N_7 = \frac{1}{2}(1-\xi^2)(1+\eta) \\
N_4 &= \frac{1}{4}(1-\xi)(1+\eta)(-\xi+\eta-1); \quad N_8 = \frac{1}{2}(1-\xi)(1-\eta^2)
\end{aligned}\right\} \tag{14.11}$$

The equilibrium equations for FGCNT-reinforced composite plate subjected to thermomechanical loading can be obtained using the principle of total minimum potential energy and expressed as

$$\left[K\right]^{e}\left\{\delta\right\}^{e} = \left\{F_{m}\right\}^{e} + \left\{F_{th}\right\}^{e} \tag{14.12}$$

where $[K]^{e}$ and $\{\delta\}^{e}$ are the stiffness matrix and displacement vector in elemental form, respectively. $\{F_{m}\}^{e}$ and $\{F_{th}\}^{e}$ are the mechanical and thermal force vectors, respectively.

14.7   RESULTS AND DISCUSSION

A homemade computer code is developed in APDL environment to compute the bending responses of FGCNT-reinforced composite perforated plate under thermomechanical loading conditions. Here FGCNT-reinforced composite plate is constituted of *PmPV* as matrix and SWCNT as reinforcement fiber materials, respectively, and the material properties are mentioned in Tables 14.1 and 14.2 as same as in Zhu et al. [12]. To examine the bending behavior, various support conditions on plate edges are considered, such as

Fully clamped support (CCCC):

$u_{0} = v_{0} = w_{0} = \phi_{x} = \phi_{y} = \phi_{z} = 0$ at $x = 0$, a and $y = 0$, b.

Fully simply supported (SSSS):

$v_{0} = w_{0} = \phi_{y} = \phi_{z} = 0$ at $x = 0$ and a; $u_{0} = w_{0} = \phi_{x} = \phi_{z} = 0$ at $y = 0$ and b.

Clamped and simply supported (SCSC):

$v_{0} = w_{0} = \phi_{y} = \phi_{z} = 0$ at $x = 0$ and a; $u_{0} = v_{0} = w_{0} = \phi_{x} = \phi_{y} = \phi_{z} = 0$ at $y = 0$ and b.

Initially, the perforated model is discretized at different mesh densities and 1,974 elements with 6,358 nodes are found adequate for further analysis, if not mentioned otherwise. To validate the present model, simply supported FGCNT-reinforced composite plate is considered at different values of volume fraction of fiber material ($V_{CNT} = 0.11, 0.14, 0.17$) and thickness ratios ($a/h = 10, 20, 50$). The nondimensional central deflection ($\bar{w} = w/h$) parameters are computed and compared with the reported results of Zhu et al. [12] and presented in Table 14.3. It is clear that the difference between the present and the previously reported results of Zhu et al. [12] is negligible. Further, to demonstrate the robustness of the developed model, a variety of numerical experiments are carried out. Here the influences of different geometrical and material parameters such as volume fraction of CNT (V_{CNT}), thickness ratio (a/h), aspect ratio (a/b), and temperature (T) and support conditions on the bending responses of FGCNT plates are demonstrated. In the forthcoming numerical examples, nondimensional central deflections of FGCNT (UD, FG-X, FG-V, FG-O) reinforced composite perforated plate are computed for the following set of parameters, if not state otherwise:

$a/b = 1$; $a/h = 10$; SSSS; $n \times n = 3 \times 3$; $V_{CNT} = 0.11, 0.14, 0.17$; temperature $T = 300K$; uniform pressure $p = 0.1$ MPa.

TABLE 14.3
Nondimensional Central Deflection of a Simply Supported FGCNT Reinforced Composite Plate Subjected to Uniformly Distributed Load

V_{CNT}	b/h	Present	Ref. [12]	% Difference
0.11	10	0.003740	0.003739	0.027
	20	0.03630	0.03628	0.055
	50	1.16000	1.15500	0.130
0.14	10	0.00330	0.00331	0.182
	20	0.03000	0.03001	0.200
	50	0.91600	0.91750	0.164
0.17	10	0.00240	0.00239	0.042
	20	0.02350	0.02348	0.085
	50	0.75400	0.75150	0.265

TABLE 14.4
Effect of Thickness Ratios (a/h) on the Nondimensional Central Deflection of Perforated FGCNT Reinforced Composite Plate

V_{CNT}	a/h	UD	FG-X	FG-V	FG-O
0.11	5	0.00073	0.00077	0.00081	0.00071
	20	0.06594	0.06988	0.05764	0.06276
	100	28.2061	30.9416	34.0696	25.8330
0.14	5	0.00069	0.00073	0.00077	0.00067
	20	0.06180	0.06524	0.06914	0.05912
	100	25.8979	28.2579	30.8997	23.8835
0.17	5	0.00046	0.00049	0.00051	0.00045
	20	0.04178	0.04434	0.04730	0.03970
	100	17.9306	19.7172	21.7350	16.3910

Table 14.4 exhibits the effect of thickness ratios (a/h = 5, 20, 100) on the deflection parameters of perforated FGCNT-reinforced composite plate at different values of V_{CNT}. It is clear that the nondimensional deflections are enhancing with the increment in thickness ratios because thick (a/h = 5) perforated plates are stiffer than the thin (a/h = 100) plates.

Table 14.5 exhibits the effect of aspect ratios (a/b = 1, 1.5, 2) on the deflection parameters of perforated FGCNT-reinforced composite plate at different values of V_{CNT}. Here the nondimensional deflections are diminishing with the increment in aspect ratios, i.e., perforated plates with a/b = 2 are stiffer compared to square (a/b = 1) plates.

Table 14.6 demonstrates the effect of support conditions (CCCC, SCSC, SSSS) on the deflection parameters of perforated FGCNT-reinforced composite plate at different values of V_{CNT}. Here perforated FGCNT-reinforced composite plates under fully clamped (CCCC) support condition exhibit the minimum deflection value, whereas maximum deflections are observed in case of fully simply supported plate.

Table 14.7 demonstrates the effect of thermal field (T = 300 K, 500 K, 700 K) on the deflection parameters of perforated FGCNT-reinforced composite plate at

TABLE 14.5

Effect of Aspect Ratios (a/b) on the Nondimensional Central Deflection of Perforated FGCNT Reinforced Composite Plate

V_{CNT}	a/b	UD	FG-X	FG-V	FG-O
	1	0.01556	0.01588	0.01636	0.01523
0.11	1.5	0.62341	0.00167	0.00176	0.0151
	2	0.000697	0.000741	0.00078	0.000674
	1	0.01478	0.01508	0.01549	0.0145
0.14	1.5	0.00149	0.00158	0.00166	0.00144
	2	0.000663	0.000704	0.000739	0.000642
	1	0.00983	0.01004	0.01034	0.00962
0.17	1.5	0.000992	0.00106	0.00112	0.000955
	2	0.000414	0.000469	0.000493	0.000426

TABLE 14.6

Effect of Support Conditions on the Nondimensional Central Deflection of Perforated FGCNT Reinforced Composite Plate

V_{CNT}	Supports	UD	FG-X	FG-V	FG-O
0.11	CCCC	0.00265	0.00285	0.00306	0.00252
	SSSS	0.01560	0.01590	0.01640	0.01520
	SCSC	0.00587	0.00619	0.00654	0.00564
0.14	CCCC	0.00249	0.00268	0.00286	0.00238
	SSSS	0.0148	0.01510	0.01550	0.0145
	SCSC	0.00554	0.00584	0.00614	0.00535
0.17	CCCC	0.00167	0.00181	0.00194	0.00159
	SSSS	0.00983	0.01000	0.01030	0.00962
	SCSC	0.00371	0.00392	0.00414	0.00357

TABLE 14.7

Effect of Temperature on the Nondimensional Central Deflection of Perforated FGCNT Reinforced Composite Plate

V_{CNT}	Temperature	UD	FG-X	FG-V	FG-O
0.11	300	0.01556	0.01588	0.01636	0.01523
	500	0.02751	0.04310	0.02865	0.02708
	700	0.13995	0.15594	0.14327	0.13881
0.14	300	0.01478	0.01508	0.01549	0.0145
	500	0.02621	0.04059	0.02722	0.02584
	700	0.13389	0.14944	0.13692	0.13286
0.17	300	0.00983	0.01004	0.13692	0.00926
	500	0.01737	0.03328	0.01808	0.01709
	700	0.0883	0.10479	0.09024	0.08757

different values of V_{CNT}. It is clear that temperature increases significantly deteriorate the stiffness of the perforated plate.

In addition to the previous observations, Tables 14.4–14.7 also reveal that in perforated FGCNT-reinforced composite plate, structural stiffness can be enhanced by increasing the volume fraction of fiber material. However, nondimensional deflections are diminishing in order of FG-V, FG-X, UD, FG-O type FGCNT composites.

14.8 CONCLUSIONS

The deformation behavior of FGCNT-reinforced composite plate with perforations is examined under thermomechanical loading. Here effective material properties of FGCNT-reinforced composite plate are evaluated using the extended rule-of-mixture for different gradation types (FG-UD, FG-X, FG-V, and FG-O). The perforated structure is discretized using FSDT-based eight-noded serendipity quadrilateral element and solved simultaneously via developed APDL code. The validation of the present model is examined by comparing present results with the previously reported results. The comprehensive parametric study reveals the robustness of developed perforated FGCNT-reinforced composite model at different sets of conditions. The impacts of thickness, aspect ratios, temperature, and support conditions on the deformation behavior of perforated FGCNT-reinforced composite plates are significant.

ACKNOWLEDGMENT

The authors would like to thank the Science and Engineering Research Board, Department of Science and Technology, Government of India (File No. ECR/2016/001829), for the financial support.

REFERENCES

1. Treacy, M. J., and Ebberse, W., "Exceptionally high young's modulus observed for individual carbon nanotubes," *Nature*, 381, pp. 678, 1996.
2. Pindera, M.-J., Aboudi, J, Glaeser, A.M., and Arnold, S. M., "Use of composites in multi-phased and functionally graded materials," *Composites, Part B*, 28, pp. 1–175, 1997.
3. Suresh, S., and Mortensen, A., *Fundamentals of Functionally Graded Materials*. 10M Communications, London, 1998.
4. Pindera, M.-J., Arnold, S. M., Aboudi, J., and Hui, D., "Use of composites in functionally graded materials," *Composites Eng.* 4, pp. 1–145, 1994.
5. Pindera, M.-J., Aboudi, J., Arnold, S. M., and Jones, W. F., "Use of composites in multi-phased and functionally graded materials," *Composites Eng.*, 5, pp. 743–974, 1995.
6. Markworth, A. J., Ramesh, K. S., and Parks, W. P.: "Review: Modeling studies applied to functionally graded materials," *J. Mater. Sci.*, 30, pp. 2183–2193, 1995.
7. Schiller, C., Siedler, M., Peters, F., and Epple, M.: "Functionally graded materials of biodegradable polyesters and bone-like calcium phosphates for bone replacement," *Functionally Graded Materials 2000*, Proceedings of the Sixth International symposium on Functionally Graded Materials, The Arnerican Ceramic Society, Westerville, OH, pp. 97–108, 2000.

8. Vodenitcharova, T., Zhang, L.C.: "Bending and local buckling of a nanocomposite beam reinforced by a single-walled carbon nanotube," *International Journal of Solid and Structures*, 43, pp. 3006–3024, 2006.
9. Shen, H. S.: "Nonlinear bending of functionally grade carbon nanotube-reinforced composite plates in thermal environments," *Composite Structures*, 91, pp. 9–19, 2009.
10. Murmu, T., Pradhan, S. C.: "Buckling analysis of a single –walled carbon nanotube embedded in an elastic medium based on nonlocal elasticity and Timoshenko beam theory and using DQM," *Physica E*, 41, pp. 1232–1239, 2009.
11. Ayatollahia, M. R., Shadlou, S., Shokrieh, M. M.: "Multiscale modelling for mechanical properties of carbon nanotube reinforced nanocomposites subjected to different types of loading," *Composite Structures*, 93, pp. 2250–2259, 2011.
12. Zhu, P., Lei, Z. X., Liew, K. M.: "Static and free vibration analysis of carbon nanotube-reinforced composite plates using finite element method with first order shear deformation plate theory," *Composite Structures*, 94, pp. 1450–1460, 2012.
13. Yas, M. H., Samadi, N.: "Free vibration and buckling analysis of carbon nanotube –reinforced composite Timoshenko beams on elastic foundation", *International Journal of Pressure Vessels and Piping*, 98, pp. 119–128, 2012.
14. Bian, Z. G., Ying, J., Chen, W. Q., Ding, H. J.: "Bending and free vibration of a smart functionally graded plate", *Structural Engineering and Mechanics*, 23, pp. 97–113, 2006.
15. Bousahla, A. A., Houari, M. S. A., Tounsi, A., Bedia, E. A. A.: "A novel higher order shear and normal deformation theory based on neutral surface position for bending analysis of advanced composite plates," *International Journal of Computational Methods*, 11, 1350082, 2014.
16. Alibeigloo, A. Liew, K. M.: "Thermoelastic analysis of functionally graded carbon nanotube reinforced composite plate using theory of elasticity," *Composite Structures*, 106, 873–881, 2013.
17. Zhang, L. W., Lei, Z. X., Liew, K. M., Yu, J. L.: "Static and dynamic of carbon nanotube reinforced functionally graded cylindrical panels," *Composite Structures*, 111, pp. 205–212, 2014.
18. Mehrabadi, S. J., Aragh, B. S.: "Stress analysis of functionally graded open cylindrical shell reinforced by agglomerated carbon nanotubes," *Thin-Walled Structures*, 80, pp. 130–141, 2014.
19. Mehar, K., Panda, S. K., Dehengia, A., Kar V. R.: "Vibration analysis of functionally graded carbon nanotube reinforced composite plate in thermal environment," *Journal of Sandwich Structures and Materials*, 18(2), pp. 2151–2173, 2015.

15 Damage Studies in Curved Hybrid Laminates under Pullout Loading

Bipin Kumar Chaurasia and Deepak Kumar

National Institute of Technology Jamshedpur,
Jamshedpur, India

Rene Roy

Research Center for Aircraft Parts Technology,
Gyeongsang National University,
Jinju, South Korea

CONTENTS

15.1   INTRODUCTION

Composite structures can provide high stiffness and strength with low weight, which can increase the payload of airplanes without losing airworthiness [1]. Carbon fiber–reinforced polymer (CFRP) is one of the commonly used composites with low density and strength-to-weight ratio equivalent to steel [2–5]. Thus, many industries, such as automobile, aerospace, and submarine, are using CFRP composites. Although its

DOI: 10.1201/9781003158813-15

application used to be limited mainly to the secondary structures of aircraft, its scope has recently expanded to include principal structures such as wing skin, tails spars, wing spars, and frames [6–8]. The spar is generally in the form of a beam of C-shaped cross-section and consists of two flanges attached to the outer skin and the web that carries the shear load. The flanges and the web meet at the curved corners. Therefore, corner radius can affect the interlaminar stresses in these areas. Delamination at the corner is one of the typical failure modes that can occur in composite spars. One of the typical failures is delamination, which can occur in composite spars. Since such delamination is mainly caused by the stresses inside the structures, the damage of the spar at the corners is closely related to the distribution of stresses in these areas [9]. Failure processes in real composite materials are very diverse, so work describing the strength of specific types of composite materials is of particular interest. Some of the published papers deal with a model to describe the deformation and failure processes of multilayer hybrid composites under a plane stress condition [10]. There are Hashin's criteria for unidirectional laminates and Puck's concept of plane of action, the physics-based LaRC02 criteria, which accurately predict matrix and fiber failure without the need for curve-fitting parameters [11]. Spottswood and Anthony used third-order shell kinematics to determine the material behavior of a thin, curved composite plate under transverse loading. Hashin failure criteria were implemented to test the shell from the first load [12]. Respecting manufacturing tolerances is a challenging issue due to the complex distortions caused by the curing process, being sometimes a major obstacle to an increasing use of composites in aeronautics. The cure-induced distortions are modeled and mitigated owing to the development of a computational model compensation strategy [13]. In the literature [14], there are simple methods for the curved laminated structure under pure bending moment. The maximum radial stress was calculated and compared with classical elasticity theory and two-dimensional finite element analysis.

The method was also verified by Cui et al. [15] when nonlinear deflection was considered in the analysis. They also developed a method to evaluate the stress in curved laminates.

Many authors [16,17] have studied the stress distribution of a curved beam under bending load using 3-D finite element analysis and also evaluated the interlaminar shear stress. In addition, they developed an analytical solution assuming plane strain and plane stress. They found that the stresses in the thickness direction differ significantly under the plane stress conditions.

The error in the interlaminar stresses was determined from the reported analytical formulas and scaled FEM in curved composite laminates under various loading conditions. Moreover, the reported error was high in the case of longitudinal loading and relatively low in the case of moment loading. Another important observation was made with respect to the maximum stress measured at one-third of the laminate thickness from the inner radius [18]. A. B. de Morais [19] had studied the mode I delamination for the composite beam using cohesive zone method. In this, he used bilinear cohesive law. Further, he observed that the present model agrees with finite element analyses of double cantilever beam and moment-loaded double cantilever beam specimens with a wide range of properties. Also, Dugdale [20] and Barenblatt [21] had investigated the failure characteristics of brittle materials by the cohesive zone model. In the Barenblatt model [21], the normal stress at the crack tip is equal to the theoretical strength of the material, and the distribution of stress in the cohesive zone depends on the distance between the crack tips. In the Dugdale model [20], the stresses in the cohesive zone are constant and equal to the yield strength of the material [22]. Geleta et al. [23] also used the cohesive zone model to simulate the delamination of plies. Further, they have inserted a zero-thickness cohesive element between the bulk finite elements for the occurrence of multiple delamination. In the cohesive zone method, various traction separation laws are implemented to know the behavior of the cohesive elements. Those traction separation laws that are proposed by different researchers over the years are triangular, exponential, cubic polynomial, and trapezoidal [23–26].

Ritesh Bhat et al. [27] had performed an experiment to check the tensile strength, flexural strength, and Barcol hardness. They found that the tensile strength and Barcol hardness increase with an increase in thickness of the specimens, but flexural strength initially increases with an increase in thickness but then begins decreasing with an increase in thickness. This phenomenon happens due to delamination behavior of GFRP. A stacking sequence was found to have a remarkable effect on delamination and in-plane tensile strength failure. Other structures that have curved regions, such as the spars of the aircraft, include a compact T-section composite. For the curved laminated composites, the transverse shear and normal stresses were the main causes of failure under different loading conditions [28]. The variation of transverse shear stresses with depth was calculated using 3-D finite element analysis. In addition, the interlaminar stresses in the curved regions were evaluated. In general, the prediction of delamination initiation in laminated composites can be based on average stress criteria, while delamination growth can be predicted by methods such as fracture mechanics, virtual crack closure technique (VCCT), and decohesive elements [29]. The normal tensile and out-of-plane shear stresses are the main cause of delamination for the stresses considered, but compressive stresses are also important and cannot be neglected in some of the cases [14]. There is various reported research available in which most of the authors have worked on the L-shaped and T-shaped composite laminates.

In this study, a finite element analysis of C-shaped hybrid laminates was carried out under pullout loading to investigate the failure behavior of these structures. The simulated laminates were taken from a C-Spar. Here we performed a damage analysis using in-built Hashin failure criteria using plane stress and plane strain model to visualize the various damages within the curved regions of the laminates. Hashin failure criteria were used to predict the final failure, and quadratic stress criteria was used to predict delamination. The failure load prediction was verified by comparing it with the results of the pullout test in the available reported literature [31].

15.2 COMPUTATIONAL DAMAGE MODEL

Progressive failure analysis of composite laminate is based on numerical damage model. For progressive failure analysis, Hashin damage criteria represent the most implemented criteria for damage initiation of composite material, whereas material degradation scheme is used for damage evolution. These methods are used for damage initiation and evolution in composite material and talk about the various failures in composite lamina in terms of damages. Hashin criterion is established in terms of the strength of composite material, whereas material property degradation is based on displacement. The damage criteria are divided into two stages of progressive failure analysis: (1) damage initiation and (2) damage evolution [32,33].

Load is acting on the composite specimen when stress increases; when it reaches the value $\sigma = \sigma_o$, then damage initation reaches critical point and complete damage initiation, after which actual damage occurs. Damage initiation is followed by Hashin damage criteria, whereas damage evolution is followed by material degradation law.

15.2.1 Damage Initiation Law

Damage initiation of composite is based on strength (as per Hashin damage initiation criteria), whereas strength is dependent on matrix and fiber, tensile or compressive stress. When any of the stress values reaches a critical point, damage initiation is complete and damage evolution begins. As per the damage initiation, four modes (Equations (15.1–15.4)) of failure are given in the Hashin criteria as follows:

Fiber tensile failure

$$F_{ft} = \left(\frac{\sigma_1}{X_T}\right)^2 + \left(\frac{\sigma_{12}}{S_{12}}\right)^2 \geq 1 \tag{15.1}$$

Fiber compressive failure

$$F_{fc} = -\left(\frac{\sigma_1}{X_c}\right) \geq 1 \tag{15.2}$$

Matrix tensile failure ($\sigma_{22} \geq 0$)

$$F_{mt} = \left(\frac{\sigma_2}{Y_t}\right)^2 + \left(\frac{\sigma_{12}}{S_{12}}\right)^2 \geq 1 \tag{15.3}$$

Matrix compressive failure ($\sigma_{22} < 0$)

$$F_{mc} = \left(\frac{\sigma_2}{2S_{12}}\right)^2 + \left(\frac{\sigma_2}{Y_c}\right)\left[\left(\frac{Y_c}{2S_{12}}\right)^2 - 1\right] + \left(\frac{\sigma_{12}}{S_{12}}\right)^2 \geq 1 \tag{15.4}$$

In the above equations, X_T and X_c are tensile and compressive strengths in the fiber direction; Y_t and Y_c are tensile and compressive strengths in the transverse direction; and S_{12} is shear strength in the 1–2 plane. If any of the failure values reaches 1, then damage initiates ($\sigma = \sigma_o$), and after that damage evolves.

15.2.2   DAMAGE EVOLUTION LAW

Damage evolution on a composite depends on material property degradation of the specimen. This material property degradation is related to displacement. The damage model with a gradual degradation scheme is applied for this evolution of damage. The equation for this damage criterion is

$$d_I = \frac{\delta_{1,eq}^f \left(\delta_{1,eq} - \delta_{1,eq}^0 \right)}{\delta_{1,eq} \left(\delta_{1,eq}^f - \delta_{1,eq}^0 \right)} \left(d_I \in [0,1] . I = ft, fc, mt, mc \right) \tag{15.5}$$

In the above equation, $\delta_{1,eq}^0$ stands for the displacement of damage initiation, $\delta_{1,eq}^f$ for final failure displacement, and $\delta_{1,eq}$ is equivalent displacement. If $\delta_{1,eq} = \delta_{1,eq}^f$, damage evolution is completed and failure occurs. In the model, a small amount of load is applied to the initial input parameters and boundary conditions, and the system is assessed for damage initiation; if damage initiation is < 1, then the stress–strain ratio is calculated and updated. The cycle is repeated until the damage initiation value reaches 1. At that point, damage evolution criteria are assessed. Here the damage due to material property is checked for failure of the composite. If the damage value d < 1, then the failure of the composite is not completed, so stress–strain and degradation of material property (because damage evolution has started) are calculated. These properties are updated and the cycle continues with the next incremental load increase. Once the damage reaches 1, the composite fails, and at that point stress–strain ratior, failure load, and type of damage are calculated and the process stopped.

15.2.3   INTERLAMINAR DAMAGE

In the cohesive zone or interface of an adjacent layer of the composite, delamination is called interlaminar damage. This interlaminar damage occurs due to traction stress and separation displacement of the node, which is governed by traction separation law (TSL). The delamination propagation takes place in the interface under mix-mode loading [32,34]. In the mix mode loading, effective relative displacement of layers is given by Equation (15.6) (Figure 15.1):

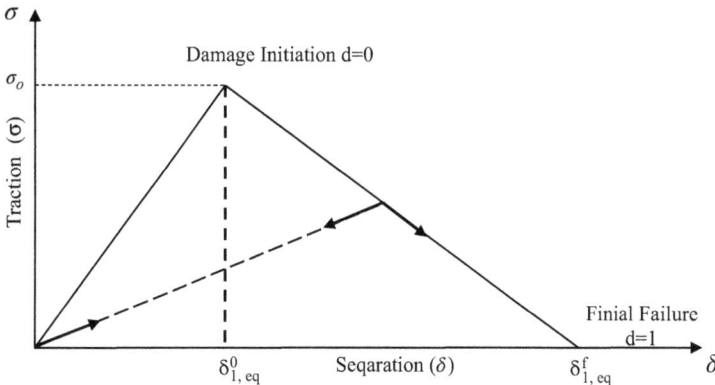

FIGURE 15.1   Traction-separation law for interlaminar damage [32].

$$\delta_m = \sqrt{\delta_1^2 + \delta_2^2} \tag{15.6}$$

For linear reduction, the damage variable (d) for evaluation of damage is

$$d = \frac{\delta_m^f\left(\delta_m^{max} - \delta_m^0\right)}{\delta_m^{max}\left(\delta_m^f - \delta_m^0\right)}\left(d \in \{0,1\}\right) \tag{15.7}$$

δ_m^0 and δ_m^f are relative displacement of the interface at initial and final damage failure. Quadratic stress criteria used to determine damage initiation (δ_m^0) and interactive power law are used for final displacement (δ_m^f); details of these expressions can be found in reference [32].

15.3 FINITE ELEMENT MODEL

15.3.1 MODELING STRATEGY

Based on the experiments of Kim et al. [31], a hybrid composite laminate under pullout loading is investigated here using 2D finite element analysis (FE). The commercially available FE-based ABAQUS software is used to perform the failure analysis.

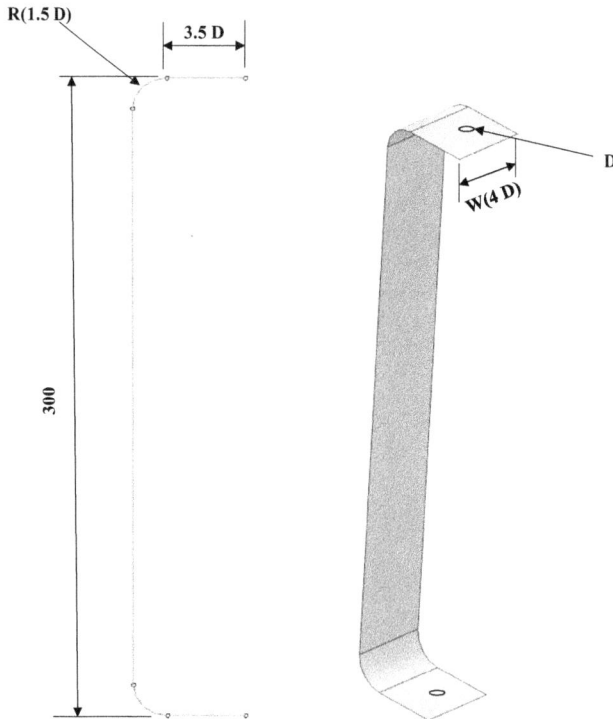

FIGURE 15.2   Section of C-Spar geometry details.

TABLE 15.1
Material Properties

Material Properties	GEP118, Glass Fabric/ Epoxy Prepreg [35]	USN125B, Unidirectional Carbon/Epoxy Prepreg [31]
E_1 (GPa)	26.5	131
E_2 (GPa)	24.8	7.6
G_{12} (GPa)	4.04	5.34
υ_{12}	0.148	0.31
X_T (MPa)	375.8	1975
Y_T (MPa)	284.4	51
Z_T (MPa)	40	37.6
S_{12} (MPa)	47.5	82.2

Hashin damage criterion is used for matrix and fiber failure, and tensile separation law is used for delamination. The geometry is created in ABAQUS as shown in Figure 15.2. The diameter (D) of the bolt (hole) is 6.35 mm and the width (W) of the specimen is 25.4 mm, four times larger than the bolt diameter. The bolt hole is located in the middle of the width of the specimen, and the total height (h) is 300 mm. The corner radius of the C-Spar section is taken as 1.5 D.

Hybrid laminates are designed considering GEP118 and unidirectional carbon/epoxy prepreg, and the stacking order is considered as $[0°/30°/-30°/0°/45°/-45°/45°/-45°/45°/-45°]_S$. The first four layers are chosen as USN125B, and the remaining layers are chosen as GEP118. The thicknesses of the glass fiber epoxy prepreg and the unidirectional carbon epoxy prepreg are 0.1455 and 0.12 mm, respectively. The total thickness of the hybrid laminate is 3.2880 mm. The material properties are listed in Table 15.1. Two types of analysis are carried out. One is the plane strain model and the other is the plane stress model.

15.3.2   PLANE STRAIN MODEL

The geometry is meshed with CPE4R elements, and the whole model has 4,325 nodes and 4,210 elements. The meshed model is shown in Figure 15.4. Here, "zero-thickness elements" (the total number of cohesive elements = 4,035) were inserted in the interface to observe the delamination behavior. The cohesive properties for the interface element were obtained from the literature and are shown in Table 15.2 (Figure 15.3).

TABLE 15.2
Material Parameters Used in the Interface Cohesive Element [36]

	Mode 1	Mode 2
Normalized elastic modulus (MPa/mm)	100,000	100,000
Interlaminar strength (MPa)	40	40
Interlaminar fracture toughness (kJ/m^2)	0.28	0.82

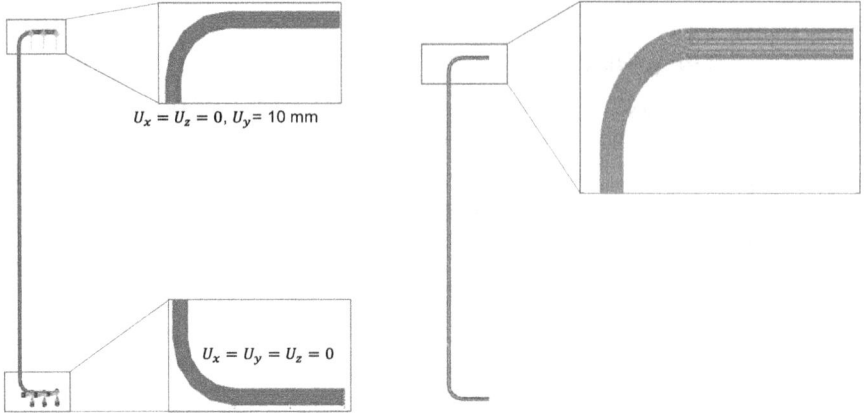

FIGURE 15.3   Boundary conditions and meshed conditions (plane strain model).

To apply the load, one end of the laminate is fixed ($U_x = U_y = U_z = 0$), and axial displacement ($U_y = 10$ mm) is applied to the other end. The model may experience excessive element distortion, as stiffness is locally reduced due to internal damage. The distortion leads to additional difficulties in numerical convergence, resulting in a slow-running or even aborted solution. Distortion control has been enabled in ABAQUS, and the damage variables have been limited to a maximum value of 0.99, which helps preserve some residual stiffness. As described in Section 15.2, the mixed-mode traction separation law is used to define the delamination evolution.

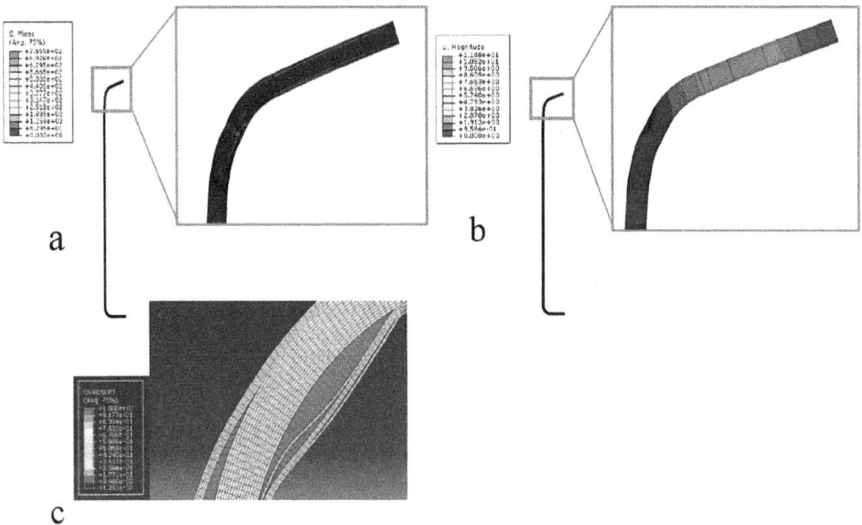

FIGURE 15.4   (a) Maximum stress obtained, 755 MPa. (b) Maximum displacement, 11.4 mm. (c) Maximum delamination failure at 0°/30° interface.

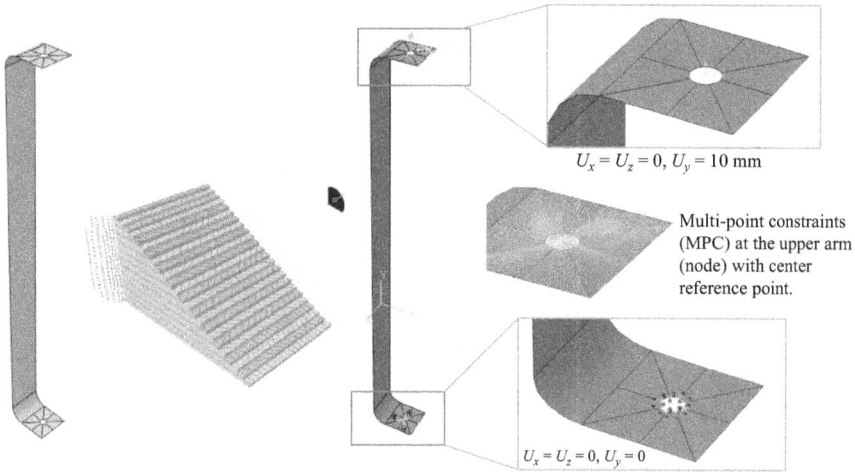

FIGURE 15.5 Fiber orientation and boundary conditions (plane stress model).

The quadratic stress criterion was used to test delamination failure at interfaces. The results show that delamination failure occurs only in the CFRP plies compared to the GFRP plies. The maximum stress is 755 MPa (Figure 15.4) and the calculated failure load is 7.90 kN.

15.3.3 PLANE STRESS MODEL

The geometry was meshed with CPS4R elements, and the whole model has 11,855 nodes and 11,440 elements. The geometrical and meshed model are shown in Figure 15.5 and 15.6, respectively. To apply the load, one end of the laminate is fixed ($U_x = U_y = U_z = 0$) and axial displacement ($U_y = 10$ *mm*) is applied to the other end. In-built Hashin failure criteria are used to predict the various damages such as fiber, matrix, and shear damages in the hybrid laminates.

The results show that the maximum stress is 597 MPa in the 0° ply (Figure 15.7), and the calculated failure load is 8.22 kN.

15.4 FAILURE LOAD PREDICTION

The pullout load applied to the flange passes through the corner radius of the C-Spar and generates the interlaminar and intralaminar stresses. Figure 15.8 shows the predicted load for the designed hybrid laminates. From Figure 15.8, peak values (load) are seen in the load-displacement diagram for the model with plane strain and for the reported experimental data [31], but in the case of plane stress, the failure load decreases continuously after reaching the maximum load. It can also be seen from the figure that the deviation of the load–displacement curve between the experiment and the plane strain model is the largest, but in the case of the plane strain model, the curve follows that of the experiment. In the

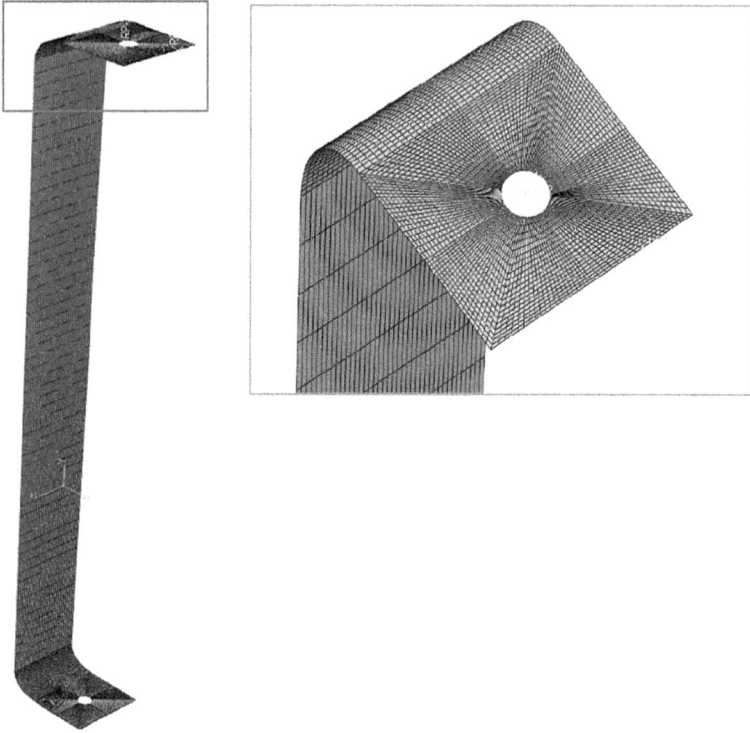

FIGURE 15.6 Meshing.

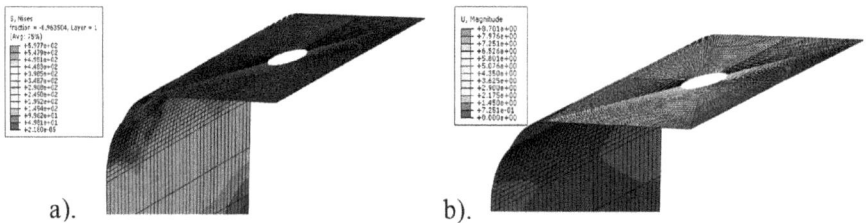

a). b).

FIGURE 15.7 (a) Maximum stress (597 MPa, 0° ply). (b) Maximum displacement.

case of plane strain model, the maximum stress before failure is 755 MPa and the deformation is 11.4 mm, while in the case of plane stress, the maximum stress is 597 MPa and observed in the 0° ply. Furthermore, the load-carrying capacity for the case of plane stress and plane strain models was found to be 8.22 kN and 7.92 kN, respectively. The computed failure load is comparable to the reported research results [31].

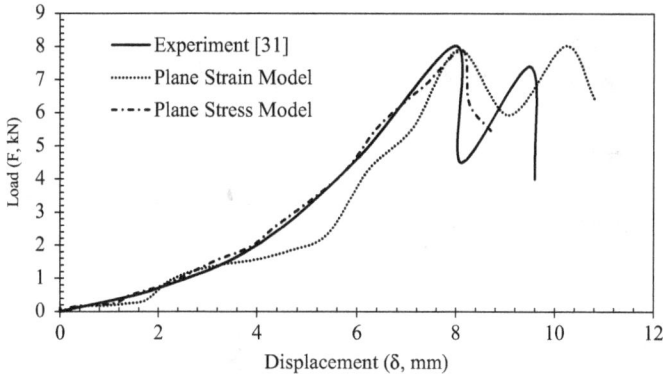

FIGURE 15.8 Load versus displacement plot.

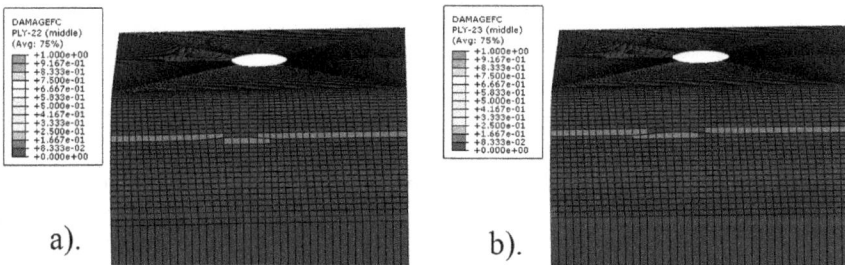

FIGURE 15.9 Predicted fiber compressive damage: (a) $-30°$ ply; (b) $30°$ ply.

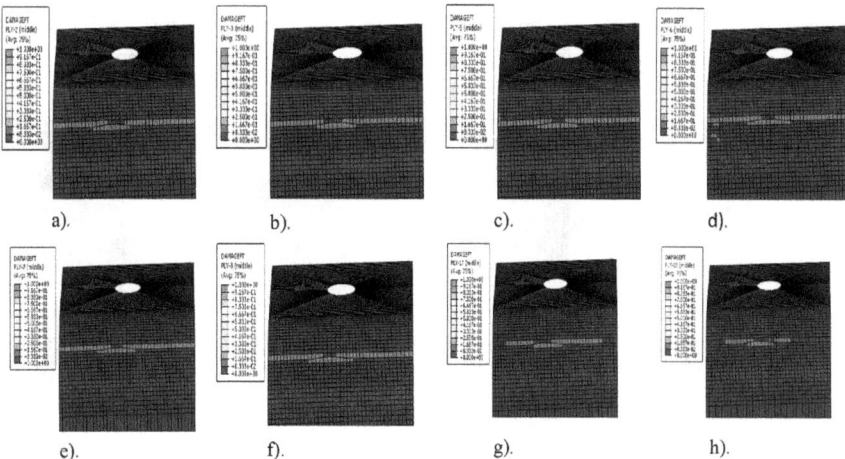

FIGURE 15.10 Predicted fiber tensile damage in various plies: (a) Ply 2 ($30°$), (b) Ply 3 ($-30°$), (c) Ply 5 ($45°$), (d) Ply 6 ($-45°$), (e) Ply 7 ($45°$), (f) Ply 8 ($-45°$), (g) Ply 17 ($-45°$) and (h) Ply 18 ($45°$).

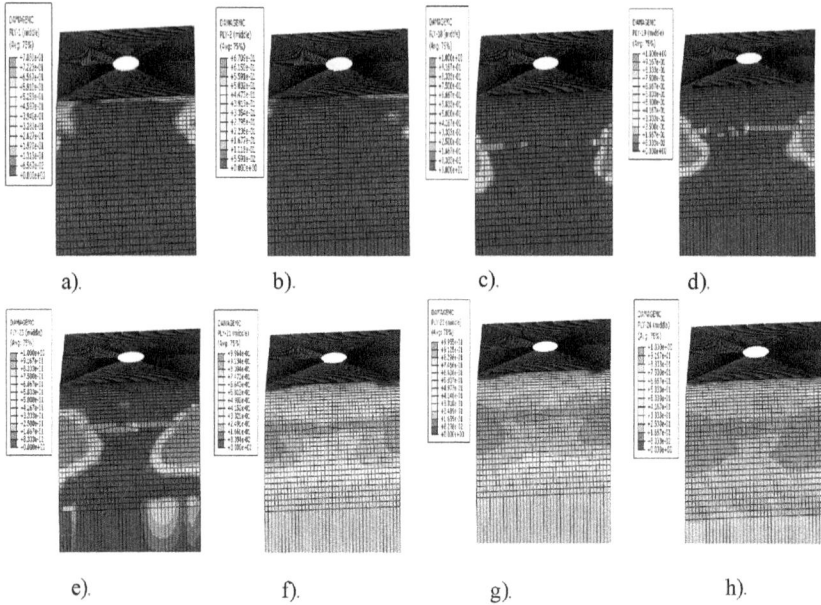

a). b). c). d).

e). f). g). h).

FIGURE 15.11 Predicted matrix compressive damage in various plies: (a) Ply 1 (0°), (b) Ply 2 (30°), (c) Ply 18 (45°), (d) Ply 19 (−45°), (e) Ply 20 (45°), (f) Ply 21 (0°), (g) Ply 22 (−30°) and (h) Ply 24 (0°).

15.5 DAMAGES PREDICTED IN CURVED REGION

The different types of damage are predicted by Hashin's failure criteria for the plane stress model. In the case of the C-Spar laminates, a folding force is applied, resulting in tensile stresses in the inner curve region, while compressive stresses occur in the upper layer of the hybrid laminates. The results show that the fibers fail in the 22nd and 23rd layers (−30°/30°). This damage (Figure 15.9) is caused by the fiber compression when

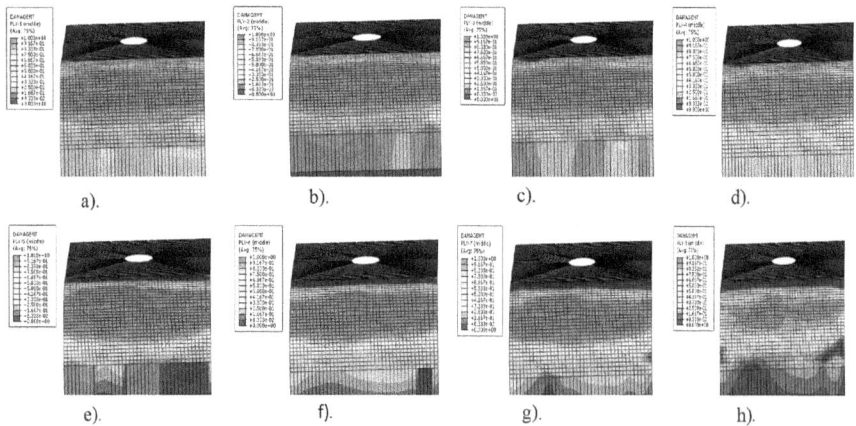

a). b). c). d).

e). f). g). h).

FIGURE 15.12 Predicted matrix tensile damage in various plies: (a) Ply 1 (0°), (b) Ply 2 (30°), (c) Ply 3 (30°), (d) Ply 4 (0°), (e) Ply 5 (45°), (f) Ply 6 (−45°), (g) Ply 7 (45°) and (h) Ply 8 (−45°).

a).   b).   c).   d).

e).   f).   g).   h).

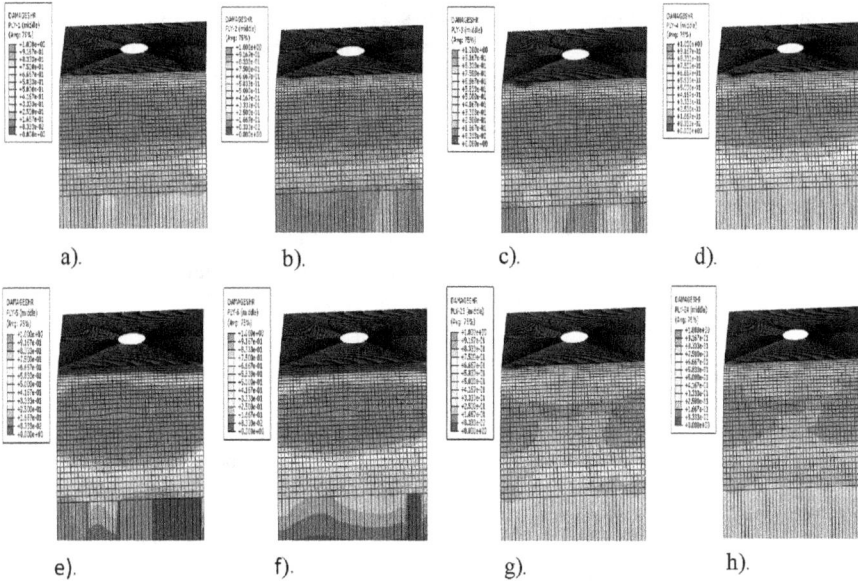

FIGURE 15.13 Predicted shear damage in various plies: (a) Ply 1 (0°), (b) Ply 2 (30°), (c) Ply 3 (−30°), (d) Ply 4 (0°), (e) Ply 5 (45°), (f) Ply 6 (−45°), (g) Ply 23 (30°) and (h) Ply 24 (0°).

the stresses in the y-direction reach the maximum. On the other hand, fiber tensile damage occurs in the inner layer of the laminates. From Figure 15.10, the effects of tensile damage decrease as we move from the inner layer of the laminate to the upper layer. Therefore, the maximum tensile damage is predicted in the 30° layer. Matrix failure due to compression is maximum in the upper layer. From Figure 15.11 (layer 24), the main failure occurs in the curve region due to the matrix failure by compression. Therefore, the damage starts in the inner region (layer 1 and 2) and then spreads. Tensile damage in the matrix is the main cause for the onset of failure in the hybrid laminates. The complete failure occurs in the inner curve section, and the damage decreases with the transition from the inner to the outer layers (Figure 15.12). From Figure 15.12, the damage predicted in the middle plies is less compared to the first plies at the applied load, and they disappear in the 24th layer. Shear damage is also a major cause of failure of designed hybrid composite laminate. The predicted damage (Figure 15.13) can be seen for the curve section of each layer of the laminate. However, it can be better predicted by the 3D model, which considers the out-of-plane interlaminar stresses.

15.6 CONCLUSION

C-shaped hybrid laminates are analyzed by considering plane stress and plane strain conditions to determine the failure load and damages. Load versus displacement is also plotted and compared with the available equivalent experimental data [31]. The failure load in the case of the plane stress model is obtained as 8.22 kN and in the case of the plane strain model as 7.94 kN. The failure load decreases continuously after reaching the maximum load in the case of plane stress. Further, it can be seen from the plot that the deviation of the load–displacement curve between the experiment and the plane strain model is the largest, but the curve follows that of the experimental data in the case of plane strain [31]. Matrix damage due to compression occurs in the upper portion of the curve, while matrix failure due to tension occurs in the inner portion of the curve. Shear damage can also have significant effects on laminate failure, but the 3D model can better predict shear damage, the main cause of failure of the model. The results lead to a conclusion that hybrid laminates fail mainly due to tensile failure of the matrix and delamination (0°/30° interface).

REFERENCES

[1] Abrate, S., 1998. *Impact on composite structures*. Cambridge University Press.
[2] Friedrich, K. and Almajid, A.A., 2013. Manufacturing aspects of advanced polymer composites for automotive applications. *Applied Composite Materials*, 20(2), 107–128.
[3] Manocha, L.M., 2003. High performance carbon-carbon composites. *Sadhana*, 28(1), 349–358.
[4] Xiao, M., Yongbo, Z., Zhihua, W. and Huimin, F., 2017. Tensile failure analysis and residual strength prediction of CFRP laminates with open hole. *Composites Part B: Engineering*, 126, 49–59.
[5] Nirbhay, M., Dixit, A., Misra, R.K. and Mali, H.S., 2014. Tensile test simulation of CFRP test specimen using finite elements. *Procedia Materials Science*, 5, 267–273.
[6] Morioka, K. and Tomita, Y., 2000. Effect of lay-up sequences on mechanical properties and fracture behaviour of CFRP laminate composites. *Materials Characterization*, 45(2), 125–136.
[7] Kumazawa, H., Aoki, T. and Susuki, I., 2006. Influence of stacking sequence on leakage characteristics through CFRP composite laminates. *Composites Science and Technology*, 66(13), 2107–2115.
[8] Awerbuch, J. and Madhukar, M.S., 1985. Notched strength of composite laminates: predictions and experiments—a review. *Journal of Reinforced Plastics and Composites*, 4(1), 3–159.
[9] Niu, M.C.Y., 1992. *Composite airframe structures*. Hong Kong: Conmilit Press.
[10] Zinoviev, P.A., Grigoriev, S.V., Lebedeva, O.V. and Tairova, L.P., 1998. The strength of multi-layered composites under a plane-stress state. *Composites Science and Technology*, 58(7), 1209–1223.
[11] Davila, C.G., Jaunky, N. and Goswami, S., 2003. *Failure criteria for FRP laminates in plane stress*. NASA/TM-2003-212663.
[12] Spottswood, S.M. and Palazotto, A.N., 2001. Progressive failure analysis of a composite shell. *Composite Structures*, 53(1), 117–131.
[13] Wucher, B., Lani, F., Pardoen, T., Bailly, C. and Martiny, P., 2014. Tooling geometry optimization for compensation of cure-induced distortions of a curved carbon/epoxy C-spar. *Composites Part A: Applied Science and Manufacturing*, 56, 27–35.
[14] Kedward, K.T., Wilson, R.S. and McLean, S.K., 1989. Flexure of simply curved composite shapes. *Composites*, 20(6), 527–536.
[15] Cui, W., Liu, T., Len, J. and Ruo, R., 1996. Interlaminar tensile strength (ILTS) measurement of woven glass/polyester laminates using four-point curved beam specimen. *Composites Part A: Applied Science and Manufacturing*, 27(11), 1097–1105.

[16] Wisnom, M.R., 1996. 3-D finite element analysis of curved beams in bending. *Journal of Composite Materials*, *30*(11), 1178–1190.

[17] Roos, R., Kress, G., Barbezat, M. and Ermanni, P., 2007. Enhanced model for interlaminar normal stress in singly curved laminates. *Composite Structures*, *80*(3), 327–333.

[18] Most, J., Stegmair, D. and Petry, D., 2015. Error estimation between simple, closed-form analytical formulae and full-scale FEM for interlaminar stress prediction in curved laminates. *Composite Structures*, *131*, 72–81.

[19] De Morais, A.B., 2013. Mode I cohesive zone model for delamination in composite beams. *Engineering Fracture Mechanics*, *109*, 236–245.

[20] Dugdale, D.S., 1960. Yielding of steel sheets containing slits. *Journal of the Mechanics and Physics of Solids*, *8*(2), 100–104.

[21] Barenblatt, G.I., 1962. The mathematical theory of equilibrium cracks in brittle fracture. In *Advances in Applied Mechanics,* Vol. 7, 55–129.

[22] Maugis, D., 1992. Adhesion of spheres: the JKR-DMT transition using a Dugdale model. *Journal of Colloid and interface Science*, *150*(1), 243–269.

[23] Geleta, T.N., Woo, K. and Lee, B., 2018. Delamination behaviour of L-shaped laminated composites. *International Journal of Aeronautical and Space Sciences*, *19*(2), 363–374.

[24] Xu, X.P. and Needleman, A., 1994. Numerical simulations of fast crack growth in brittle solids. *Journal of the Mechanics and Physics of Solids*, *42*(9), 1397–1434.

[25] Tvergaard, V., 1990. Effect of fibre debonding in a whisker-reinforced metal. *Materials Science and Engineering: A*, *125*(2), 203–213.

[26] Tvergaard, V. and Hutchinson, J.W., 1992. The relation between crack growth resistance and fracture process parameters in elastic-plastic solids. *Journal of the Mechanics and Physics of Solids*, *40*(6), 1377–1397.

[27] Bhat, R., Mohan, N., Sharma, S., Pratap, A., Keni, A.P. and Sodani, D., 2019. Mechanical testing and microstructure characterization of glass fiber reinforced isophthalic polyester composites. *Journal of Materials Research and Technology*, *8*(4), 3653–3661.

[28] Zimmermann, K., Zenkert, D. and Siemetzki, M., 2010. Testing and analysis of ultra-thick composites. *Composites Part B: Engineering*, *41*(4), 326–336.

[29] Camanho, P.P., Davila, C.G. and De Moura, M.F., 2003. Numerical simulation of mixed-mode progressive delamination in composite materials. *Journal of Composite Materials*, *37*(16), 1415–1438.

[30] Li, X., Hallett, S.R. and Wisnom, M.R., 2008. Predicting the effect of through-thickness compressive stress on delamination using interface elements. *Composites Part A: Applied Science and Manufacturing*, *39*(2), 218–230.

[31] Kim, J.H., Nguyen, K.H., Choi, J.H. and Kweon, J.H., 2016. Experimental and finite element analysis of curved composite structures with C-section. *Composite Structures*, *140*, 106–117.

[32] Liu, P.F. and Islam, M.M., 2013. A nonlinear cohesive model for mixed-mode delamination of composite laminates. *Composite Structures*, *106*, 47–56.

[33] Tekalur, S.A., Shivakumar, K. and Shukla, A., 2008. Mechanical behaviour and damage evolution in E-glass vinyl ester and carbon composites subjected to static and blast loads. *Composites Part B: Engineering*, *39*(1), 57–65.

[34] Zhang, C., Duodu, E.A. and Gu, J., 2017. Finite element modeling of damage development in cross-ply composite laminates subjected to low velocity impact. *Composite Structures*, *173*, 219–227.

[35] HexPly Prepreg Technology. Publication No. FGU 017c, January 2013.

[36] Truong, V.H., Nguyen, K.H., Park, S.S. and Kweon, J.H., 2018. Failure load analysis of C-shaped composite beams using a cohesive zone model. *Composite Structures*, *184*, 581–590.

16 Dynamic Behavior of Laminated Composites with Internal Delamination

Chetan K. Hirwani

National Institute of Technology Patna,
Patna, India

*Rishabh Pal, Mrinal Chaudhury
and Subrata Kumar Panda*

National Institute of Technology Rourkela,
Rourkela, India

Nitin Sharma

Kalinga Institute of Industrial Technology Bhubaneswar,
Bhubaneswar, India

CONTENTS

DOI: 10.1201/9781003158813-16

16.1   INTRODUCTION

The application of modern and advanced laminated composite material structures in various sustainable engineering structures is increasing nowadays due to their relatively superior attributes compared to metallic counterparts. Apart from their superior structural properties such as high specific strength and stiffness and excellent fatigue damage resistance, the large library of constitutive materials (fiber and matrix), their different combination, and a variety of ways of fiber placement allow the design and development of composite structure with desired properties for a specific application. Despite several advantageous attributes, the peak of the application of laminated composite has not been achieved yet due to its complex modes of failure. The major reason for the complex failure is its inherent propensity to damage such as delamination, matrix cracking, and fiber breakage. Out of these, delamination is the cause of the majority of composite failures.

Delamination, also called interlayer fracture, is a separation of two continuous/adjacent lamina of a laminated composite. The separation/delamination arises at the free edge as well as inside the laminate. The delamination that arises at the free edge is called free edge delamination, and the delamination that arises inside the laminate is called internal delamination. Both types of delamination may appear in composite laminate anytime during the life span, i.e., during manufacturing and functioning/service life. During manufacturing, they appear due to undesirable processes such as inadequate curing, bubble formation (due to air entrapment), and the introduction of foreign elements and irregular geometry of the product. Additionally, during functioning, they appear as the composite structure subjected to the various type of operational loading such as impact loading, eccentric loading, low-velocity impact, out-of-plane stresses, and unfavorable environmental conditions (high/low temperature, humidity, and combined temperature and humidity). In general, the edge delamination can be identified/seen by the naked eye or by some regular means and can be remedied during application, while the internal delamination cannot be identified easily due it generally occurring away from top/bottom surfaces.

The presence of delamination is significant and causes severe damage to the laminated composite structures. Their presence breaks the material continuum and disintegrates the material, which leads to the significant reduction in strength and stiffness, fatigue life, and load-bearing capacity. The reduction in structural attributes (strength, stiffness, etc.) may cause catastrophic failure of the structures in service life span, with consequent serious reliability and safety issues. For example, the reduced frequency of the laminated structure may meet the working frequency, and the condition of resonance may arise. In this regard, to increase design safety and reliability in future applications, the in-depth understanding of structural dynamic behavior (fundamental frequency) of the layered composite is of great importance, because the flat composites (beam and plate) are mostly used in various engineering structures as a primary component. Over the years, owing to delamination severity, modeling (using analytical and numerical methods) and analysis of laminated/delaminated composite have been attempted by the research community, and a summary of the most important of them is presented in the subsequent paragraphs.

Band and Desai (2015) investigated both flexural and frequency responses of the laminated composite beam structure using a transition element that combines the subdomain of the higher-order equivalent single layers to two-dimensional mixed layer-wise. Zhen and Wanji (2016) presented the novel higher-order zig-zag theory for the investigation of the modal characteristics (free and forced) of the laminated composite beam structure having different thicknesses of each lamina. Sayyad and Ghugal (2017) presented the critical review of the flexural, vibration, and buckling behavior of the laminated and sandwich composite beam structure using different benchmark solutions and numerical and analytical methods. Talekar and Kotambkar (2020) predicted the influence of the layup sequence and thickness variation effect on the modal behavior of the composite beam by using the FE model. Brunelle and Robertson (1976) investigated the frequency behavior of the nonuniform primarily stressed thick flat panel with simply supported edge constraints and show that in some cases the frequency of higher modes can be lower than that of the lower modes. Qatu (1991) presented Ritz method and algebraic polynomial (as a displacement function) based mathematical model for the analysis of the influence of fiber orientation, material, and edge condition of modal characteristics of the flat composite structure. Ghosh and Dey (1994) derived the FE model of the constant-thickness laminated flat panel structure using higher-order theory and analyzed the frequency responses of the same. Niyogi et al. (1999) proposed a FE model of the composite panel using Lagrangian element and first-order shear deformation theory (FSDT) kinematics for investigating the free as well as forced vibration characteristics. Ganapathi and Makhecha (2001) employed the effective approach by considering the realistic in-plane as well as transverse displacement variation through the thickness for modeling and analysis of the layered composite flat structure. An analytical solution in conjunction with higher-order refined plate theory is proposed by Kant and Swaminathan (2001) for the modal response prediction of the layered and sandwich composite plate structure. The rectangular layered structure is also analyzed by Liew and Huang (2003) by using a mathematical model based on FSDT along with the moving least-squares differential quadrature method. The arbitrarily layered flat panel with clamped edges is investigated by Shi et al. (2004) for frequency characteristics using the Galerkin method. Ashour (2004) studied the dynamic characteristics of variable-thickness plate structures by using the finite strip transition matrix technique. Topal and Uzman (2007) analyzed the fundamental frequency of the laminated composite plate by using a mathematical model (derived from FSDT) and FSDT and finite element method (FEM)-based simulation software ANSYS. Zhang and Yang (2009) presented the development in numerical modeling techniques (different theories along with FEM) for analyzing the dynamic behavior of composite material for various structural applications. The new technique called *radial basis function collocation* along with FSDT is utilized by Ngo-Cong et al. (2011) for dynamic behavior investigation of the layered composite plate. The strain smoothing technique based on the newly developed smoothed quadrilateral flat element is utilized by Nguyen-Van et al. (2011) to analyze the dynamic behavior of flat as well as curved panel structure. Sharma et al. (2011) and Sharma and Mittal (2013) analyzed different modes of vibration frequency of the laminated plate structure by using the numerical model based on differential quadrature method and FEM along with FSDT, respectively. Carrera Unified Formulation and isogeometric analysis are combined by Natarajan et al. (2014) to derive the numerical model of plate structure and to investigate the dynamic characteristics of the same. Mantari et al. (2014) obtained the fundamental frequency responses of the advanced composite plate using the numerical model based on higher-order shear deformation theory (HSDT). Sayyad and Ghugal (2015) presented a review of different theories and techniques used in the analysis of multilayered composite and sandwich panel structure. Pingulkar and Suresha (2016) utilized the FSDT and FEM-based software ANSYS and investigated the free vibration responses and mode shapes of the glass and carbon fiber polymer composite. Belarbi et al. (2017) proposed a new higher-order layerwise model by combining the concept of first-order as well as higher-order theories for the investigation of the fundamental frequency responses of the layered composite and sandwich flat-panel structures. Sharma (2017) investigated the modal frequency characteristics of shear-deformable multilayered composite plates by using FEM. Alesadi et al. (2017) employed the IGA and CFU for the development of the numerical model of cross-ply multilayered composite plate and analyzed the frequency responses of the same. Carrera et al. (2017) utilized the revised beam element and CUF for the modeling and analysis of the thin beam structure. Mandal et al. (2020) performed the experimental modal behavior investigation on glass/epoxy composite and verified with that of the numerical responses derived from the FSDT based FE model. Chanda and Sahoo (2021) presented the fundamental frequency characteristics and time-dependent transverse deflection characteristics of the multilayer composite and sandwich flat panel structure using a model based on zigzag theory along with FEM.

In continuation, the laminated structure with delamination has also been studied and discussed in the following lines. Chang et al. (1998) investigated the frequency responses of the layered composite structure considering the influence of different shapes, sizes, and locations of delamination as well as the effect of discontinuous elastic foundation. Parhi et al. (1999) analyzed the eigenfrequency characteristics of the laminated composite twisted plate structure using the FSDT based FE model. Della and Shu (2007) presented an extensive review of the analytical and numerical model used to analyze the fundamental eigenfrequency behavior of the delaminated composite structure. Park et al. (2009) analyzed the frequency characteristics of the skew laminated composite flat panel structure considering the effect of delamination around cut using the mathematical model based on HSDT kinematics. The HSDT model in association with the FE has also been utilized by Dey and Karmakar (2013) to investigate the dynamic behavior of debonded conical shells subjected to low-velocity impact. Das et al. (2013) utilized the ANSYS platform to model the edge delaminated composite and investigate the influence of same on stress behavior. Marjanović and Vuksanović (2014) developed the layerwise theory-based FE model of the layered and sandwich flat panel for the investigation of the frequency and buckling characteristics. Ganesh et al. (2016) also utilized the FE model with FSDT kinematics for the investigation of fundamental frequency responses of the damaged (debonded) layered composite panel. The behavior of through width delamination has been studied by the Chermoshentseva et al. (2015) experimentally by fabricating the synthetic fiber and hydrophobic nanopowder composite and testing them in

Universal Testing Machine (UTM). Szekrenyes (2016) performed the dynamic (frequency and buckling) analysis of the layered beam structure by taking the influence of the presence of delamination. Kapuria and Ahmed (2019) proposed a mathematical model based on combining the zigzag theory and FE concept for the analysis of layered/sandwich composite flat panel structure with multiple delaminations. Tiwari et al. (2019) proposed a three-dimensional generated elements and Heaviside step functions-based mathematical model for analyzing the laminated plate with internal damage (delamination). Kamaloo et al. (2019) derived the mathematical model of conical shell structure with throughout circumferential delamination and analyzed the delaminated shell structure for the frequency behavior. Yang and Oyadiji (2020) fabricated the advanced fiber-reinforced polymer composite and created the artificial delamination by inserting the polytetrafluoroethylene (PTFE) at the predefined place of certain interfaces. They analyzed the frequency characteristics of the same experimentally and compared the results with the FE model. Chen et al. (2020) modeled the initiation as well as propagation behavior of delamination in mode-I for high strain rate loading using analytical theory and investigated the energy release rate. They also compared the analytical results with that of the experimental results too. Mondal and Ramachandra (2020) utilized the layerwise theory and jump discontinuity for the modeling and analysis (dynamics) of the delaminated composite flat structure subjected to the dynamic pulse loading. Li (2020) presented the micromechanical temperature-dependent (TD) mathematical model considering the effect of damping debonding, fiber breaking, and crack spacing to predict the vibration behavior of the fiber-reinforced ceramic matrix composite. Kim and Park (2021) derived the mathematical steps using Ritz method and investigated the free eigenfrequency characteristics of the composite plate with a hole.

From the extensive survey of the literature, it is understood that dynamic analysis of the laminated/delaminated panel structure has already been analyzed by using various numerical models and experimental modal analysis. Additionally, it has also been understood that most of the work studied the effect of the presence of delamination and its size and positions on frequency characteristics. Also, the delamination shape is mostly assumed as square. However, in reality, the panel structures may be of various shapes. In view of the above, to bridge the research gap, the present work aims to investigate the influence of different shapes of delamination on the dynamic behavior of layered plate structure. In order to perform such analysis, a model of the delaminated structure will be developed in FEM-based software ABAQUS. The model accuracy will be checked by solving the numerical illustration and validating the current responses with that of the published literature responses. Later on, the effect of the various shape of delamination, including the geometrical parameters, will be analyzed on modal responses and discussed in detail.

16.2 THEORETICAL FORMULATION

16.2.1 Displacement Kinematics

The layered composite structure as in Figure 16.1 is defined in a coordinate system $(0, x, y, z)$. The span length of the panel is defined as a, b, and h along x, y, and z axes, respectively.

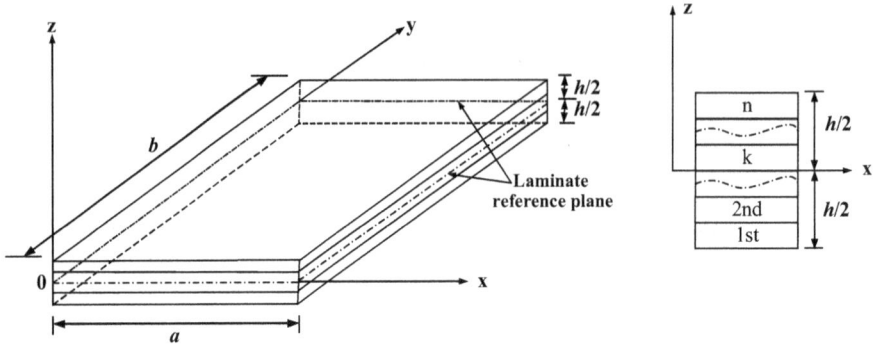

FIGURE 16.1   Laminated composite structure.

The displacement equation is assumed per FSDT and presented as

$$
\begin{aligned}
u(x,y,z,t) &= u_0(x,y) + z\phi_x(x,y) \\
v(x,y,z,t) &= v_0(x,y) + z\phi_y(x,y) \\
w(x,y,z,t) &= w_0(x,y) + z\phi_z(x,y)
\end{aligned}
\tag{16.1}
$$

where u, v, and w are the displacement field along x, y, and z-axes, respectively. t denotes the time. u_0, v_0, and w_0 are the displacements of the midplane, ϕ_x and ϕ_y are the rotations of normal to the reference plane (i.e., $z = 0$) about y and x-axes, respectively.

16.2.2 STRESS-STRAIN RELATION

The constitutive (stress-strain) relation of the lamina can be expressed as

$$
\{\sigma_x\,\sigma_y\,\sigma_z\,\tau_{yz}\,\tau_{xz}\,\tau_{xy}\}^T = [\overline{Q}]\{\varepsilon_x\,\varepsilon_y\,\varepsilon_z\,\gamma_{yz}\,\gamma_{xz}\,\gamma_{xy}\}^T
\tag{16.2}
$$

where $\{\sigma_x\,\sigma_y\,\sigma_z\,\tau_{yz}\,\tau_{xz}\,\tau_{xy}\}$ and $\{\varepsilon_x\,\varepsilon_y\,\varepsilon_z\,\gamma_{yz}\,\gamma_{xz}\,\gamma_{xy}\}$ represents linear stress and strain components, respectively, in laminate coordinate axes. $\overline{Q}$ is the elastic constants matrix for an orthotropic lamina, and the expansion of the same can be seen in Equation (16.3).

$$
[\overline{Q}] =
\begin{bmatrix}
\overline{Q}_{11} & \overline{Q}_{12} & \overline{Q}_{13} & 0 & 0 & \overline{Q}_{16} \\
\overline{Q}_{12} & \overline{Q}_{22} & \overline{Q}_{23} & 0 & 0 & \overline{Q}_{26} \\
\overline{Q}_{13} & \overline{Q}_{23} & \overline{Q}_{33} & 0 & 0 & 0 \\
0 & 0 & 0 & \overline{Q}_{44} & \overline{Q}_{45} & 0 \\
0 & 0 & 0 & \overline{Q}_{45} & \overline{Q}_{55} & 0 \\
\overline{Q}_{16} & \overline{Q}_{26} & 0 & 0 & 0 & \overline{Q}_{66}
\end{bmatrix}
\tag{16.3}
$$

where $\bar{Q}_{ij} = \bar{Q}_{11}, \bar{Q}_{12} \dots \bar{Q}_{66}$ denotes the transformed reduced stiffness whose expansion in terms of elastic moduli can be seen in Reddy (2003). And the strain–displacement equation can be represented as:

$$
\begin{Bmatrix} \varepsilon_x \\ \varepsilon_y \\ \varepsilon_z \\ \gamma_{yz} \\ \gamma_{xz} \\ \gamma_{xy} \end{Bmatrix} = \begin{Bmatrix} \dfrac{\partial u}{\partial x} \\[1mm] \dfrac{\partial v}{\partial y} \\[1mm] \dfrac{\partial w}{\partial z} \\[1mm] \dfrac{\partial v}{\partial z} + \dfrac{\partial w}{\partial y} \\[1mm] \dfrac{\partial u}{\partial z} + \dfrac{\partial w}{\partial x} \\[1mm] \dfrac{\partial u}{\partial y} + \dfrac{\partial v}{\partial x} \end{Bmatrix}
\tag{16.4}
$$

Now, Equation (16.4) can be modified after substituting the values from Equation (16.1), as:

$$
\{\varepsilon\} = [Z]\{\bar{\varepsilon}\}
\tag{16.5}
$$

where $[Z]$ is a matrix of thickness coordinate and $\{\bar{\varepsilon}\}$ is the strain (mid-plane) terms.

16.2.3 ENERGY RELATION

The strain energy and kinetic energy relation for the laminate can be expressed as

$$
U = \frac{1}{2} \int_V \varepsilon^T \sigma \, dV
\tag{16.6}
$$

$$
T = \frac{1}{2} \int_V \rho \{\dot{d}\}^T \{\dot{d}\} \, dV
\tag{16.7}
$$

were ρ is the density and $\{\dot{d}\}$ is the derivative of displacement vector with respect to time.

16.2.4 FINITE ELEMENT FORMULATION

The displacement of the laminated plate with six degrees-of-freedom at each node using FE steps is expressed in the following form:

$$
\{d_0\} = \sum_{i=1}^{6} [N_i]\{d_{0i}\}
\tag{16.8}
$$

where

$$\{d_0\} = [u_0 \quad v_0 \quad w_0 \quad \phi_x \quad \phi_y \quad \phi_z]^T$$

$$\{d_{0i}\} = [(u_0)_i \quad (v_0)_i \quad (w_0)_i \quad (\phi_x)_i \quad (\phi_y)_i \quad (\phi_y)_i]^T$$

Now mid-plane strain vector for the finite element model can be written as

$$\{\bar{\varepsilon}\} = [B]\{d_{0i}\} \tag{16.9}$$

where $[B]$ is the matrix for strain–displacement relation.

Now the elemental mass $[m]$ and stiffness matrix $[k]$ can be written as in Equation (16.10) and (16.11), respectively.

$$[m] = \int_A [N]^T \rho [N] dA \tag{16.10}$$

$$[k] = \int_A [B]^T [D][B] dA \tag{16.11}$$

To derive the equation for the delaminated structure, internal delamination (Area abcd) is assumed in the laminate as in Figure 16.2. In Figure 16.2, both laminated/integrated (part A) and delaminated (part B and C) sections can be seen. The mathematical steps for the laminated case are already discussed above. Now, similar to the laminated part, the

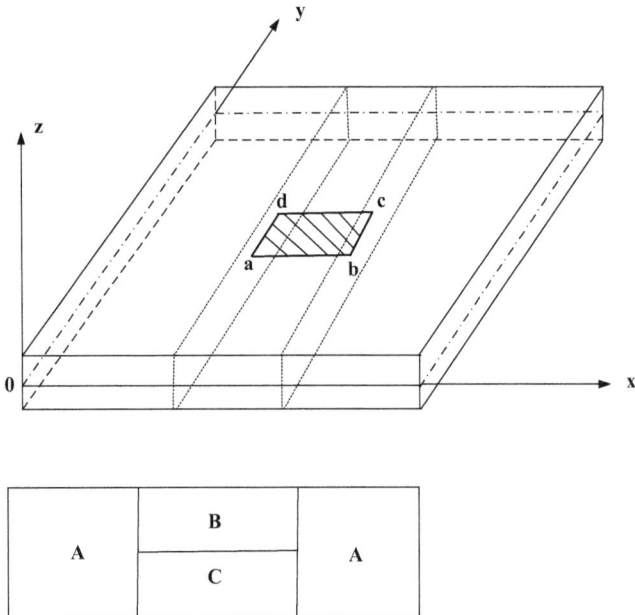

FIGURE 16.2 Laminated composite with delamination.

equation can be developed for the delaminated part also. And the mass and stiffness matrix for the delaminated element can be written as

$$[m_1] = \int_A [N]^T \rho [N] dA \qquad (16.12)$$

$$[k_1] = \int_A [B]^T [D][B] dA \qquad (16.13)$$

Now, using the appropriate continuity condition, the equation of the laminated and delaminated segment can be joined together.

16.2.5 System Governing Equation

The regulating equation of motion for free vibration of the composite laminated plate can be written using Hamilton's principle:

$$d \int_{t_1}^{t_2} (T - U) dt = 0 \qquad (16.14)$$

Now, substituting the values of U and T and assembling the elemental equation, the governing equation can be written as

$$[M]\{\ddot{d}\} + [K]\{d\} = 0 \qquad (16.15)$$

where $[M]$ and $[K]$ are the mass and stiffness matrix in the global form. $\{\ddot{d}\}$ and $\{d\}$ are representing the acceleration and global displacement vector in the global form.

Further, Equation (16.15) can be modified in the eigenvalue form represented as

$$([K] - \lambda^2 [M]) d = 0 \qquad (16.16)$$

The following boundary conditions have been used to get the final solution.
Simply supported:

$$v_0 = w_0 = \phi_y = \phi_z = 0 \quad at \; x = 0 \quad and \; a$$
$$u_0 = w_0 = \phi_x = \phi_z = 0 \quad at \; y = 0 \quad and \; b \qquad (16.17)$$

Clamped:

$$u_0 = v_0 = w_0 = \phi_x = \phi_y = \phi_z = 0$$
$$at \; x = 0 \; and \; a, \; at \; y = 0 \; and \; b \qquad (16.18)$$

16.3 ABAQUS MODEL DEVELOPMENT

In order to develop the finite element model of the delaminated composite flat panel structure, sublaminate approach has been utilized. The detailed description of the modeling steps has been discussed in the following lines:

- First, the layered composite structure has been divided into two sublaminates from the delamination plane, i.e., the number of layers below the delamination plane is considered as one sublaminate and the number of layer above the delamination plane is considered as another sublaminate.
- Both the sublaminate are modeled separately by the 2D shell approach, i.e., two shells are created for each sublaminate.
- Now a partition is created on each shell as per the shape and size of the delamination, and the required number of layers are assigned to each shell part.
- Both shell parts are joined together using the tie constraints and offset option in the ABAQUS software modeling tree.
- In order to introduce the FE concept, the discretization of the physical model has been done using the four-noded doubly curved quadrilateral elements available in the element library.
- Further, the necessary edge constraint condition has been applied and the solution has been obtained for the frequency responses.

16.4 RESULTS AND DISCUSSION

In this section, different numerical examples have been solved using the current FE model. In the entire example, a square ($a = b = 25$ m and $h = 0.0125$m) simply supported eight-layer symmetric ($0°/90°/45°/90°/90°/45°/90°/0°$) laminated composite plate have examined (Figure 16.3). The delamination has been considered at the center location of the mid-plane else stated otherwise. Different sizes of delamination say 0%, 6.25%, and 25% of the total area of the plate are considered and given the names DS-A, DS-B, and DS-C, respectively. The shape of the delamination has also been varied say square, rectangular, elliptical, and circular as shown in Figure 16.4. The description of the geometrical parameter of different shapes of delamination is also provided in Table 16.1. The material property used in the entire example is provided in Table 16.2, unless stated otherwise.

16.4.1 VALIDATION STUDY

For the comparison study, a square plate structure with all side-clamped configurations has been analyzed for frequency responses. The plate structure is composed of four symmetric plies with square delamination ($0°/90°/90°$-del-$0°$). The input parameters related to geometrical configuration and material properties are taken as same as Vuksanović et al. (2016). The frequency characteristics (six modes) are obtained using the current FE model compared with the reference results and presented in

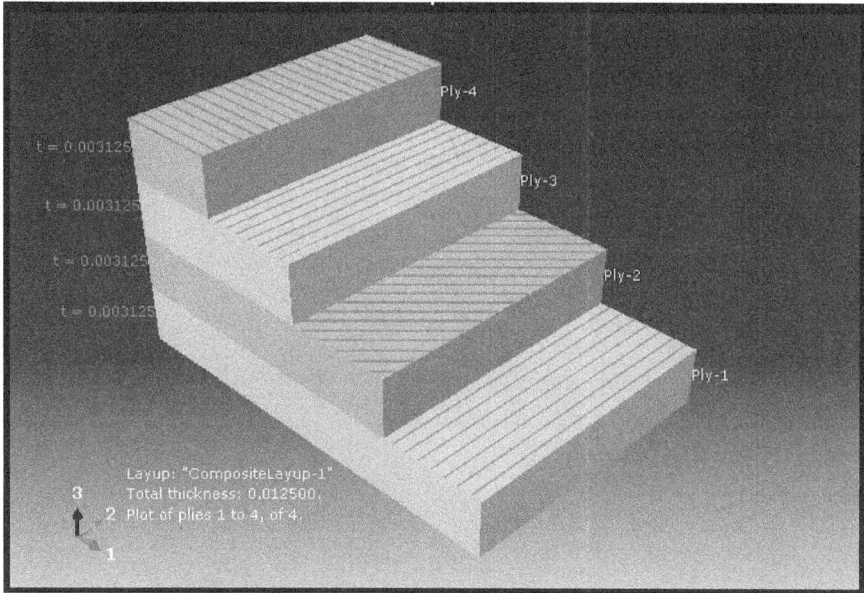

FIGURE 16.3   Stacking sequence.

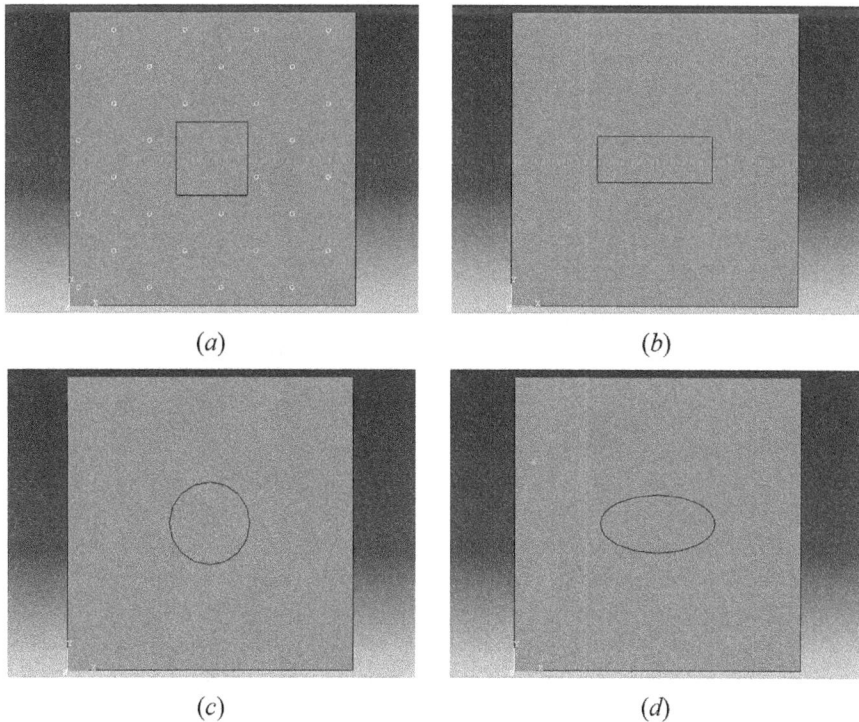

FIGURE 16.4   Different shapes of delamination: (a) square, (b) rectangular, (c) circular and (d) elliptical.

TABLE 16.1
Geometrical Parameter of Different Shape of Delamination

Delamination Type	Delamination Area	Delamination Shape Dimension (m)			
		Square (Length)	Rectangular (Length × Width)	Circular (Radius)	Elliptical (Major & Minor Diameter)
DS-A	0	0	0 × 0	$r = 0$	0 & 0
DS-B	6.25%	0.0625	0.1 × 0.0391	$r = 0.0355$	0.05 & 0.0249
DS-C	25%	0.1250	0.1 × 0.1563	$r = 0.0705$	0.1 & 0.0497

TABLE 16.2
Material Property Used

Young's modulus	E_1 (N/m^2)	1.32×10^{11}
	$E_2 = E_3$ (N/m^2)	5.35×10^9
Shear modulus	$G_{12} = G_{13}$ (N/m^2)	2.79×10^9
	G_{23} (N/m^2)	1.40×10^9
Poisson's ratio	$v_{12} = v_{13}$	0.291
	v_{23}	0.3
Density	ρ (kg/m^3)	1446.2

TABLE 16.3
Comparison Study of Frequency (Hz) Responses of Four-Layered (0°/90°/90°-del-0°) Delaminated Composite Plate Structure

Mode	Vuksanović et al. (2016)	Present	% Difference
1	239.53	240.40	0.36
2	336.24	344.51	2.46
3	382.50	380.40	−0.55
4	450.49	452.71	0.49
5	576.27	592.48	2.81
6	656.51	653.40	−0.47

Table 16.3. From the comparison table, it is seen that the current responses are close to the reference results, i.e., the maximum deviation of the responses is ~3%. Hence the current model can be utilized for the dynamic behavior analysis of delaminated composite plate structure.

After successful validation of the presently developed FE model, it is now used to explore the influence of delamination shape, size, and geometrical parameter on frequency characteristics of the plate structure. For the said exploration, different numerical examples have been solved and presented in the following lines.

TABLE 16.4

Natural Frequency (Hz) of Laminated Composite with Different Size of Delamination

	Intact	Rectangular		Elliptical		Square		Circular	
				Delamination Shape and Size					
Mode	DS-A	DS-B	DS-C	DS-B	DS-C	DS-B	DS-C	DS-B	DS-C
1	403.8	380.7	353.2	380.3	352.5	380.1	348.9	380.0	340.7
2	753.9	674.0	557.0	666.4	555.4	659.4	517.8	656.9	514.5
3	905.6	830.5	612.6	815.2	621.0	810.9	617.0	805.1	585.4
4	1134.7	1053.7	846.7	1054.8	850.6	1056.6	659.2	1059.4	614.2
5	1373.3	1187.4	869.3	1194.2	879.0	1196.3	813.4	1203.1	809.2
6	1634.7	1483.0	892.3	1473.3	951.6	1468.3	895.8	1461.7	858.7

16.4.2 NEW NUMERICAL ILLUSTRATIONS

16.4.2.1 Delamination Size Effect on Different Modes of Flat Composite

This example presents the influence of the size of delamination on different modes of eigenfrequency responses of a layered composite flat structure. Three different sizes of delamination (DS-A, DS-B, and DS-C) have been incorporated at the center mid-plane of the layered composite. The dynamic responses are obtained via the FE model and shown in Table 16.4. From the table data, irrespective of different shapes of delamination, the decrease in frequency characteristics with the increase in delamination size has been observed. This indicates that there may be a gradual drop in the overall stiffness of the panel, which further decreases the frequency characteristics with increasing the delamination size.

16.4.2.2 Effect of Different Shapes of Delamination

In this example, four different shapes of delamination—square, rectangular, circular, and elliptical—have been considered at the center mid-plane of the layered composite plate structure. The size of all shapes of delamination is taken as the same, i.e., 25% of the total plate area (DS-C). Next the frequency responses are obtained using the current FE model and presented in Table 16.5. In most cases, the highest-frequency characteristics are obtained for the plate with rectangular delamination and the lowest frequency is obtained for the plate with circular delamination. In between the frequency the following trends with different shapes of delamination is observed: rectangular > elliptical > square > circular. As the frequency responses are directly proportional to the overall stiffness, it is clear from the current investigation that the presence of circular delamination affects (decreases) the overall stiffness of the plate more compared to the other shapes of delamination. Additionally, frequency characteristics are decreases irrespective of different shapes of delamination with increasing the length to thickness ratio.

TABLE 16.5

Natural Frequency (Hz) of Laminated Composite with Different Shapes of Delamination

Span to Thickness Ratio (a/h)	Mode	Different Shape of Delamination			
		Rectangular	Elliptical	Square	Circular
10	1	1971.2	1965.0	1965.8	1957.1
	2	2629.2	2639.4	2580.9	2580.3
	3	3280.6	3308.8	3239.9	3222.6
	4	3579.2	3604.8	3625.1	3571.3
	5	3920.0	3949.9	3828.3	3629.7
	6	4619.0	4684.7	3955.3	3941.5
30	1	1054.3	1049.8	1046.2	1029.1
	2	1556.0	1553.0	1474.0	1469.9
	3	1802.9	1825.6	1801.2	1789.2
	4	2238.0	2278.0	1985.3	1789.4
	5	2382.1	2386.8	2211.3	2203.3
	6	2610.1	2765.9	2442.5	2375.0
50	1	681.7	679.4	674.9	661.0
	2	1046.9	1043.3	980.3	975.1
	3	1176.3	1191.9	1180.6	1139.8
	4	1556.6	1575.9	1276.9	1173.4
	5	1619.6	1627.0	1512.6	1506.1
	6	1711.0	1821.0	1666.9	1606.0
70	1	497.9	496.6	492.3	481.3
	2	777.1	774.5	724.4	719.9
	3	861.9	873.6	866.8	828.0
	4	1171.5	1180.6	930.6	862.1
	5	1208.3	1218.1	1130.4	1124.8
	6	1255.0	1337.5	1244.8	1195.3
100	1	353.2	352.5	348.9	340.7
	2	557.0	555.4	517.8	514.5
	3	612.6	621.0	617.0	585.4
	4	846.7	850.6	659.2	614.2
	5	869.3	879.0	813.4	809.2
	6	892.3	951.6	895.8	858.7

16.4.2.3 Mode Shapes of Different Shape of Delamination

This illustration presents the different modes shapes of the delaminated plate structure. The plate with various shapes of delamination (DS-B) at the mid-plane of the laminated is assumed for the analysis. The mode shape of six modal frequencies has been extracted and presented in Table 16.6.

TABLE 16.6

Mode Shape of Delaminated Composite Plate

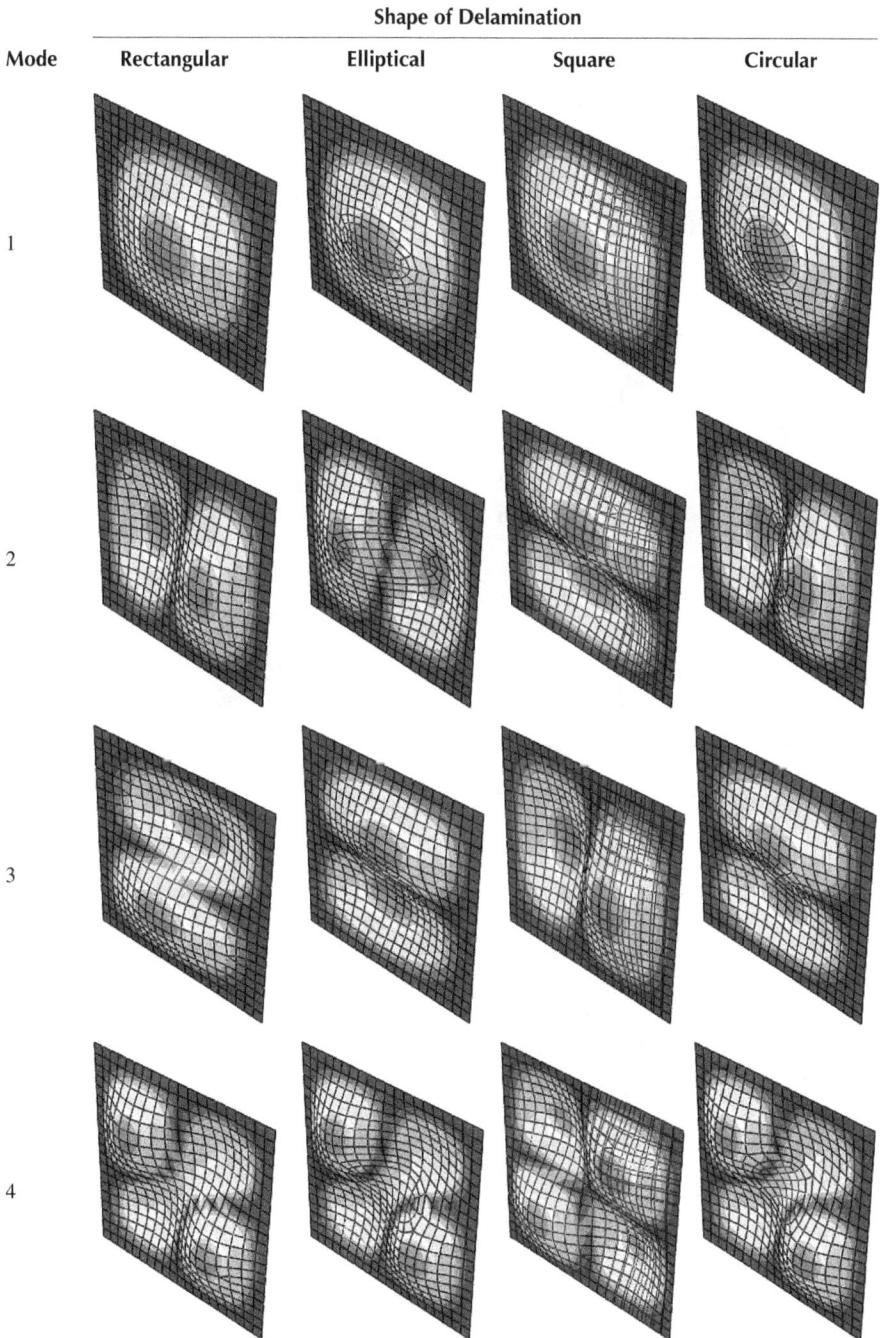

	Shape of Delamination			
Mode	Rectangular	Elliptical	Square	Circular
1				
2				
3				
4				

(Continued)

TABLE 16.6 (CONTINUED)
Mode Shape of Delaminated Composite Plate

Shape of Delamination

Mode	Rectangular	Elliptical	Square	Circular
5				
6				

16.5 CONCLUSIONS

In this work, the influence of different shapes and sizes of delamination on various modes of eigenfrequency responses of the flat composite panel structure has been analyzed. For the analysis, an FE model of the intact and damaged (delaminated) composite panel has been prepared by using the ABAQUS platform. Four different shapes of delamination (square, rectangular, elliptical, and circular) have been considered at the center mid-plane of the laminated structure. From the investigation, it is concluded that the circular shape of delamination is more critical than the other shapes. Additionally, it is also concluded that the increase in the size of delamination reduces the various mods of frequency characteristics, and with the increase in the span-to-thickness ratio the different modes of frequency responses decrease.

REFERENCES

Askaripour K, Zak A. A survey of scrutinizing delaminated composites via various categories of sensing apparatus. *Journal of Composites Science* 2019;3:95. https://doi.org/10.3390/jcs3040095.

Band UN, Desai YM. Multi-model finite element scheme for static and free vibration analyses of composite laminated beams. *Latin American Journal of Solids and Structures* 2015;12:2061–77. https://doi.org/10.1590/1679-78251743.

Belarbi MO, Tati A, Ounis H, Khechai A. On the free vibration analysis of laminated composite and sandwich plates: A layerwise finite element formulation. *Latin American Journal of Solids and Structures* 2017;14:2265–90. https://doi.org/10.1590/1679-78253222.

Chang TP, Hu CY, Jane KC. Vibration analysis of delaminated composite plates under axial load. *Mechanics of Structures and Machines* 1998;26:195–218. https://doi. org/10.1080/08905459808945427.

Chen T, Harvey CM, Wang S, Silberschmidt V v. Delamination propagation under high loading rate. *Composite Structures* 2020;253:112734. https://doi.org/10.1016/j. compstruct.2020.112734.

Chermoshentseva AS, Pokrovskiy AM, Bokhoeva LA. The behavior of delaminations in composite materials - Experimental results. *IOP Conference Series: Materials Science and Engineering*, vol. 116, Institute of Physics Publishing; 2016, p. 012005. https://doi. org/10.1088/1757-899X/116/1/012005.

Das S, Choudhury P, Halder S, Sriram P. Stress and free edge delamination analyses of delaminated composite structure using ANSYS. *Procedia Engineering*, vol. 64, Elsevier Ltd; 2013, p. 1364–73. https://doi.org/10.1016/j.proeng.2013.09.218.

Della CN, Shu D. Vibration of delaminated composite laminates: A review. *Applied Mechanics Reviews* 2007;60:1–20. https://doi.org/10.1115/1.2375141.

Dey S, Karmakar A. Dynamic analysis of delaminated composite conical shells under low velocity impact. *Journal of Reinforced Plastics and Composites* 2013;32:380–92. https://doi.org/10.1177/0731684412465663.

Ganapathi M, Makhecha DP. Free vibration analysis of multi-layered composite laminates based on an accurate higher-order theory. *Composites Part B: Engineering* 2001;32:535–43. https://doi.org/10.1016/S1359-8368(01)00028-2.

Ganesh S, Kumar KS, Mahato PK. Free vibration analysis of delaminated composite plates using finite element method. *Procedia Engineering*, vol. 144, Elsevier Ltd; 2016, p. 1067–75. https://doi.org/10.1016/j.proeng.2016.05.061.

Guha Niyogi A, Laha MK, Sinha PK. Finite element vibration analysis of laminated composite folded plate structures. *Shock and Vibration* 1999;6:273–83.

Imran M, Khan R, Badshah S. *A Review on the Vibration Analysis of Laminated Composite Plate*. vol. 62. 2019.

Ju F, Lee HP, Lee KH. Free vibration of composite plates with delaminations around cutouts. *Composite Structures* 1995;31:177–83. https://doi.org/10.1016/0263-8223(95)00016-X.

Kamaloo A, Jabbari M, Tooski MY, Javadi M. Nonlinear free vibrations analysis of delaminated composite conical shells. *International Journal of Structural Stability and Dynamics* 2019. https://doi.org/10.1142/S0219455420500108.

Kant T, Swaminathan K. Analytical solutions for free vibration of laminated composite and sandwich plates based on a higher-order refined theory. *Composite Structures* 2001;53:73–85. https://doi.org/10.1016/S0263-8223(00)00180-X.

Kapuria S, Ahmed A. An efficient zigzag theory based finite element modeling of composite and sandwich plates with multiple delaminations using a hybrid continuity method. *Computer Methods in Applied Mechanics and Engineering* 2019;345:212–32. https:// doi.org/10.1016/j.cma.2018.10.035.

Kim Y, Park J. A theory for the free vibration of a laminated composite rectangular plate with holes in aerospace applications. *Composite Structures* 2020;251:112571. https://doi. org/10.1016/j.compstruct.2020.112571.

Li L. A micromechanical temperature-dependent vibration damping model of fiber-reinforced ceramic-matrix composites. *Composite Structures* 2021;261:113297. https://doi. org/10.1016/j.compstruct.2020.113297.

Marjanović M, Vuksanović D. Layerwise solution of free vibrations and buckling of laminated composite and sandwich plates with embedded delaminations. *Composite Structures* 2014;108:9–20. https://doi.org/10.1016/j.compstruct.2013.09.006.

Mondal S, Ramachandra LS. Nonlinear dynamic pulse buckling of imperfect laminated composite plate with delamination. *International Journal of Solids and Structures* 2020;198:170–82. https://doi.org/10.1016/j.ijsolstr.2020.04.010.

Nguyen-Van H, Mai-Duy N, Karunasena W, Tran-Cong T. Buckling and vibration analysis of laminated composite plate/shell structures via a smoothed quadrilateral flat shell element with in-plane rotations. *Computers and Structures* 2011;89:612–25. https://doi.org/10.1016/j.compstruc.2011.01.005.

Parhi PK, Bhattacharyya SK, Sinha PK. Dynamic analysis of multiple delaminated composite twisted plates. *Aircraft Engineering and Aerospace Technology* 1999;71:451–61. https://doi.org/10.1108/00022669910296891.

Park T, Lee SY, Voyiadjis GZ. Finite element vibration analysis of composite skew laminates containing delaminations around quadrilateral cutouts. *Composites Part B: Engineering* 2009;40:225–36. https://doi.org/10.1016/j.compositesb.2008.11.004.

Qatu MS. Free vibration of laminated composite rectangular plates. *International Journal of Solids and Structures* 1991;28:941–54. https://doi.org/10.1016/0020-7683(91)90122-V.

Reddy J N. *Mechanics of Laminated Composite Plates and Shells*. CRC Press. 2003. https://doi.org/10.1201/b12409.

Sayyad AS, Ghugal YM. Bending, buckling and free vibration of laminated composite and sandwich beams: A critical review of literature. *Composite Structures* 2017;171:486–504. https://doi.org/10.1016/j.compstruct.2017.03.053.

Szekrényes A. Vibration and parametric instability analysis of delaminated composite beams. *International Journal of Materials and Metallurgical Engineering* 2016;10:821–38. https://doi.org/doi.org/10.5281/zenodo.1125055.

Talekar N, Kotambkar M. Modal analysis of four layered composite cantilever beam with lay-up sequence and length-to-thickness ratio. *Materials Today: Proceedings*, vol. 21, Elsevier Ltd; 2020, p. 1176–94. https://doi.org/10.1016/j.matpr.2020.01.068.

Tiwari P, Barman SK, Maiti DK, Maity D. Free vibration analysis of delaminated composite plate using 3D degenerated element. *Journal of Aerospace Engineering* 2019;32:04019070. https://doi.org/10.1061/(ASCE)AS.1943-5525.0001053.

Vuksanović Đ, Marjanović M, Kovačević D. Finite element modeling of free vibration problem of delaminated composite plates using abaqus cae. *6th international conference civil engineering - science and practice, žabljak*, 7–11 March 2016.

Yang C, Olutunde Oyadiji S. A new delamination making technique for cured glass/epoxy samples and validation using vibration characteristics. *Composite Structures* 2020;257:113019. https://doi.org/10.1016/j.compstruct.2020.113019.

Zhen W, Wanji C. Free and forced vibration of laminated composite beams. *AIAA Journal* 2018;56:2877–86. https://doi.org/10.2514/1.J055506.

Index

A

ABAQUS, 180, 235, 266, 267, 268, 281, 286, 292
Admissible functions, 96
Analytical solution, 36, 49, 55, 58, 74, 89, 95, 96, 106, 123, 202, 262, 279
Anisotropy, 174, 180
ANSYS, 23, 116, 175, 195, 199
Auxetic, 174, 192

B

Bending, 58, 65, 91, 92, 187, 189, 203, 207, 225
Biocomposites, 139, 163
Blast load, 122, 123, 126
Buckling, 2, 14, 58, 70, 88, 140, 168, 169
Bulk modulus, 7, 18, 143

C

Carbon nanotubes (CNTs), 88, 90, 110, 111, 114, 246, 247, 251
Classical laminate theory, 202
Coherent potential approximation, 7
Cohesive elements, 263, 267
Cohesive zone, 263, 265
Compressive stress, 264
Continuum Theory, 190
Corrugation, 60, 202, 218, 219
C-shaped, 262
C-Spar, 266

D

Damage criterion, 265
Damage, 110, 175, 245, 263, 264, 265, 268, 271, 272, 273, 278, 280, 292
Degradation, 263, 265
Delamination, 2, 58, 218, 262, 265, 278, 284, 286
Differential transform method (DTM), 36, 38
Direct iterative method, 207
Dirichlet boundary condition, 43
Doubly curved, 286

E

Effective material properties, 6, 60, 114, 142, 164, 219, 251
Efficiency parameters, 91, 114
Electrical conductivity, 76, 83
Electrical resistivity, 77, 83

Energy principle, 64, 167
Equations of motion, 93
Extended rule, 114, 251
Extensional stiffness matrix, 187

F

Failure load prediction, 269
Finite element, 19, 115, 120, 146, 165, 194, 222, 266, 283
First-order shear deformation theory (FSDT), 116, 184, 190
Flexural stiffness matrix, 187
Force vectors, 255
Functionally graded materials (FGMs), 2, 58, 140, 218, 247

G

Gasik-Ueda model, 9
Generalized Hooke's law, 188, 236

H

Hamilton's principle, 20, 93, 285
Hashin damage criteria, 262, 264
Higher-order, 15, 20, 147, 204, 220
Hirano model, 8
Homogenization, 4, 142, 235
Honeycomb, 174, 175, 190
Hybrid composites, 262
Hydroxyapatite, 142, 147
Hygrothermal load, 207

I

Implant, 140, 162
Interlaminar, 2, 265, 267
Isoparametric, 203, 215

K

Kerner model, 7

L

Lagrangian elements, 22, 146

M

Maclaurin series, 37
Macroscale, 236

295